FRONTIERS OF
4d- AND 5d-TRANSITION
METAL OXIDES

Gang Cao • Lance DeLong

University of Kentucky, USA

FRONTIERS OF
4d- AND **5d**-TRANSITION
METAL OXIDES

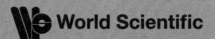

World Scientific

NEW JERSEY • LONDON • SINGAPORE • BEIJING • SHANGHAI • HONG KONG • TAIPEI • CHENNAI

Published by

World Scientific Publishing Co. Pte. Ltd.

5 Toh Tuck Link, Singapore 596224

USA office: 27 Warren Street, Suite 401-402, Hackensack, NJ 07601

UK office: 57 Shelton Street, Covent Garden, London WC2H 9HE

British Library Cataloguing-in-Publication Data
A catalogue record for this book is available from the British Library.

FRONTIERS OF 4d- AND 5d-TRANSITION METAL OXIDES

ISBN 978-981-4374-85-9

Printed in Singapore by World Scientific Printers.

Preface

The systematic study of 4d and 5d transition metal oxides began roughly in the 1990's, and now represents a fertile and highly active field of condensed matter physics. However, no broad overview of the experimental situation concerning the ruthenates and iridates currently exists. The immense growth in publications addressing the physical properties of these highly intriguing materials underlines the need to document recent advances and the current state of this field.

The contributing authors for this volume are among the pioneers of research on the 4d and 5d oxides, and we thank them for their substantial efforts in bringing this work to fruition. The following persons were of great help in processing the manuscript: Alina Oleksandrova and Jasminka Terizc oversaw the initial formatting of the text; and Jasminka Terizc compiled the index for the book and proof-read all captions and figures of the book. We also wish to thank Ms. Julia Huang and Mr. Yeow-Hwa Quek of *World Scientific Publishing* for their professional assistance and patience in completing the manuscript, and Dr. K. K. Phua, Editor-in-Chief of *World Scientific Publishing*, for his initial encouragement of our efforts. Professor Michael Cavagnero, Chair of the Department of Physics and Astronomy at the University of Kentucky, deserves thanks for his constant support of this endeavor. GC wishes to thank the College of Arts and Sciences of the University of Kentucky for its support, and LD wishes to thank the College of Arts and Sciences of the University of Kentucky for its support in the form of a Sabbatical Leave during the completion of the manuscript.

Finally, we wish to thank our dear wives, Qi Zhou and Janie, for their inspiration, counsel, and support.

Gang Cao
Lance E DeLong
Lexington, Kentucky, USA

CONTENTS

Contents

Chapter 1

INTRODUCTION

Gang Cao and Lance E. DeLong

Department of Physics and Astronomy and Center for Advanced Materials,
University of Kentucky, Lexington, Kentucky 40506, USA
E-mail: cao@uky.edu; delong@pa.uky.edu

Transition metal oxides have recently been the subject of enormous activity within both the applied and basic science communities. However, the overwhelming balance of interest was devoted to 3d-elements and their binary compounds for many decades, extending from the 19th to the late 20th Century. The strong magnetic and elastic properties of these materials were key to various technologies, although the relatively robust superconducting properties of several Nb intermetallic compounds (e.g., Nb_3Sn, NbTi) shifted some attention toward the 4d-elements during the 1960's.

A sea change occurred with the discovery of "high temperature" superconductivity in ternary and more complex copper oxides in 1986. The ongoing explosion of interest in 3d oxides produced further breakthroughs with the discovery of "colossal magnetoresistance" (CMR) in ternary Mn oxides in 1990's, which rapidly led to successful commercial applications of CMR in data storage devices. These advances have been accompanied by a remarkable shortening of the time lag between the initial discoveries of novel materials with surprising fundamental properties, and the development of derivative materials and hybrid structures for the marketplace. Furthermore, an intensified interplay between the basic research and information technologies has led to the creation of whole new fields of investigation and commercial development, including "spintronics", "magnonics", and "nano-science" or "nano-engineering". These remarkable developments have also been accompanied by nothing less than an explosion of publications (including professional and popular works), which has completely outstripped the steady increases in journal publications on condensed matter physics and chemistry which took place in the 1960's and 1970's.

It is now apparent that *novel materials*, which often exhibit surprising or even revolutionary physical properties, are necessary for critical advances in technologies that affect the everyday lives of average people. For example, a number of discoveries involving superconducting and magnetic materials, polymers and thin-film processing have underpinned the development of novel medical diagnostic tools, personal electronic devices, advanced computers, and powerful motors and actuators for automobiles.

Fundamental scientists, confronted with ever-increasing pace and competition in research, are beginning to examine the remaining "unknown territories" located in the lower rows of the periodic table of the elements. Although the rare earth and light actinide elements have been aggressively studied for many decades, the heavier 4d- and 5d-elements and their oxides have largely been ignored until recently. The reduced abundance and increased production costs for many of these elements have certainly discouraged basic and applied research into their properties. What has not been widely appreciated, however, is that 4d- and 5d-elements and their compounds exhibit unique competitions between fundamental interactions that result in physical behaviors and empirical trends that markedly differ from their 3d counterparts.

The unique opportunities offered by the heavier transition elements are exemplified by recent discoveries concerning Ru oxides. Evidence for exotic superconducting pairing in Sr_2RuO_4 was discovered in 1994, and this work was no doubt stimulated by the central importance of the analogous cuprate, La_2CuO_4, to understanding the earliest examples of high-T_C superconductors. Initial research on Sr_2RuO_4 has been followed by a burgeoning body of work on other ruthenates that exhibit a variety of novel and perplexing magnetic, dielectric and elastic properties. More recently, Ir oxides or iridates have attracted growing attention, due to the influence of even stronger spin-orbit interactions on their physical properties; these effects were largely ignored in theoretical treatments of 3d and other materials of interest (excepting the rare earth and actinide classes). Specifically, the existence of several overlapping energy scales and competing interactions present in the 4d- and 5d-oxides, including spin-orbit, coulomb and exchange interactions, offers wide-ranging opportunities for the discovery of new physics and, ultimately, new device paradigms.

Below, we briefly overview the contributions of the Chapter Authors for this monograph, and their relationship to current progress in the field.

Soon Jae Moon and Tae Won Noh describe recent infrared and optical spectroscopy data and supporting theoretical studies of the effects of spin-orbit coupling in iridates in Chapter 2. A new signature effect of spin-orbit coupling is the "$J_{eff} = 1/2$ insulating state", identified by Noh and collaborators in iridates. Moon and Noh thoroughly discuss the electronic structure of Sr_2IrO_4, which has come to represent an early archetype for new physics driven by spin-orbit coupling.

I. Zegkinoglou and B. Keimer give a brief overview in Chapter 3 of the most recent advances in X-ray scattering studies of 4d- and 5d-transition metal compounds They emphasize three ruthenates with thoroughly studied magnetic and orbital properties: the single-layer Mott system, Ca_2RuO_4, the bilayer magnetic analogue, $Ca_3Ru_2O_7$, and the rutheno-cuprate $RuSr_2GdCu_2O_8$. The results of these studies underline the versatility and power of resonant x-ray scattering for revealing the role of fundamental electronic structure in physical properties.

Lance Cooper reviews the methods used and results obtained in Raman scattering studies of the Ruddelsen-Popper ruthenates in Chapter 4. Cooper discusses how structural changes associated with the RuO_6 octahedra—induced variously by temperature, atomic substitution, magnetic field, and pressure—play a definitive role in many of the exotic properties observed in the layered ruthenates, especially in their magnetic and orbitally polarized phases. Raman scattering data reveal the anomalous behavior of the B_{1g} symmetry RuO_6 octahedral phonon mode in all the ruthenate materials studied by Cooper. After a brief description of the theory of Raman scattering in solids, the low-temperature, high-pressure, and high-magnetic-field Raman methods used in these studies are discussed. Raman scattering results are reviewed in detail for single-layer $Ca_{2-x}Sr_xRuO_4$, bilayered $Ca_3Ru_2O_7$, and triple-layered $Sr_4Ru_3O_{10}$, including evidence for orbital order in $Ca_3Ru_2O_7$.

Rongying Jin discusses the electronic properties of single-layer Ruddlesden-Popper ruthenates with emphasis on their metal-insulator transitions in Chapter 5. Such transitions can be induced by varying either temperature, pressure, or chemical doping. Jin discusses the existing experimental properties and relevant theoretical calculations to assess the complex couplings between charge, spin, lattice and orbital degrees of freedom.

Gang Cao, Lance DeLong and Pedro Schlottmann briefly overview transport, magnetic and thermal properties of Ca ruthenates in Chapter 6. The balance of the Chapter focuses on the bilayered Ruddelsen-Popper phase, $Ca_3Ru_2O_7$, first reported by Cao and collaborators in 1997. Experimental results for the remarkable physical properties of single-crystal $Ca_3Ru_2O_7$ are discussed as functions of crystal orientation, magnetic induction as high as 45 T, and temperature as low as 0.3 K. Depending on the measuring field direction, $Ca_3Ru_2O_7$ exhibits signatures of all possible electronic or magnetic ground states except superconductivity. In particular, $Ca_3Ru_2O_7$ simultaneously exhibits behaviors that are conventional hallmarks of mutually exclusive metallic and insulating states. This peculiar material presents a conundrum for condensed matter theory, and a profound challenge to understand the complex interplay between spin, orbital, charge and lattice degrees of freedom present in 4d-oxides.

Ward Plummer reviews recent research on the exotic behavior of correlated electrons in Chapter 7, with emphasis on surface properties. Better experimental techniques, higher quality samples, the discovery of new functional materials, and a focus on complexity have all driven the current flurry of research on ruthenates. Plummer describes how the unique environment of a sample surface, with its inherently broken translational symmetry, provides an exquisite playground for probing the physics of transition metal compounds. The relationships between surface structure and physical properties are demonstrated, especially where small structural changes strongly affect a variety of physical properties, including the metal-to-insulator transition temperature, the occurrence of superconductivity, magnetic ordering, electron-phonon coupling, spin-phonon coupling, and even quantum critical behavior.

The most profound result of the spin-orbit interaction on the iridates is the $J_{eff} = 1/2$ insulating state. In Chapter 8, Gang Cao and Lance DeLong review the underlying physical properties of layered iridates and results of transport and thermodynamic studies of $Sr_{n+1}Ir_nO_{3n+1}$ (n = 1 and 2) that emphasize spin-orbit-tuned ground states stabilized by chemical doping, application of pressure and magnetic field. These weak perturbations are capable of directly reducing the spin-orbit interaction. This will rebalance comparable interactions and generate a rich phase diagram of strongly competing ground states that remain largely controlled by the spin-orbit interaction.

Chapter 2

SPECTROSCOPIC STUDIES OF STRONG SPIN–ORBIT COUPLING IN 4D AND 5D TRANSITION METAL OXIDES

Soon Jae Moon[1,2] and Tae Won Noh[1]

[1]*Department of Physics and Astronomy, Seoul National University, Seoul 151-747, Korea*
[2]*Department of Physics, University of California, San Diego, La Jolla, California 92093, USA*
E-mail: sjmoon@physics.ucsd.edu

In $4d$ and $5d$ transition metal oxides, the magnitude of spin-orbit coupling becomes comparable to those of other fundamental interactions, such as electron-phonon and on-site Coulomb interactions. So it can either compete with or cooperate with other interactions. Recent investigations of $4d$ and $5d$ transition metal oxides revealed that the spin-orbit coupling can modify electronic and magnetic structures significantly. Moreover, it can produce a novel quantum state that has never been observed in $3d$ transition metal oxides. In this chapter, we introduce spectroscopic and theoretical studies on the roles of spin-orbit coupling and the associated novel physical phenomena in $4d$ and $5d$ transition metal oxides.

Contents

2.1. Introduction

Transition metal oxides with $3d$ or $4d$ electrons exhibit numerous intriguing phenomena, such as unconventional superconductivity, Mott transitions, colossal magnetoresistance, and spin/orbital orderings.[1] These phenomena originate from various interactions, including electron correlation, hopping, electron–phonon coupling, and spin–orbit coupling.

The majority of studies on transition metal oxides have focused on $3d$ electron systems.[1] In $3d$ transition metal oxides, the electronic states are reasonably well localized due to strong Coulomb interactions. The localized electronic states can be described in terms of crystal-field states, such as triply degenerate t_{2g} (xy, yz, and zx) and doubly degenerate e_g (x^2-y^2 and $3z^2-r^2$) orbital states for octahedral environments. The degeneracy between these crystal-field states can be further lifted by electron–phonon interactions (e.g., Jahn–Teller effect). The energy scale of the level splitting of t_{2g} (e_g) orbital states is typically ~0.1 eV (~1 eV), which is much larger than that of spin–orbit coupling (~20 meV).[2] Therefore, in $3d$ transition metal oxides, the orbital angular momentum is completely quenched by degeneracy lifting, and the electronic state can be described in terms of the spin-only Hamiltonian.

The magnitude of spin–orbit coupling increases approximately as the fourth power of the atomic number. The spin–orbit coupling of $5d$ elements is an order of magnitude larger than that of $3d$ transition metals;[3] thus, it can be particularly important in $5d$ transition metal oxides. Indeed, a theoretical study showed that spin–orbit coupling induced formation of a distinct and narrow band at the Fermi level, with profound implications in the charge-density wave formation and Mott transition in $1T$-TaS_2.[4] When spin–orbit coupling is very strong, the spin and orbital angular momentum can become entangled, producing an electronic state that can be described by total angular momentum. Therefore, in some $5d$ transition metal oxides with strong spin–orbit coupling, the physics can be drastically different from systems in which spin–orbit coupling acts as a minor perturbation.

The role of spin–orbit coupling has recently attracted attention from the condensed matter physics community. In particular, the recent discovery of a novel spin–orbit coupling-induced $J_{eff} = 1/2$ Mott state in iridate compounds has stimulated intense research on novel quantum behavior in $5d$ transition metal oxides.[5,6] A theoretical study predicts that the $J_{eff} = 1/2$ Mott state leads to the low-energy quantum Hamiltonian of the Kitaev model, relevant for quantum computations.[7] Additionally, spin–orbit coupling is believed to play a crucial role in achieving the robust conducting edge state on the boundary of two-dimensional (2D) band insulators, so-called topological insulators.[8] Indeed, theoretical predictions have shown that $5d$ Na_2IrO_3 with a honeycomb lattice could exhibit the quantum spin Hall effect at room temperature due to the large spin–orbit coupling of $5d$ Ir ions.[9] In this chapter, we review recent spectroscopic and theoretical results that have revealed the critical roles of spin–orbit coupling in $4d$ and $5d$ transition metal oxides.

2.2. Spin–orbit Coupling-induced Fermi Surface Modification in $4d$ Sr_2RuO_4 and Sr_2RhO_4

The electron correlation is expected to be weaker in $4d$ transition metal oxides compared with $3d$ transition metal oxides, and thus, the Fermi surfaces of metallic $4d$ transition metal oxides should be well described by simple band theory calculations. Therefore, we considered two $4d$ transition metal oxides with layered perovskite structures, Sr_2RuO_4 and Sr_2RhO_4.

Sr_2RuO_4 is the one of the most extensively studied $4d$ transition metal oxides due to its unconventional superconductivity.[10] The formal valence of the Ru ion is 4+; it has four $4d$ electrons. Under octahedral crystal-field energy of 10Dq, the $4d$ orbital states split into the t_{2g} and e_g orbital states. In $4d$ and $5d$ transition metal oxides, the magnitude of 10Dq is ~3 eV, which is sufficient to yield low-spin states.[11,12] Therefore, four d electrons of Sr_2RuO_4 are expected to occupy the lower lying t_{2g} orbital states, resulting in a metallic ground state with partially filled, wide t_{2g} bands.

Comprehensive information on the Fermi surface of Sr_2RuO_4 was initially obtained by the de Haas–van Alphen experiment.[13] Later, angle-resolved photoemission spectroscopy (ARPES) confirmed the de Haas–van Alphen results.[14] These experimental results also agreed well with density-functional band-structure calculations.[15,16] Figure 2.1(c) shows

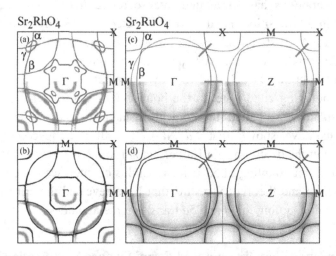

Figure 2.1. Fermi surfaces of (a) Sr_2RhO_4 and (c) Sr_2RuO_4 from local-density approximation (LDA) calculations without spin–orbit coupling. Fermi surfaces of (b) Sr_2RhO_4 and (d) Sr_2RuO_4 from LDA calculations with spin–orbit coupling. Solid lines correspond to the theoretical Fermi surface. The experimental Fermi surfaces obtained by angle-resolved photoemission spectroscopy (ARPES, thick gray line) are overlaid.[17]

the Fermi surface of Sr_2RuO_4 obtained by local-density approximation (LDA) calculations without spin–orbit coupling.[17] The LDA Fermi surface agreed relatively well with that from ARPES for Sr_2RuO_4. The Fermi surface of Sr_2RuO_4 consisted of three sheets, labeled α, β, and γ. The α and β sheets were composed of one-dimensional (1D)-like yz and zx orbitals, which exhibit anticrossing behavior along the zone diagonal. The γ sheet was mainly composed of the xy orbital and was highly 2D, indicating that spin–orbit coupling played a minor role in determining the Fermi surface of Sr_2RuO_4.

Rh is located to the right of Ru in the periodic table. Thus, the Ru ion in Sr_2RhO_4 should have five $4d$ electrons, all of which are expected to partially fill the lower lying t_{2g} orbital states. Considering the relative unimportance of spin–orbit coupling in describing the Fermi surface of Sr_2RuO_4, one might expect it to play a minor role in describing the Fermi surface of Sr_2RhO_4. However, the experimental Fermi surface obtained by ARPES and de Haas–van Alphen measurements contradicts this suggestion.[18,19] Figure 2.1(a) shows the Fermi surface of Sr_2RhO_4 obtained by LDA calculations without spin–orbit coupling.[17] This theoretical result was inconsistent with the experimental Fermi surface obtained by ARPES, marked by the thick gray line in Fig. 2.1(a).

The spin–orbit coupling constant is $\zeta_{SO} = 191$ meV (161 meV) for Rh^{4+} (Ru^{4+});[17] thus, the effect of spin–orbit coupling should not be neglected. Indeed, LDA calculations that included spin–orbit coupling (LDA+SOC) showed closer agreement with experimental results, as shown in Fig. 2.1(b): the γ sheet was completely absent, and α and β sheets were significantly smaller. This calculation demonstrated the importance of spin–orbit coupling in determining the electronic structure of Sr_2RhO_4.

Note that the effect of spin–orbit coupling is much weaker in Sr_2RuO_4 than in Sr_2RhO_4, although the magnitudes of ζ_{SO} for the two compounds are comparable. Closer inspection of the band dispersion provides more information on the effect of spin–orbit coupling in Sr_2RhO_4 and Sr_2RuO_4. Figure 2.2(b) and 2.2(d) show band dispersions from LDA and LDA+SOC calculations for Sr_2RhO_4, respectively. Note that the LDA band dispersion near the Fermi level is much weaker in Sr_2RhO_4 than in Sr_2RuO_4 due to structural distortion of Sr_2RhO_4 (*i.e.*,

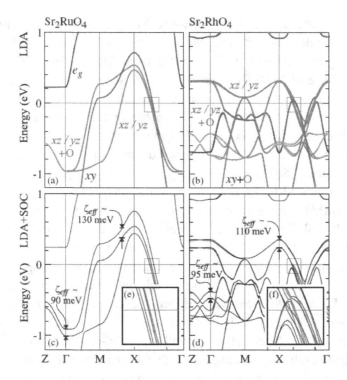

Figure 2.2. (a) Local-density approximation (LDA) and (c) local-density approximation with spin–orbit coupling (LDA+SOC) band dispersions of Sr_2RuO_4. (b) LDA and (d) LDA+SOC band dispersions of Sr_2RhO_4. (e), (f) Enlarged plot of the marked regions near E_F in (c),(d).[17]

rotation of RhO_6 octahedra). The Rh–O–Rh bond angle of Sr_2RhO_4 is about 160°,[20] which is much smaller than the Ru–O–Ru bond angle, 180°, of Sr_2RuO_4.[21] The rotation induced hybridization of xy and x^2-y^2 bands, forming a gap between bonding and antibonding bands.[19] Due to this lattice distortion, the hybridized d-bands were weakly dispersed. Therefore, the splitting due to spin–orbit coupling may have substantially modified the Fermi surface of Sr_2RhO_4, as shown in Fig. 2.2(f). On the other hand, because the band dispersion in Sr_2RuO_4 was much steeper, the energy shift due to spin–orbit coupling did not significantly change the Fermi surface of Sr_2RuO_4.

An excellent agreement between the experimental and theoretical Fermi surfaces of Sr_2RhO_4 is reported by Liu *et al.*[22] Figure 2.3 illustrates

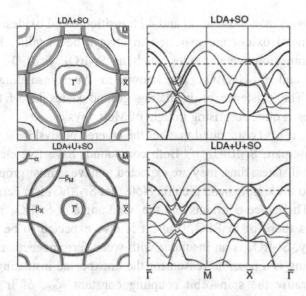

Figure 2.3. Fermi surfaces and band dispersions of Sr_2RhO_4 from local-density approximation + spin–orbit coupling (LDA+SO) and local-density approximation + Coulomb interaction + spin–orbit coupling (LDA+U+SO) calculations. Thin lines denote calculated Fermi surfaces. The thick gray line in the left panel is the experimental Fermi surface obtained by angle-resolved photoemission spectroscopy (ARPES).[22]

the significant improvement when both spin–orbit coupling and Coulomb interactions were included with proper structural information. These results suggest that there is an interplay between spin–orbit coupling and Coulomb interactions. The calculated band dispersions in Fig. 2.3 show that the band splitting in LDA+SO calculations is further enhanced when Coulomb interactions were included.

2.3. Spin–orbit Coupling-induced $J_{eff} = 1/2$ Mott State of $5d$ Sr_2IrO_4

Mott physics has been the foundation for understanding numerous intriguing physical properties of strongly correlated electron systems, including $3d$ and $4d$ transition metal oxides.[1] In the Hubbard model, the electronic ground state of correlated electron systems can be classified based on the ratio of the on-site Coulomb interaction, U, to the bandwidth, W. Because $5d$ orbitals are more spatially extended than $3d$ and $4d$ orbitals, the U (W) values of $5d$ transition metal oxides should be smaller (larger) than those of $3d$ and $4d$ transition metal oxides. Thus, $5d$ transition metal oxides are expected to always be metallic. However, some $5d$ transition metal oxides, such as Sr_2IrO_4, $Sr_3Ir_2O_7$, $Y_2Ir_2O_7$, $Cd_2Os_2O_7$, and Ba_2NaOsO_6, are known to have insulating ground states.[23–28] The existence of insulating ground states in $5d$ transition metal oxides is quite surprising in light of Mott physics.

Sr$_2$IrO$_4$ is a $5d$ compound that has the layered perovskite structure as its $4d$ counterpart, Sr_2RhO_4.[29,30] Both compounds have five electrons in their d orbital states; thus, they are expected to have similar ground states according to the simple band picture. Note that Sr_2RhO_4 is a Fermi-liquid metal.[18,19] The lattice constants and the bond angle of Sr_2IrO_4 are nearly the same as those of Sr_2RhO_4; thus, it is also expected to be metallic. Surprisingly, Sr_2IrO_4 is an insulator with weak ferromagnetism.[23] Spin–orbit coupling is a possible reason for the unexpected insulating state of Sr_2IrO_4 because the spin–orbit coupling constant, ζ_{SO}, of Ir^{4+} ions is larger than that of Rh^{4+} ions by a factor of two.[6,17] In this section, we discuss the crucial role of strong spin–orbit coupling in the Sr_2IrO_4 $J_{eff} = 1/2$ Mott state and the associated phenomena. Transport, magnetic, thermal and dielectric properties of Sr_2IrO_4 and related systems are discussed in Chapter 8.

2.3.1. *Theoretical Description of the $J_{eff} = 1/2$ Mott State of Sr_2IrO_4*

2.3.1.1. *Schematic Model for the $J_{eff} = 1/2$ Mott State in the Atomic Limit*

Figure 2.4 schematically illustrates the novel $J_{eff} = 1/2$ Mott state induced by the cooperative interaction between strong spin–orbit coupling and on-site Coulomb repulsion, U.[5] Due to the large value of 10Dq, the five d electrons of Sr_2IrO_4 occupy the lower lying t_{2g} orbital states, and the system is expected to be a metal with partially filled, wide t_{2g} bands. As in $3d$ or $4d$ transition metal oxides, large U is required to yield a typical $S = 1/2$ Mott insulating state, as shown in Figs. 2.4(a) and 2.4(b). However, the larger W and smaller U values of $5d$ transition metal oxides alone cannot yield the $S = 1/2$ Mott state based on conventional Mott physics.

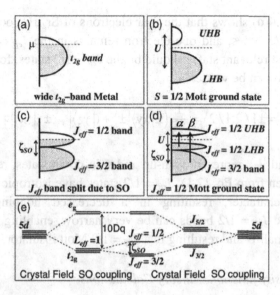

Figure 2.4. Schematic illustration of the $J_{eff} = 1/2$ Mott insulating state with $5d$ t_{2g}^5 electronic configuration. Band diagrams (a) without spin–orbit coupling and U, (b) with unrealistically large U but no spin–orbit coupling, (c) with spin–orbit coupling but no U, and (d) with both spin–orbit coupling and U. Possible optical transitions α and β are denoted by arrows. (e) Energy-level splitting of the $5d$ orbital states due to crystal-field energy and spin–orbit coupling.[5]

When the spin–orbit coupling is included, the partially filled t_{2g} orbital states split; when the spin–orbit coupling is very strong, the resulting electronic state can be described by total angular momentum, J. For the d^5 electronic configuration without crystal-field splitting, J can be either 5/2 or 3/2. However, in the low-spin states of Sr_2IrO_4, the crystal field is very large, and only triply degenerate t_{eg} orbitals are involved. Then, t_{2g} orbital states can be mapped into effective angular momentum $L_{eff} = 1$ states. More explicitly, $(|yz\rangle \pm i|zx\rangle)/\sqrt{2}$ states correspond to $L_{eff}^z = \pm 1$ and the $|xy\rangle$ state corresponds to $L_{eff}^z = 0$. As schematically shown in Fig. 2.4(e), with strong spin–orbit coupling, t_{2g} orbital states split into the effective total angular momentum doubly degenerate $J_{eff} = 1/2$ and quadruply degenerate $J_{eff} = 3/2$ states. It should be noted that the energy levels of $J_{eff} = 3/2$ states are lower than those of the $J_{eff} = 1/2$ states because the $J_{eff} = 1/2$ states split off from the $J = 5/2$ states.

Figure 2.4(d) shows that the four electrons of Sr_2IrO_4 occupy lower lying $J_{eff} = 3/2$ states, and one electron remains in the $J_{eff} = 1/2$ states. Thus, the most relevant states should be the $J_{eff} = 1/2$ states, for which the wavefunctions can be written as

$$\left| J_{eff} = \pm 1/2, 1/2 \right\rangle = \frac{1}{\sqrt{3}} \left(\pm \left| xy \right\rangle \left| \pm \right\rangle + \left(\left| yz \right\rangle \left| \mp \right\rangle \pm i \left| zx \right\rangle \left| \mp \right\rangle \right) \right), \quad (2.1)$$

where $\left| + \right\rangle$ and $\left| - \right\rangle$ represent spin-up and spin-down states, respectively. As can be seen from Eq. (2.1), $J_{eff} = 1/2$ states have isotropic orbital and mixed-spin character, resulting in a decreased hopping integral. Therefore, the $J_{eff} = 1/2$ bands can be very narrow, enabling a Mott gap opening of small U and resulting in a $J_{eff} = 1/2$ Mott insulator.

2.3.1.2. *Density-functional-theory Calculations*

Density-functional-theory calculations within the LDA scheme support the $J_{eff} = 1/2$ Mott state of Sr_2IrO_4. Figure 2.5 shows the calculated Fermi surfaces and band structures from the LDA and LDA+U calculations on Sr_2IrO_4 with and without including spin–orbit coupling. The LDA calculation predicted that the ground state is metallic with wide bands of t_{2g} orbital character crossing the Fermi level, E_F (Fig. 2.5(a)). The calculated Fermi surface was nearly the same as that of its $4d$ counterpart, Sr_2RhO_4,[18,19] which was expected from the d^5 configuration and nearly identical structural distortion (i.e., rotation of RhO_6 and IrO_6 octahedra).

When the spin–orbit coupling was included, the band structure changed drastically. At high-symmetry points in k-space, such as M, the highly degenerate bands in LDA calculations were split by spin–orbit coupling; the t_{2g} bands split into two narrow $J_{eff} = 1/2$ bands and broader $J_{eff} = 3/2$ bands. In the LDA band structure, the contribution of xy components above the Fermi level were strongly suppressed relative to those of the yz and zx states, whereas the Ir $5d$ bands ranging from −2.5 to 0.5 eV were still dominated by t_{2g} orbital components. On the other hand, for spin–orbit coupling-induced $J_{eff} = 1/2$ states, all three t_{2g} orbital components were almost equally distributed throughout the LDA+SO band structure.

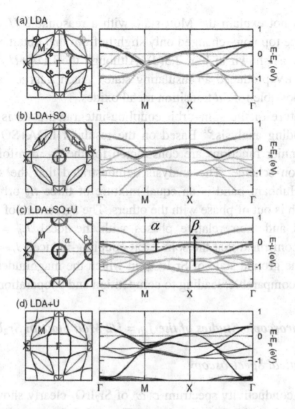

Figure 2.5. Theoretical Fermi surfaces and band dispersions of Sr_2IrO_4 from (a) local-density approximation (LDA), (b) local-density approximation + spin–orbit coupling (LDA+SO), (c) local-density approximation + spin–orbit coupling + Coulomb interaction (LDA+SO+U), and (d) local-density approximation + Coulomb interaction (LDA+U). The value of U was 2 eV. In (c), the topology of the valence band maxima (binding energy $E_B = 0.2$ eV) are shown instead of the Fermi surface.[5] Possible optical transitions labeled α and β are indicated by the arrows in (c). The transition α corresponds to that from the lower Hubbard band (LHB) to the upper Hubbard band (UHB) of the $J_{eff} = 1/2$ states. The transition β corresponds to that from the $J_{eff} = 3/2$ bands to the UHB of the $J_{eff} = 1/2$ states. Because the LHB and UHB have little dispersion and are parallel, the optical transition α should be very narrow.

The narrowness of the $J_{eff} = 1/2$ bands in Fig. 2.5(b) implies that Mott instability can be generated with U smaller than the LDA W value. Including $U = 2$ eV caused the $J_{eff} = 1/2$ bands to split into lower and upper Hubbard bands (LHB and UHB), which were nondispersive and parallel. It should be noted that LDA+U calculations without spin–orbit coupling do not explain the Mott state with a reasonable U value. The Fermi surface topology changed only slightly from that obtained by LDA calculations, as shown in Fig. 2.5(d). Although the LDA+U calculation with $U = 6$ eV produced an insulating state,[31] such a large value is not physically possible for $5d$ transition metal oxides.

The nature of the spin–orbit coupling-integrated state is identified by tight-binding analysis.[31] Based on the result of LDA+SO+U calculations, Wannier functions are constructed for the t_{2g} manifold through the projection scheme. The analysis demonstrated that the spin–orbit-integrated Hubbard band is an equal mixture of three t_{2g} orbital states, one of which is out of phase with the others. The agreement of the orbital components and their relative phases with the ideal $J_{eff} = 1/2$ state (Eq. (2.1)) confirmed that the spin–orbit coupling-induced $J_{eff} = 1/2$ state should be the ground state of Sr_2IrO_4, in which the magnitudes of U, W, and ζ_{SO} are comparable, leading to competition and cooperation.

2.3.2. *Spectroscopic Studies of the $J_{eff} = 1/2$ Mott State in Sr_2IrO_4*

2.3.2.1. *Optical Spectroscopy*

The optical conductivity spectrum $\sigma(\omega)$ of Sr_2IrO_4 clearly shows a novel feature of the $J_{eff} = 1/2$ Mott insulator.[6] Figure 2.6 shows $\sigma(\omega)$ of Sr_2IrO_4 at room temperature. Sr_2IrO_4 exhibited insulating behavior with an optical gap of about 0.3 eV. More interestingly, $\sigma(\omega)$ of Sr_2IrO_4 displayed unique spectral features that have not been observed in other correlation-induced Mott insulators: a double-peak structure with a sharp α peak and a broader β peak. Note that the α peak is narrower than the correlation-induced peaks in $\sigma(\omega)$ of $3d$ and $4d$ transition metal oxides by a factor of 3–5, as shown in Fig. 2.6. Considering the large spatial extent of $5d$ orbitals, it is impossible to explain such a narrow peak using the conventional Mott picture.

The peculiar spectral feature in $\sigma(\omega)$ of Sr_2IrO_4 is a natural consequence of the spin–orbit coupling-induced $J_{eff} = 1/2$ Mott insulating state. Figure 2.5(c) shows that the LHB and the UHB displayed little dispersion and were parallel. Therefore, the optical transition between the two Hubbard bands should be very narrow, as the α peak. Another

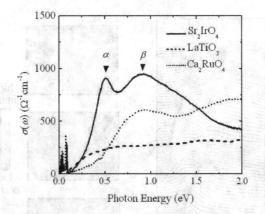

Figure 2.6. Optical conductivity $\sigma(\omega)$ of Sr_2IrO_4 at room temperature. Peak α corresponds to the transition from the lower Hubbard band (LHB) to the upper Hubbard band (UHB) of the $J_{eff} = 1/2$ states. Peak β corresponds to the transition from the $J_{eff} = 3/2$ bands to the UHB of the $J_{eff} = 1/2$ states. For comparison, $\sigma(\omega)$ of other Mott insulators, such as $3d$ LaTiO$_3$ and $4d$ Ca$_2$RuO$_4$, are shown.[6, 32, 33]

possible optical transition occurs from the $J_{eff} = 3/2$ bands to the UHB of the $J_{eff} = 1/2$ states, *i.e.*, peak β. Due to the broader $J_{eff} = 3/2$ bands, the β peak should be broader than the α peak.

It should also be noted that the LDA+SO+U calculation with $U = 2$ eV, which is a reasonable value for $5d$ transition metal oxides, reproduced the energies of the two optical transitions quite well. The separation between the Hubbard bands was about 0.5 eV and that between the $J_{eff} = 3/2$ bands and the UHB was about 0.9 eV, in agreement with the positions of the α and β peaks in experimental $\sigma(\omega)$. Based on these peak assignments, the values of the effective U and spin–orbit coupling constant, ζ_{SO}, were estimated to be about 0.5 and 0.4 eV, respectively. Note that the values of U and ζ_{SO} for $4d$ ruthenates were estimated to be about 1 eV and 0.16 eV in previous spectroscopic studies.[17,33] As expected, U was lower and ζ_{SO} was greater for the $5d$ electrons compared with the $4d$ electrons.

2.3.2.2. *Angle-Resolved Photoemission Spectroscopy*

The band structure of the $J_{eff} = 1/2$ Mott state of Sr_2IrO_4 is also experimentally confirmed using ARPES.[5] The energy distribution curves

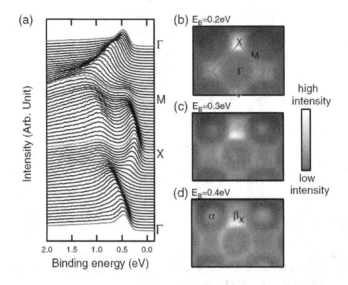

Figure 2.7. (a) Energy distribution curves of Sr_2IrO_4 up to $E_B = 2.0$ eV along high-symmetry lines. Angle-resolved photoemission spectroscopy (ARPES) intensity maps at (b) $E_B = 0.2$ eV, (c) 0.3 eV, and (d) 0.4 eV. The square in (b) represents the Brillouin zone that was reduced due to rotation of IrO_6 octahedra.[5]

in Fig. 2.7(a) show that none of the bands crossed the Fermi level, as expected for an insulator. Figure 2.7(b)–2.7(d) show the intensity maps at $E_B = 0.2$, 0.3, and 0.4 eV, respectively. The first valence band maximum, denoted as β_X, appeared at the X points. As E_B increased, another maximum (α) appeared at the Γ points. The topmost valence band, corresponding to the LHB of the $J_{eff} = 1/2$ states, had a small dispersion of about 0.5 eV, despite the large spatial extent of $5d$ orbitals. These experimental observations agreed quite well with the band dispersion predicted by LDA+SO+U calculations, as shown in Fig. 2.5(c).

2.3.2.3. *X-ray Absorption Spectroscopy*

Oxygen $1s$ X-ray absorption spectroscopy (XAS) can provide information on the formation of the $J_{eff} = 1/2$ state because it characterizes the orbital components.[5] O $1s$ XAS detects the transition from the O $1s$ orbital to the O $2p$ orbital. XAS measurements can probe unoccupied density of d orbital states that are strongly hybridized with O $2p$ orbitals.

Under normal incidence of light (*E//ab*), the XAS spectra show only *d* orbital states that are strongly coupled with O $2p_x$ and $2p_y$ orbitals. Under incident light that is mostly polarized along the out-of-plane direction (*E//c*), the corresponding XAS spectra detect *d* orbital states that are hybridized with the O $2p_z$ orbital.

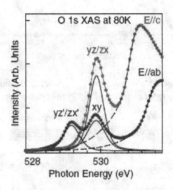

Figure 2.8. Polarization-dependent O $1s$ x-ray absorption spectroscopy (XAS) spectra of Sr_2IrO_4 measured at 80 K (solid circles). *xy* and *yz/zx* represent the transition from in-plane oxygen. *yz'/zx'* denote the transition from apical oxygen. The solid lines show calculated XAS spectra under an assumption of *xy*:*yz*:*zx* = 1:1:1.[5]

Figure 2.8 shows polarization-dependent O $1s$ XAS spectra of Sr_2IrO_4 measured at 80 K (solid circles). The two peaks at about 529 and 530 eV in *E//ab* spectra were assigned to *yz/zx* orbital states hybridized with apical O (*yz'/zx'*) and *xy* orbital states hybridized with planar O (*xy*), respectively. The peak at about 530 eV in *E//c* spectra was assigned to *yz/zx* orbital states hybridized with planar O (*yz/zx*).[34] XAS spectra (solid and dashed lines) calculated using an orbital ratio of *xy*:*yz*:*zx* = 1:1:1 reproduced the experimental data quite well within an estimation error of less than 10% for unoccupied t_{2g} orbital states. Note that the three t_{2g} orbital components are equally mixed in the unoccupied *d* bands, confirming the importance of the $J_{eff} = 1/2$ state in Eq. (2.1).

2.3.2.4. *Resonant X-ray Scattering*

The $J_{eff} = 1/2$ state should be a characteristic equal mixture of *xy*, *yz*, and *zx* orbitals with complex number *i* involved in one of the orbital

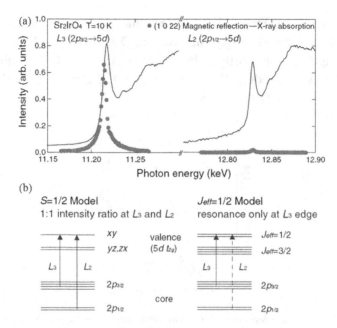

Figure 2.9. (a) Resonant enhancement of the magnetic reflection (1 0 22) at the Ir L edge. Solid lines are x-ray absorption spectroscopy (XAS) spectra at the Ir L_3 ($2p_{3/2}$) and L_2 ($2p_{1/2}$) edges. The solid circles represent the intensity of the magnetic (1 0 22) peak. (b) For the $S = 1/2$ state, the XAS intensities at the L_3 and L_2 edges are expected to be equal. On the other hand, resonant enhancement occurs only for the L_3 edge for the $J_{eff} = 1/2$ state.[35]

components, as shown in Eq. (2.1). Indeed, the complex phase in the wavefunction of the $J_{eff} = 1/2$ state was identified by resonant X-ray scattering (RXS) experiments.[35] Figure 2.9 shows the resonant enhancement of the magnetic reflection (1 0 22) at the L edge of Sr_2IrO_4. Interestingly, although the L_3 edge exhibited drastic enhancement of the magnetic reflection, resonance at the L_2 edge was very small.[35]

The scattering amplitude, f_{ab}, in the RXS process is expressed by

$$f_{ab} = \sum_m \frac{m_e \omega_{im}^3}{\omega} \frac{\langle i|R_b|m\rangle\langle m|R_a|i\rangle}{\hbar\omega - \hbar\omega_{im} + i\Gamma/2},\qquad(2.2)$$

where m_e, Γ, and R_a (R_b) are the electron mass, broadening factor, and position operators with polarization a (b), respectively. The incident photo energy is ω, and RXS can occur by second-order processes with intermediate states. In Eq. (2.2), ω_{im} is the energy difference between the initial and intermediate states. In the RXS process, the interference between various scattering channels provides information on the scattering intensity with phase sensitivity.

Calculation of the scattering amplitude, f_{ab}, for the L_3 and L_2 edges revealed that the $J_{eff} = 1/2$ state can explain the experimental observation. The general wavefunction for the t_{2g} states can be written as

$$c_1|xy\rangle|+\sigma\rangle + c_2|yz\rangle|-\sigma\rangle + c_3|zx\rangle|-\sigma\rangle, \tag{2.3}$$

where σ represents the spin state. When the distortion of IrO_6 octahedra becomes dominant, the resulting electronic state will be a pure real state with one hole in the highest energy t_{2g} state. In this case, the insulating ground state is the conventional $S = 1/2$ Mott insulator. On the other hand, when the magnitude of spin–orbit coupling is larger than the distortion-induced degeneracy lifting in the t_{2g} states, the magnitude of the c_i values will equal the c_1 real, c_2 real, and c_3 imaginary values, as in Eq. (2.1). The calculated f_{ab} for the L_2 edge using Eq. (2.3) is

$$\begin{pmatrix} (c_1+ic_3)(c_1^*-ic_3^*) & (ic_1^*+c_3)(c_1-c_2) & 0 \\ (-ic_1+c_3)(c_1^*-c_2^*) & (c_1-c_2)(c_1^*-c_2^*) & 0 \\ 0 & 0 & (c_2+ic_3)(c_2^*-ic_3^*) \end{pmatrix} \tag{2.4}$$

Because the magnetic intensity comes from the imaginary part of the off-diagonal components,[35] the intensity of the L_2 edge is zero only when $c_1 = c_2$ or $c_3 = ic_1$. Note that this condition rules out the $S = 1/2$ model. Under this condition, the scattering intensities for the L_3 edge were calculated to be

$$I_{L_3} = \left(Im(c_1 c_3^*)\right)^2 / 4 \ \text{ or } \ \left(Re(c_1^* c_2)\right)^2 / 4 \tag{2.5}$$

The large enhancement of the L_3 edge implies that the yz orbital was out of phase with the zx orbital. Additionally, the intensity difference between the L_3 and the L_2 edges was largest when the relative phase was $\pi/2$, confirming the complex phase of the $J_{eff} = 1/2$ states, shown in Eq. (2.1).

2.3.3. Temperature-dependence of the Electronic Structure of the $J_{eff} = 1/2$ Mott State

Thus far, we established the discovery of the $J_{eff} = 1/2$ Mott state of Sr_2IrO_4 using numerous spectroscopic methods. It is important to investigate the physical properties of the $J_{eff} = 1/2$ Mott state and compare them with those of the $S = 1/2$ Mott state. In this section, the temperature dependence of the optical conductivity, $\sigma(\omega)$, of Sr_2IrO_4 is discussed.

Figure 2.10(a) shows the temperature-dependent change in $\sigma(\omega)$.[36] The two peaks experienced significant spectral changes with varying temperature. As temperature increased, the α peak broadened and shifted to lower energy. At the same time, the spectral weight of the α peak increased and that of the β peak decreased with increasing temperature. These spectral changes indicate that the spin–orbit coupling-induced Mott gap decreased gradually, consistent with its resistivity behavior.[35] The temperature dependence of the resistivity followed the variable-range hopping model, and the increase in the hopping rate with increasing temperature might have been associated with the decreased optical gap.

Previous optical studies on $3d$ $S = 1/2$ Mott insulators, such as $LaTiO_3$, $YTiO_3$, and $Yb_2V_2O_7$, showed that their $\sigma(\omega)$ were weakly temperature dependent.[37–39] More interestingly, the electronic structure and Mott gap in the $\sigma(\omega)$ spectrum of $LaTiO_3$ changed slightly with temperature, although it underwent structural and magnetic transitions.[40,41] This behavior is in sharp contrast to the strong variation in the $\sigma(\omega)$ of Sr_2IrO_4 without a structural transition. The strong temperature dependence in $\sigma(\omega)$ for Sr_2IrO_4 was likely related to the large hybridization of orbitals and the resulting sensitivity to other interactions.

Interestingly, the temperature-dependent evolution of the $\sigma(\omega)$ spectrum of Sr_2IrO_4 was quite similar to that of La_2CuO_4. A narrow charge-transfer peak observed in the optical spectra of La_2CuO_4 broadened and shifted to lower energy.[42] At temperatures below 100 K, the peak did not show discernible changes, similar to the behavior of the α peak in the $\sigma(\omega)$ of Sr_2IrO_4. These temperature-induced changes were explained in terms of the electron–hole interaction and electron–phonon coupling. The similarities between the changes in the optical spectra of La_2CuO_4 and Sr_2IrO_4 suggest that electron–phonon coupling might play an important role in the changes in the optical conductivity of Sr_2IrO_4.

The temperature dependence of the optical gap, Δ_{opt}, also suggests that the electronic structure and magnetic ordering might be closely coupled. Sr_2IrO_4 shows weak ferromagnetic ordering below 240 K, originating from the canted antiferromagnetic ordering of the $J_{eff} = 1/2$ moments in the in-plane.[23, 31] As shown in Fig. 2.10(b), Δ_{opt} changed little below 200 K and decreased quickly above 200 K, which is near the magnetic transition temperature. The $\Delta_{opt}(T)$, differentiated with respect to temperature, in the inset of Fig. 2.10(b) revealed that the change in Δ_{opt} was largest near the magnetic transition temperature, as indicated by the gray rectangle.

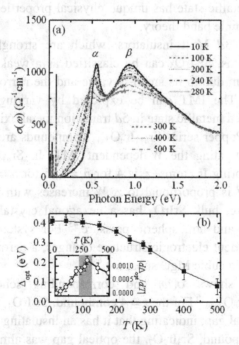

Figure 2.10. (a) Temperature-dependent optical conductivity, $\sigma(\omega)$, of Sr_2IrO_4. As temperature increased, the α and β peaks broadened, and the Mott gap decreased. The sharp spikes below the optical gap energy are due to optical phonon modes. (b) Temperature-dependent optical gap Δ_{opt}. The inset shows the value of Δ_{opt} differentiated with respect to temperature.[36]

2.4. Correlated Metallic State of 5*d* Iridates

2.4.1. *Dimensionality-controlled Insulator–Metal Transition in Ruddlesden–Popper Series* $Sr_{n+1}Ir_nO_{3n+1}$ *(n = 1, 2, and ∞)*

The effects of electron correlation, U, which give rise to various intriguing physical phenomena in strongly correlated electron systems, have been a central issue in condensed-matter physics.[1] In particular, the insulator–metal transition (IMT) and anomalous properties of the resulting metallic states are very prominent features of correlated electron systems. A simple way to induce the IMT is through bandwidth control. As W increases, becoming comparable to U, the system will change from an insulator to a metal. Near the IMT, the conducting carriers should have a large effective mass due to electron correlation. This correlated metallic state has unique physical properties that cannot be explained by simple band theory.

Compared to 3*d* Mott insulators, which are strongly correlated narrow band systems, Sr_2IrO_4 can be classified as a weakly correlated narrow band system due to its small U value and the narrowness of the J_{eff} = 1/2 bands. The IMT can be explored by varying W and the associated correlated metallic state in 5*d* transition metal oxides.

Ruddlesden–Popper series $Sr_{n+1}Ir_nO_{3n+1}$ compounds are good model systems for investigating the W-dependent IMT. In $Sr_{n+1}Ir_nO_{3n+1}$, the number of neighboring Ir atoms, z, is 4 for $n = 1$, 5 for $n = 2$, and 6 for $n = ∞$. Because W is proportional to z, W increases with increasing n. However, in nature, bulk $SrIrO_3$ has a hexagonal crystal structure at room temperature and atmospheric pressure.[43] For systematic investigation of W-dependent electronic structure changes, a perovskite $SrIrO_3$ film was grown on a cubic MgO substrate.[6,44]

Figure 2.11 shows $\sigma(\omega)$ and corresponding schematic band diagrams of $Sr_{n+1}Ir_nO_{3n+1}$.[6] For the $z = 4$ compound, Sr_2IrO_4, $\sigma(\omega)$ showed a finite-sized optical gap, indicating that it has an insulating ground state. For the $z = 5$ compound, $Sr_3Ir_2O_7$, the optical gap was almost zero. For the $z = 6$ compound, $SrIrO_3$, the optical gap disappeared, and a Drude-like response due to conducting carriers appeared in the low-frequency region. These spectral changes indicate that an IMT occurred between $Sr_3Ir_2O_7$ and $SrIrO_3$.

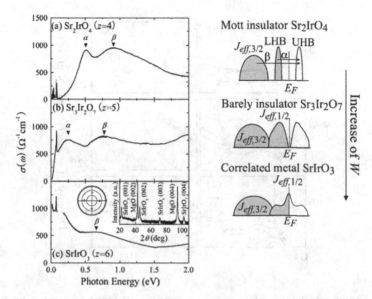

Figure 2.11. Optical conductivity, $\sigma(\omega)$, and schematic band diagrams of (a) Sr_2IrO_4, (b) $Sr_3Ir_2O_7$, and (c) $SrIrO_3$. Because of the strong phonon absorption of the MgO substrate, $\sigma(\omega)$ of $SrIrO_3$ cannot be obtained between 0.09 and 0.15 eV. The insets of (c) show the x-ray pole figure measured around the (011) reflection (left) and diffraction pattern (right) of $SrIrO_3$. The diffraction pattern shows that the $SrIrO_3$ film grew epitaxially as a perovskite structure. The pole figure shows that the film had the fourfold symmetry of a perovskite structure with (001) orientation.[6]

The LDA+U calculation, including spin–orbit coupling, demonstrated that W could be varied by changing the dimensionality, *i.e.*, z.[6] Figure 2.12(a)–2.12(c) show the band dispersions of Sr_2IrO_4, $Sr_3Ir_2O_7$, and $SrIrO_3$ from the LDA+U calculations with spin–orbit coupling, respectively. The Ir $5d$ states were the main contributors in the energy region between −1.5 and 1.0 eV. The light and dark lines represent the $J_{eff} = 1/2$ bands and the $J_{eff} = 3/2$ bands, respectively. For $Sr_3Ir_2O_7$ and $SrIrO_3$, the $J_{eff} = 1/2$ bands were split due to increasing interlayer coupling. When z increases, the neighboring Ir ions along the c-axis develop stronger hybridization of the d bands through the apical O ions. The hybridization can split the bands into bonding and antibonding states, increasing W.

Figure 2.12(d)–2.12(f) show the calculated total density of states (DOS) of Sr_2IrO_4, $Sr_3Ir_2O_7$, and $SrIrO_3$, respectively. The DOS between

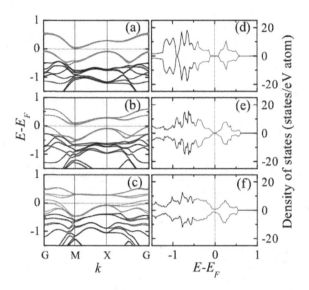

Figure 2.12. Results from local-density approximation + Coulomb interaction (LDA+U) calculations, including spin–orbit coupling: band structures of (a) Sr_2IrO_4, (b) $Sr_3Ir_2O_7$, and (c) $SrIrO_3$, and total density of states of (d) Sr_2IrO_4, (e) $Sr_3Ir_2O_7$, and (f) $SrIrO_3$. A U value of 2 eV was used, producing electronic structures consistent with experimental $\sigma(\omega)$. The positive and negative densities of states represent spin-up and spin-down bands, respectively. The light and dark lines represent the $J_{eff,1/2}$ and the $J_{eff,3/2}$ bands, respectively.[6]

−0.5 and −1.5 eV had strong contributions from the J_{eff} = 3/2 states. The DOS between −0.5 and 0.5 eV were from the J_{eff} = 1/2 states. For Sr_2IrO_4, the J_{eff} = 1/2 bands were split into the LHB and UHB due to electron correlation, U. The separation between the centers of the UHB and LHB (J_{eff} = 3/2 bands) was approximately 0.5 eV (0.9 eV), consistent with the position of the α (β) peak in Fig. 2.11(a). As z increased, the J_{eff} = 1/2 and J_{eff} = 3/2 bands clearly broadened. As the $J_{eff,1/2}$ bands broadened, the Mott gap closed; therefore, the IMT in $Sr_{n+1}Ir_nO_{3n+1}$ was likely controlled by W.

Extended Drude analysis has been commonly used to obtain electrodynamic parameters, such as the effective mass and scattering rate of the carriers in correlated materials.[45,46] In particular, near the IMT of an S = 1/2 Mott insulator, the free carriers at the metallic side have high effective mass, which is considered the hallmark of Mott instability.

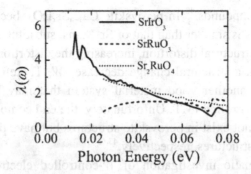

Figure 2.13. Frequency-dependent mass enhancement, $\lambda(\omega)$, of 5d SrIrO$_3$. The $\lambda(\omega)$ of 4d SrRuO$_3$ and Sr$_2$RuO$_4$ at room temperature.[6]

Then, the free carriers in the metallic state near the IMT should be highly correlated, and the corresponding compound should be in a correlated metallic state. Mott instability in 5d Sr$_{n+1}$Ir$_n$O$_{3n+1}$ was confirmed by extended Drude analysis on metallic SrIrO$_3$.[6] Figure 2.13 shows the mass enhancement, $\lambda(\omega)$, of conducting carriers in SrIrO$_3$, corresponding to the effective mass normalized by the band mass. As shown in Fig. 2.13, the $\lambda(\omega)$ of SrIrO$_3$ reached about 6 at the lowest energy. The $\lambda(\omega)$ of 3d correlated metal (Ca,Sr)VO$_3$ was reported to be approximately 3–4.[47] Note that the $\lambda(\omega)$ of SrIrO$_3$ is comparable to or larger than those of the 3d and 4d transition metal oxides,[46,47] implying that SrIrO$_3$ should be a correlated metal near the Mott transition.

These unique results originated from the large spin–coupling of 5d transition metal ions. Theoretical studies on the 4d counterpart, Sr$_2$RhO$_4$, showed that spin–coupling could only modify the Fermi surface within the metallic phase.[17,22] On the other hand, larger spin–coupling in 5d systems drastically enhanced the effect of electron correlation, and this cooperative interaction drives some 5d systems much closer to Mott instability.

2.4.2. Electronic Structure Evolution in the Bandwidth-controlled Ca$_{1-x}$Sr$_x$IrO$_3$ System

The W-controlled electronic structure change in the spin–orbit coupling-induced $J_{eff} = 1/2$ state was also found in perovskite-type

$Ca_{1-x}Sr_xIrO_3$ compounds.[48] In perovskite $Ca_{1-x}Sr_xIrO_3$, because the ionic size of Ca^{2+} ions is smaller than that of Sr^{2+} ions, substitution of Sr^{2+} with Ca^{2+} results in structural distortion, increasing the distortion angle of IrO_6 octahedra. Such a structural change decreases W. Therefore, perovskite $Ca_{1-x}Sr_xIrO_3$ is another good material system that may be capable of tuning the W-controlled IMT. Unfortunately, the end compounds, $CaIrO_3$ and $SrIrO_3$, do not exist in perovskite structures but have post-perovskite and hexagonal structures, respectively.[43,49]

For systematic investigation of W-controlled electronic structure changes, perovskite-type $Ca_{1-x}Sr_xIrO_3$ compounds were artificially synthesized using an epitaxial stabilization technique.[48] Figure 2.14 shows the temperature-dependent resistivity, $\rho(T)$, of perovskite $CaIrO_3$, $Ca_{0.5}Sr_{0.5}IrO_3$, and $SrIrO_3$ thin films grown on $GdScO_3(110)$ substrates. Interestingly, $d\rho/dT$ changed from positive to negative as x decreased. Although both $SrIrO_3$ and $Ca_{0.5}Sr_{0.5}IrO_3$ exhibited metallic behavior ($d\rho/dT>0$), the absolute value of ρ of $Ca_{0.5}Sr_{0.5}IrO_3$ was slightly larger than that of $SrIrO_3$. On the other hand, $CaIrO_3$ exhibited insulator-like behavior ($d\rho/dT<0$), and the absolute value of the conductivity was above the Mott minimum value. Compared with the highly metallic character of $4d$ perovskite $Ca_{1-x}Sr_xRuO_3$ and $Ca_{1-x}Sr_xRhO_3$ compounds,[50–52] $5d$ $Ca_{1-x}Sr_xIrO_3$ compounds have much higher resistivity values. Therefore, the $5d$ $Ca_{1-x}Sr_xIrO_3$ compounds should be located near Mott instability.

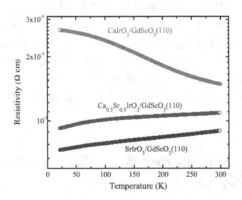

Figure 2.14. Resistivity, $\rho(T)$, of $CaIrO_3$, $Ca_{0.5}Sr_{0.5}IrO_3$, and $SrIrO_3$ thin films grown on $GdScO_3$ (110) substrates.[46]

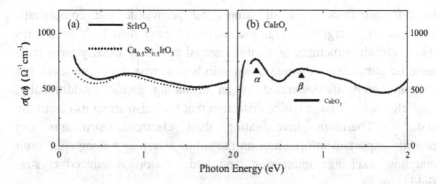

Figure 2.15. (a) Optical conductivity, $\sigma(\omega)$, of $SrIrO_3$ and $Ca_{0.5}Sr_{0.5}IrO_3$, and (b) $CaIrO_3$ thin films grown on $GdScO_3(110)$ substrates. In (b), $\sigma(\omega)$ below 0.1 eV was obtained from the fit of reflectivity of $CaIrO_3$ thin films on $SrTiO_3(001)$ substrates.[48]

Figure 2.15(a) and 2.15(b) show $\sigma(\omega)$ of $Ca_{1-x}Sr_xIrO_3$ compounds. For $SrIrO_3$, a distinct Drude-like response appeared due to conducting carriers. For $Ca_{0.5}Sr_{0.5}IrO_3$, a Drude-like response was observed, but its intensity was suppressed. In contrast, in $\sigma(\omega)$ of $CaIrO_3$, a peak (α) at about 0.2 eV appeared instead of the Drude-like response. The spectral feature for $CaIrO_3$ was qualitatively similar to that of Sr_2IrO_4. Indeed, LDA+U+SO calculations on $CaIrO_3$ demonstrated that the insulator-like electronic state was achieved by the cooperative interaction between U and spin–orbit coupling.[48] The α (β) peaks were assigned to the transition from the LHB of the $J_{eff} = 1/2$ states ($J_{eff} = 3/2$ bands) to the UHB of the $J_{eff} = 1/2$ states, as in Sr_2IrO_4. These findings suggest that $Ca_{1-x}Sr_xIrO_3$ was located near the Mott transition due to strong spin–orbit coupling of $5d$ Ir ions.

2.5. Roles of Spin–orbit Coupling in Double Perovskite Rhenates and Other Iridates

2.5.1. *Double perovskite A_2FeReO_6 (A = Ba, Ca)*

Double perovskite A_2FeReO_6 (A = Ba and Ca) is another class of interesting materials for investigating the roles of spin–orbit coupling because these materials contain $5d$ Re ions. Because these compounds have both $3d$ Fe and $5d$ Re ions, the effect of spin–orbit coupling could be different from those of simple $5d$ perovskite-type compounds. Moreover, the large exchange interaction of more than 1 eV dominates the electronic structures of double perovskites; they usually have half-metallic ground states where down-spin bands cross the Fermi level, but up-spin bands are separated by an insulating gap.[53,54] Additionally, Ca_2FeReO_6 has a large lattice distortion that may also affect its electronic state.[55,56] Therefore, investigating their electronic structures can provide important information on the subtle interplay among spin–orbit coupling, exchange interaction, and lattice-distortion-induced crystal-field splitting.

Figure 2.16 shows $\sigma(\omega)$ of A_2FeReO_6 at room temperature.[57] Although the $\sigma(\omega)$ spectra of Ba_2FeReO_6 and Ca_2FeReO_6 show common charge-transfer peaks (β and B), their low-energy spectral features are quite different. $\sigma(\omega)$ of Ba_2FeReO_6 shows a Drude-like peak centered at zero frequency and a peak (α) at about 0.7 eV. In $\sigma(\omega)$ of Ca_2FeReO_6, the low-energy spectral weight was significantly suppressed, with a peak (A) at about 1 eV.

The LDA+U+SO calculations reproduced the prominent features revealed in $\sigma(\omega)$ and demonstrated the important role of spin–orbit coupling.[57] For Ba_2FeReO_6, spin–orbit coupling-induced large changes in the electronic structure, *i.e.*, split-off of two nearly parallel bands (thick lines in Fig. 2.17(b)) near E_F, as shown in Figs. 2.17(a) and

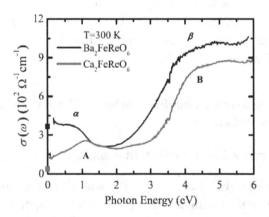

Figure 2.16. (Color online) $\sigma(\omega)$ of Ba_2FeReO_6 (black) and Ca_2FeReO_6 (red). Dark and light squares correspond to direct-current (DC) conductivity.[56]

2.17(b). The optical transition between these two parallel bands corresponds to the distinct α peak in $\sigma(\omega)$. The strong α peak cannot be explained without spin–orbit coupling. Therefore, the results of LDA+U+SO calculations indicate that spin–orbit coupling played an important role in determining the electronic structure of Ba_2FeReO_6.

On the other hand, structural distortion played very important roles in determining general aspects of the electronic structure of Ca_2FeReO_6. Note that the experimental value of the Fe–O–Re bond angle is about $153°$.[56] By including these structural properties in the LDA+U calculation, a nearly open gap was predicted, as shown in Fig. 2.17(c). The

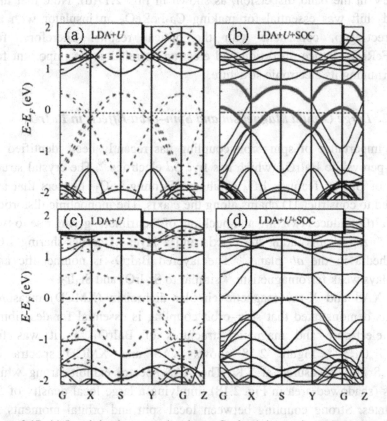

Figure 2.17. (a) Local-density approximation + Coulomb interaction (LDA+U) and (b) local-density approximation + Coulomb interaction + spin–orbit coupling (LDA+U+SO) for Ba_2FeReO_6. (c) LDA+U and (d) LDA+U+SO for Ca_2FeReO_6. The values of $U_{eff}=5$ and 3 eV were used for Fe and Re ions, respectively.[57]

prediction of a near gap-like structure in the LDA+U calculation
originated from band narrowing due to significant structural distortion.
Even without spin–orbit coupling, the lattice distortion should play a
major role in determining electronic structure.

Although the effect of spin–orbit coupling in Ca_2FeReO_6 is not as
pronounced as in Ba_2FeReO_6, it is still important in describing the
detailed optical properties. As shown in Fig. 2.17(c), both the valence
and conduction bands touch the Fermi level, E_F, at different k positions.
Thus, without spin–orbit coupling, the predicted ground state should be
metallic. The inclusion of spin–orbit coupling caused a shift of about
0.2 eV in the band dispersion, as shown in Fig. 2.17(d). Note that this
band shift was essential for making Ca_2FeReO_6 an insulator with an
indirect gap, consistent with the optical results. Therefore, for
Ca_2FeReO_6, lattice distortion and electron correlation are important for
determining its electronic structure.

2.5.2. *Large Orbital Magnetism and Spin–orbit Effects in BaIrO₃*

The importance of spin–orbit coupling has recently been identified in
non-perovskite $BaIrO_3$, which has five $5d$ electrons.[58] The crystal structure of $BaIrO_3$ features IrO_6 octahedra forming Ir_3O_{12} clusters that are
linked to construct 1D chains along the c axis. The monoclinic distortion
in $BaIrO_3$ induces twisting and buckling of the trimers, giving rise to two
1D zigzag chains along the c axis and a layer of corner-sharing IrO_6
octahedra in the ab plane.[59] The layered $BaIrO_3$ is nonmetallic and
displays weak ferromagnetism,[59] similar to Sr_2IrO_4 and $Sr_3Ir_2O_7$.

XAS and X-ray magnetic circular dichroism (XMCD) measurements demonstrated that spin–orbit coupling is essential for describing
the electronic and magnetic structures of $BaIrO_3$, as it was for
$Sr_{n+1}Ir_nO_{3n+1}$.[58] Figure 2.18 shows XAS and XMCD spectra of
$Ba_{1-x}Sr_xIrO_3$ measured at 5 K. The XAS spectra exhibit strong white
lines (shadowed area in Fig. 2.18), implying a large local density of $5d$
Ir states. Strong coupling between local spin and orbital moments in
$Ba_{1-x}Sr_xIrO_3$ is required to analyze the branching ratio, $BR = I_{L_3}/I_{L_2}$,
where I_{L_2,L_3} is the integrated intensity at a particular edge. The BR for
$Ba_{1-x}Sr_xIrO_3$ is about 4. This value is larger than the BR of 2 for metallic

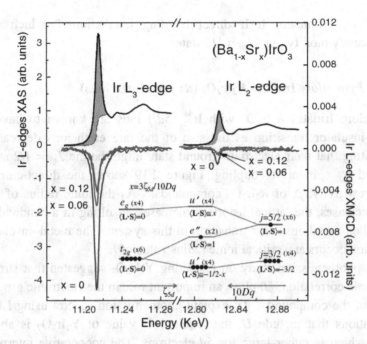

Figure 2.18. X-ray absorption spectroscopy (XAS, left axis) and X-ray magnetic circular dichroism (XMCD, right axis) spectra at Ir $L_{3,2}$ edges for $Ba_{1-x}Sr_xIrO_3$. The inset shows a schematic energy diagram of how the d levels split under crystal-field and spin–orbit coupling.[58]

Ir compounds.[60–62], indicating the local nature of spin and orbital moments and their coupling in $Ba_{1-x}Sr_xIrO_3$.

Further insight into the importance of spin–orbit coupling can be obtained from the expectation value of the spin–orbit coupling $\langle \vec{L} \cdot \vec{S} \rangle$. The number of holes, n_h, can be estimated from the relationship $BR = (2 + r)/(1-r)$ and $r = \langle \vec{L} \cdot \vec{S} \rangle / \langle n_h \rangle$.[63] Because the value of n_h is about 5 for Ir^{4+} ions, the resulting $\langle \vec{L} \cdot \vec{S} \rangle$ value was estimated to be 2. Note that the $\langle \vec{L} \cdot \vec{S} \rangle$ value for pure $J_{eff} = 1/2$ states with one hole should be 1. This discrepancy can be reconciled by including the interactions among spin–orbit coupling, Coulomb interactions, and octahedral crystal fields; configuration interaction calculations including these interactions indeed reproduced the experimental value of $\zeta_{so} = 0.3$ eV.[59] It is thought that both $J_{eff} = 1/2$ and $J_{eff} = 3/2$ states contribute to the magnetic moment and provide $\langle \vec{L} \cdot \vec{S} \rangle$ of about 2. This result is not surprising because there should be prominent Ir–Ir direct bonding interactions,[64] which can significantly modify pure $J_{eff} = 1/2$ states.

2.5.3. *Pyrochlore Iridates $R_2Ir_2O_7$ (R: rare earth ions)*

Pyrochlore Iridates $R_2Ir_2O_7$ with Ir^{4+} ($5d^5$) ions are known to have a metal–insulator transition as the size of the rare earth ions decreases, suggesting that their electronic ground state might be the $J_{eff} = 1/2$ state induced by spin–orbit coupling. Figure 2.19 shows the direct-current (DC) resistivity, ρ, of $R_2Ir_2O_7$ compounds.[25] As the ionic radius of R^{3+} ions decreases, the lattice distortion increases, resulting in a bandwidth-controlled metal–insulator transition in this system. The metal–insulator transition occurs at a critical ionic radius value of R^{3+}.

A photoemission study on insulating $Y_2Ir_2O_7$ suggested that strong electronic correlation, U, plays an important role in the electronic ground states of the compound.[65] To explain photoemission spectra using LDA calculations that include U, the required U value of $Y_2Ir_2O_7$ is about 4 eV, which is rather large for $5d$ electrons. The cooperative interplay

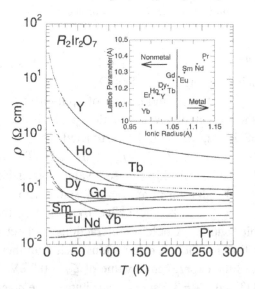

Figure 2.19. Direct-current (DC) resistivities of $R_2Ir_2O_7$ (R^{3+}: rare earth ions). Inset shows the relationship between the ionic radius of R^{3+} and the cubic lattice constants.[21]

between strong spin–orbit coupling and electron correlation might explain the insulating ground state of $Y_2Ir_2O_7$ with a reasonable U value for a system with $5d$ electrons, similar to $Sr_{n+1}Ir_nO_{3n+1}$ compounds.[6]

Indeed, a recent theoretical study demonstrated the importance of spin–orbit coupling in pyrochlore iridates.[66] Interestingly, the interplay between spin–orbit coupling and electron correlation can give rise to a distinct phase of matter, topological insulators, in the pyrochlore lattice. Figure 2.20 shows the predicted phase diagram of pyrochlore iridates. For weak electron correlation and weak spin–orbit coupling, the system has a metallic ground state. When spin–orbit coupling is large, the system becomes a topological band insulator. Theoretical calculations also predict that increasing electron correlation will lead to the appearance of a new topological Mott insulator (TMI) phase.

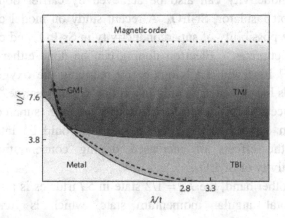

Figure 2.20. Phase diagram as a function of electron correlation, U, and spin–orbit coupling, λ. The four main phases with moderately strong U are the metal, the topological band insulator (TBI), the topological Mott insulator (TMI), and the gapless Mott insulator (GMI). The dashed line represents a zero-gap semiconductor state resulting from an accidental gap closing. The dotted line denotes the large-U region, where magnetic ordering is expected.[66]

2.6. Future Studies

The role of strong spin–orbit coupling and the associated novel physical phenomena in several $4d$ and $5d$ transition metal oxides were reviewed in this chapter. Spin–orbit coupling is an essential component in determining electronic and magnetic ground states of $4d$ and $5d$ transition metal oxides. In $4d$ Sr_2RuO_4 and Sr_2RhO_4, spin–orbit coupling modified their Fermi surface topologies. In $5d$ Sr_2IrO_4, the effect of spin–orbit coupling was much more dramatic. The large spin–orbit coupling in $5d$ Ir ions drastically enhanced the effect of electron correlation by inducing a very narrow $J_{eff} = 1/2$ band near the Fermi level and placing the iridates close to Mott instability.

Strong spin–orbit coupling applies the physics of strongly correlated $3d$ transition metal oxides to weakly correlated $5d$ transition metal oxides, i.e., Mott physics. Note that high-temperature superconductivity in $3d$ cuprates was induced by carrier doping in single-band Mott insulators. The spin–orbit coupling-induced single-band $J_{eff} = 1/2$ state can be regarded as a t_{2g} analogue of the cuprates. In this regard, one may expect that superconductivity can also be achieved by carrier doping of the $J_{eff} = 1/2$ Mott insulator, Sr_2IrO_4. A recent study on model calculations discussed the possibility of superconductivity in Sr_2IrO_4 and compared it with that of cuprates.[67] Electron doping can be done either by substitution of Sr^{2+} ions with La^{3+} ions or by reducing the oxygen content. Both methods have been used only in a dilute doping regime.[68,69] Doping greatly reduced the electrical resistivity, but the superconductivity has not been experimentally reported yet. It would be interesting to investigate the effects of increased doping concentration on the superconductivity of Sr_2IrO_4.

On the other hand, the $J_{eff} = 1/2$ state in $5d$ iridates is a spin–orbital entangled total angular momentum state, which is fundamentally different from the spin-only state of $3d$ transition metal oxides. This new quantum state is predicted to induce the quantum spin Hall effect in Na_2IrO_3, which has a honeycomb lattice, even at room temperature.[9] Additionally, a dissipationless conducting edge state is also predicted to appear when the $J_{eff} = 1/2$ state is on the pyrochlore lattice.[66] These novel topological states exist on the surface of 2D systems. Recently, single-crystalline Na_2IrO_3 has been synthesized.[70] Electrical and magnetic measurements indicate that it is an antiferromagnetic Mott insulator. Experiments probing the surface state should confirm whether the quantum spin Hall effect could exist in this compound. A study on polycrystalline pyrochlore iridates also showed that they might be able to achieve the $J_{eff} = 1/2$ state.[25] Synthesis of high-quality single crystals of insulating pyrochlore iridates is still anticipated. Experiments on single-crystalline pyrochlore iridates should prove the existence of the $J_{eff} = 1/2$ state and possibly the robust conducting edge state of topological insulators.

Acknowledgments

We appreciate the careful reading of this manuscript by Professor Young-June Kim and Mr. Da Woon Jeong. This work was supported by a National Research Foundation of Korea (NRF) grant funded by the Korea government (MEST) (No. 2009-0080567 and No. 2010-0020416).

References

1. M. Imada, A. Fujimori, and Y. Tokura, *Rev. Mod. Phys.* **70**, 1039 (1998).
2. S. Maekawa, T. Tohyama, S. E. Barnes, S. Ishihara, W. Koshibae, and G. Khaliullin, *Phyiscs of Transition Metal Oxides* (Springer, Berlin, 2004).
3. L. F. Mattheiss, *Phys. Rev.* **B13**, 2433 (1976).
4. K. Rossnagel and N. V. Smith, *Phys. Rev.* **B73**, 073106 (2006).
5. B. J. Kim, H. Jin, S. J. Moon, J.-Y. Kim, B.-G. Park, C. S. Leem, J. Yu, T. W. Noh, C. Kim, S.-J. Oh, J.-H. Park, V. Durairaj, G. Cao, and E. Rotenber, *Phys. Rev. Lett.* **101**, 076402 (2008).
6. S. J. Moon, H. Jin, K. W. Kim, W. S. Choi, Y. S. Lee, J. Yu, G. Cao, A. Sumi, H. Funakubo, C. Bernhard, and T. W. Noh, *Phys. Rev. Lett.* **101**, 226402 (2008).
7. G. Jackeli and G. Khaliullin, *Phys. Rev. Lett.* **102**, 017205 (2009).
8. D. Hsieh, D. Qian, L. Wray, Y. Xia, Y. S. Hor, R. J. Cava, and M. Z. Hasan, *Nature* **452**, 970 (2008).
9. A. Shitade, H. Katura, J. Kuneš, X.-L. Qi, S.-C. Zhang, and N. Nagosa, *Phys. Rev. Lett.* **102**, 256403 (2009).
10. Y. Maeno, H. Hashimoto, K. Yoshida, S. Nishizaki, T. Fujita, J. G. Bednorz, and F. Lichtenberg, *Nature* **372**, 532 (1994).
11. S. J. Moon, M. W. Kim, K. W. Kim, Y. S. Lee, J.-Y. Kim, J.-H. Park, B. J. Kim, S.-J. Oh, S. Nakatsuji, Y. Maeno, I. Nagai, S. I. Ikeda, G. Cao, and T. W. Noh, *Phys. Rev.* **B74**, 113104 (2006).
12. Y. S. Lee, J. S. Lee, T. W. Noh, D. Y. Byun, K. S. Yoo, K. Yamamura, and E. Takayama-Muromachi, *Phys. Rev.* **B67**, 113101 (2003).
13. A. P. Mackenzie, S. R. Julian, A. J. Diver, G. J. McMullan, M. P. Ray, G. G. Lonzarich, Y. Maeno, S. Nishizaki, and T. Fujita, *Phys. Rev. Lett.* **76**, 3786 (1996).
14. A. Damascelli, D. H. Lu, K. M. Shen, N. P. Armitage, F. Ronning, D. L. Feng, C. Kim, Z.-X. Shen, T. Kimura, Y. Tokura, Z. Q. Mao, and Y. Maeno, *Phys. Rev. Lett.* **85**, 5194 (2000).

15. T. Oguchi, *Phys. Rev* **B51**, 1385 (1995).
16. D. J. Singh, *Phys. Rev.* **B52**, 1358 (1995).
17. M. W. Haverkort, I. S. Elfimov, L. H. Tjeng, G. A. Sawatzky, and A. Damascelli, *Phys. Rev. Lett.* **101**, 026406 (2008).
18. F. Baumberger, N. J. C. Ingle, W. Meevasana, K. M. Shen, D. H. Lu, R. S. Perry, A. P. Mackenzie, Z. Hussain, D. J. Singh, and Z.-X. Shen, *Phys. Rev. Lett.* **96**, 246402 (2006).
19. B. J. Kim, J. Yu, H. Koh, I. Nagai, S. I. Ikeda, S.-J. Oh, and C. Kim, *Phys. Rev. Lett.* **97**, 106401 (2006).
20. T. Vogt and D. J. Buttrey, *J. Solid State. Chem.* **123**, 186 (1996).
21. M. Braden, A. H. Moudden, S. Nishizaki, Y. Maeno, and T. Fujita, *Physica* **C273** 248 (1997).
22. G.-Q. Liu, V. N. Antonov, O. Jepsen, and O. K. Andersen, *Phys. Rev. Lett.* **101**, 026408 (2008).
23. G.Cao, Y. Xin, C. S. Alexander, J. E. Crow, P. Schottmann, M. K. Crawford, R. L. Harlow, and W. Marshall, *Phys. Rev.* **B66**, 214412 (2002).
24. G.Cao, J. Bolivar, S. McCall, J. E. Crow, and R. P. Guertin, *Phys. Rev.* **B57**, R11039 (1998).
25. D. Yanagisima and Y. Maeno, *J. Phys. Soc. Jpn*, **70**, 2880 (2001).
26. D. Mandrus, J. R. Thomson, R. Gaal, L. Forro, J. C. Bryan, B. C. Chakoumakos, L. M. Woods, B. C. Sales, R. S. Fishman, and V. Keppens, *Phys. Rev.* **B63**, 195104 (2001).
27. W. J. Padilla, D. Mandrus, and D. N. Basov, *Phys. Rev.* **B66**, 035120 (2002).
28. A. S. Erickson, S. Misra, G. J. Miller, R. R. Gupta, Z. Schlesinger, W. A. Harrison, J. M. Kim, and I. R. Fisher, *Phys. Rev. Lett.* **99**, 016404 (2007).
29. Q. Huang, J. L. Soubeyroux, O. Chmaissem, I. Natali Sora, A. Santoro, R. J. Cava, J. J. Krajewski, and W. F. Peck, Jr, *J. Solid State. Chem.* **112**, 355 (1994).
30. M. K. Crawford, M. A. Subramanian, R. L. Halrlow, J. A. Fernandez-Baca, Z. R. Wang, and D. C. Johnston, *Phys. Rev.* **B49**, 9198 (1994).
31. H. Jin, H. Jeong, T. Ozaki, and J. Yu, *Phys. Rev.* **B80**, 075112 (2009).
32. T. Arima, Y. Tokura, and J. B. Torrance, *Phys. Rev.* **B48**, 17006 (1993).
33. J. S. Lee, Y. S. Lee, T. W. Noh, S.-J. Oh, J. Yu, S. Nakatsuji, H. Fukazawa, and Y. Maeno, *Phys. Rev. Lett.* **89**, 257402 (2002).
34. T. Mizokawa, L. H. Tjeng, G. A. Sawatzky, G. Ghiringhelli, O. Tjernberg, N. B. Brookes, H. Fukazawa, S. Nakatsuji, and Y. Meano, *Phys. Rev. Lett.* **87**, 077202 (2001).
35. B. J. Kim, H. Ohsumi, T. Komesu, S. Sakai, T. Morita, H. Takagi, and T. Arima, *Science* **323**, 1329 (2009).
36. S. J. Moon, H. Jin, W. S. Choi, J. S. Lee, S. S. A. Seo, J. Yu, G. Cao, T. W. Noh, and Y. S. Lee, *Phys. Rev.* **B80**, 195110 (2009).
37. P. Lukenheimer, T. Rudolf, J. Hemberger, A. Pimenov, S. Tachos, F. Lichtenberg, and A. Loidl, *Phys. Rev.* **B68**, 245108 (2003).
38. N. N. Kovaleva, A. V.Boris, P. Yordanov, A. Maljuk, E. Brücher, J. Strempfer, M. Konuma, I. Zegkinoglou, C. Bernhard, A. M. Stoneham, and B. Keimer, *Phys. Rev.* **B76**, 155125 (2006).

39. K. Waku, T. Suzuki, and T. Katsufuji, *Phys. Rev.* **B74**, 024402 (2006).
40. M. Cwik, T. Lorenz, J. Baier, R. Müller, G. André, F. Bourée, F. Lichtenberg, A. Freimuth, R. Schmitz, E. Müller-Hartmann, and M. Braden, *Phys. Rev.* **B68**, 060401(R) (2003).
41. J. Hemberger, H.-A. Krug von Nidda, V. Fritsch, J. Deisenhofer, S. Lobina, T. Rudolf, P. Lukenheimer, F. Lichtenberg, A. Loidl, D. Bruns, and B. Büchner, *Phys. Rev. Lett.* **91**, 066403 (2003).
42. J. P. Falck, A. Levy, M. A. Kastner, and R. J. Birgeneau, *Phys. Rev. Lett.* **69**, 1109 (1992).
43. J. M. Longo, J. A. Kafalas, and R. J. Arnott, *J. Solid State Chem.* **3**, 174 (1971).
44. Y. K. Kim, A. Sumi, K. Takahashi, S. Yokoyama, S. Ito, T. Watanabe, K. Akiyama, S. Kaneko, K. Saito, and H. Funakubo, *Jpn. J. Appl. Phys.* **45**, L36 (2006).
45. D. N. Basov and T. Timusk, *Rev. Mod. Phys.* **77**, 721 (2005).
46. J. S. Lee, S. J. Moon, T. W. Noh, S. Nakatsuji, and Y. Maeno, *Phys. Rev. Lett.* **96**, 057401 (2006).
47. H. Makino, I. H. Inoue, M. J. Rozenberg, I. Hase, Y. Aiura, and S. Onari, *Phys. Rev.* **B58**, 4384 (1998).
48. S. Y. Jang, H. Kim, S. J. Moon, W. S. Choi, B. C. Jeon, J. Yu, and T. W. Noh, *J. Phys. Condens. Matter* **22**, 485602 (2010).
49. K. Ohgushi, H. Gotou, T. Yagi, Y. Kiuchi, F. Sakai, and Y. Ueda, *Phys. Rev.* **B74**, 241104(R) (2006).
50. K. Yamamura, E. Takayama-Muromachi, *Phys. Rev.* **B64**, 224424 (2001).
51. K. Yamamura, E. Takayama-Muromachi, *Physica* **C445**, 54 (2006).
52. C. B. Eom, R. J. Cava, R. M. Fleming, J. M. Phillips, R. B. van Dover, J. H. Marshall, J. W. P. Hsu, J. J. Krajevski, W. F. Jr. Peck, *Science* **258**, 1766 (1992).
53. K.-I. Kobayashi, T. Kimura, H. Sawada, K. Terakura, and Y. Tokura, *Nature (London)* **395**, 677 (1998).
54. Z. Szotek, W. M. Temmerman, A. Svane, L. Petit, and H. Winter, *Phys. Rev. B* **68**, 104411 (2003).
55. H. Kato, T. Okuda, Y. Okimoto, Y. Tomioka, K. Oikawa, T. Kamiyama, and Y. Tokura, *Phys. Rev.* **B69**, 184412 (2004).
56. K. Oikawa, T. Kamiyama, H. Kato, and Y. Tokura, *J. Phys. Soc. Jpn.* **72**, 1411 (2003).
57. B. C. Jeon, C. H. Kim, S. J. Moon, W. S. Choi, H. Jeong, Y. S. Lee, J. Yu, C. J. Won, J. H. Jung, N. Hur, and T. W. Noh, *J. Phys.: Condens. Matter* **22**, 345602 (2010).
58. M. A. Laguna-Marco, D. Haskel, N. Souza-Neto, J. C. Lang, V. V. Krishnamurthy, S. Chikara, G. Cao, and M. van Veenendaal, *Phys. Rev. Lett.* **105**, 216407 (2010).
59. G. Cao, X. N. Lin, S. Chikara, V. Durairaj, and E. Elhami, *Phys. Rev.* **B69**, 174418 (2004).
60. Y. Jeon, B. Qi, F. Lu, and M. Croft, *Phys. Rev.* **B40**, 1538 (1989).
61. G. Schütz, R. Wienke, W. Wilhelm, W. Wagner, P. Kienle, R. Zeller, and R. Frahm, *Z. Phys. B: Condens. Matter* **75**, 495 (1989).
62. F. Wilhelm, P. Poulopoulos, H. Wende, A. Scherz, K. Baberschke, M. Angelakeris, N. K. Flevaris, and A. Rogalev, *Phys. Rev. Lett.* **87**, 207202 (2001).

63. G. van der Laan and B. T. Thole, *Phys. Rev. Lett.* **60**, 1977 (1988).
64. G. Mihály, I. Kézsmárki, F. Zámborsky, M. Miljak, K. Penc, P. Fazekas, H. Berger, and L. Forró, *Phys. Rev.* **B61**, R7831 (2000).
65. R. S. Singh, V. R. R. Medicherla, Kalaboran. Maiti, and E. V. Sampathkumaran, *Phys. Rev.* **B77**, 201102(R) (2008).
66. D. Pesin and L. Balents, *Nat. Phys.* **6**, 376 (2010).
67. F. Wang and T. Senthil, *arXiv:1011.3500* (unpublished).
68. Y. Klein and I. Terasaki, *J. Phys.: Condens. Matter* **20**, 295201 (2008).
69. O. B. Korneta, T. Qi, S. Chikara, S. Parkin, L. E. De Long, P. Schlottmann, and G. Cao, *Phys. Rev.* **B82**, 115117 (2010).
70. Y. Singh and P. Gegenwart, *Phys. Rev.* **B82**, 064412 (2010).

Chapter 3

X-RAY SCATTERING STUDIES OF 4D- AND 5D-ELECTRON TRANSITION METAL OXIDES

Ioannis Zegkinoglou

Department of Physics, University of Wisconsin, Madison
Advanced Light Source, Lawrence Berkeley National Laboratory

Bernhard Keimer

Max Planck Institute for Solid State Research, Stuttgart

Contents

3.1. Introduction

X-rays have been widely used as a probe for the investigation of condensed matter down to the atomic level for almost a century. Since the first pioneering experiments by W.L. Bragg on crystalline sodium chloride (NaCl) in 1913, x-ray diffraction has been the main tool for structure determination and material analysis for virtually every kind of material. Meanwhile, the great possibilities opened by the advances in synchrotron radiation facilities in the last three decades have enabled the use of x-ray diffraction for applications far beyond lattice structure determination, such as the investigation of magnetic properties, the observation of orbital ordering and the study of short-range charge order, to name only a few of them. This has contributed enormously to the progress of research on strongly correlated electron systems and has helped in the understanding of many of their fascinating properties.

Transition metal oxides with partially occupied $4d$ and $5d$ electron orbitals have attracted significant attention in recent years due to their intriguing electronic properties. Magnetic and/or orbital order, electronic liquid-crystal phases, unusual metal-insulator transitions, non-Fermi-liquid-behavior and superconductivity are only a few examples from the large manifold of exciting properties they exhibit. Many of them originate in a competition between many-body states with different spin, orbital and charge ordering patterns. The study of the interplay between the active degrees of freedom is of crucial importance for understanding the behavior of the systems. Resonant x-ray diffraction has the unique advantage of directly probing all active degrees of freedom – spin, orbital, charge – while being element-specific and valence-state-sensitive. It is therefore a powerful tool for investigating the physics of $4d$ and $5d$ electron systems.

This chapter will give a brief overview of the most recent advances in x-ray scattering studies of $4d$ and $5d$ transition metal compounds, with emphasis on the investigations of three ruthenium-based electron systems with long-discussed magnetic and orbital properties: the single-layered Mott transition system Ca_2RuO_4, the bilayered magnetically ordered counterpart $Ca_3Ru_2O_7$, and the ruthenocuprate $RuSr_2GdCu_2O_8$. The results of these studies underline the versatility and power of resonant x-ray diffraction.

3.2. Non-Resonant X-Ray Scattering

X-ray diffraction is the elastic, coherent scattering of x-rays by the bound electrons of a target material. In the classical picture, the electrons of the material's ions are accelerated by the electric and magnetic fields of the electromagnetic radiation and are set into oscillation. The oscillating electrons emit radiation, according to the classical electromagnetic theory, which has the same frequency as the primary x-ray beam. The emitted radiation is what we call scattered radiation.

For small crystals (so small relative to their distance from the x-ray source, that the primary x-ray beam can be treated by the plane-wave approximation), the kinematical theory of diffraction can be used for its description.[1] The theory is adequate if the diffracted beam is so weak, that the interaction of the incoming and the diffracted beams can be neglected. This means that multiple scattering effects, caused by the the re-scattering of the diffracted beam back into the direction of the incident beam, are not taken into consideration. The reduction of the amplitude of the incident wave due to the scattering of a small fraction of it into the exit beam at every atomic plane in the crystal (extinction) is also ignored.[2] The fulfillment of this condition is guaranteed by the small size of the crystals, i.e. by the small number of atomic planes in the sample.

The cross-section of elastic scattering is determined by the diagonal matrix elements of the total scattering amplitude operator **G**, averaged over the polarization states of the primary x-ray beam. In kinematical theory, the cross-section for the elastically scattered signal observed in an element of solid angle $d\Omega$ is given by the trace operation:

$$\frac{d\sigma}{d\Omega} = r_e^2 \cdot Tr\{\mu| < \mathbf{G} > |^2\} \tag{3.1}$$

where r_e is the classical radius of the electron and μ is a matrix describing the density of the photon polarization states.[3] The off-diagonal matrix elements of **G** determine the cross-sections of inelastic scattering processes.

The average value of the scattering amplitude operator multiplied by the classical electron radius r_e is called scattering length g:

$$g = r_e < \mathbf{G} > \tag{3.2}$$

The scattering length, which contains in general both charge and magnetic scattering contributions, can be expressed as a Fourier expansion in reciprocal space, in which the coefficients are unit cell structure factors:

$$g = r_e < \mathbf{G} >= -r_e \sum_{\tau} \delta(\mathbf{Q} - \tau)[(\epsilon' \cdot \epsilon)F_c(\mathbf{Q}) - i\tau \mathbf{F_s}(\mathbf{Q}) \cdot \mathbf{B}] \qquad (3.3)$$

where ϵ, ϵ' are the polarization vectors of the incoming and the scattered beams, respectively, τ is a reciprocal lattice vector, and $F_c(\mathbf{Q})$, $\mathbf{F_s}(\mathbf{Q})$ are the Fourier transforms of the time-averaged electron charge and spin densities per unit cell, respectively, for the scattering vector $\mathbf{Q}$. The vector $\mathbf{B}$ is given by:

$$\mathbf{B} = (\epsilon' \times \epsilon) - (\hat{\mathbf{k}}' \times \epsilon') \times (\hat{\mathbf{k}} \times \epsilon) + (\hat{\mathbf{k}}' \cdot \epsilon)(\hat{\mathbf{k}}' \times \epsilon') - (\hat{\mathbf{k}} \cdot \epsilon')(\hat{\mathbf{k}} \times \epsilon) \qquad (3.4)$$

where the unit vectors $\hat{\mathbf{k}}$ and $\hat{\mathbf{k}}'$ have directions along the incoming and scattered beams, respectively.

The first term of (3.3) is the contribution of the Thomson charge scattering to the total scattering length. The corresponding structure factor F_c is equal to:

$$F_c(\mathbf{Q}) = \sum_j f_c^j(\mathbf{Q})e^{i\mathbf{Q}\cdot\mathbf{R}_j} \qquad (3.5)$$

where the sum runs over all the ions j of the unit cell, and $f_c^j(\mathbf{Q})$ is the charge form factor of the ion at position $\mathbf{R}_j$.[3] The latter form factor is given by the Fourier transform of the electronic charge distribution $\rho_j(r)$ within the ion j:[4]

$$f_c^j(\mathbf{Q}) = -\frac{1}{e} \int \rho_j(\mathbf{r})e^{i\mathbf{Q}\cdot\mathbf{r}}dr \qquad (3.6)$$

The second term of (3.3) is the contribution of magnetic scattering to the total scattering length. While the charge density structure factor (3.5) is a scalar quantity, expressing (for $\mathbf{Q} = 0$) the total amount of charge in a unit cell, the spin density structure factor $\mathbf{F_s}(\mathbf{Q})$ is a vector quantity, expressing both the configuration and the orientation of the spin moments in the unit cell. In many materials it can be expressed as a sum of individual contributions from each ion of the unit cell:

$$\mathbf{F_s}(\mathbf{Q}) = \sum_j < \mathbf{S_j} > f_s^j(\mathbf{Q})e^{i\mathbf{Q}\cdot\mathbf{R}_j} \qquad (3.7)$$

where $< \mathbf{S_j} >$, $f_s^j(\mathbf{Q})$ are the spin moment and the spin form factor, respectively, of the ion located at position $\mathbf{R}_j$.

Away from resonance, the magnetic scattering amplitude is much smaller than the one of charge scattering. Non-resonant magnetic scattering intensity is typically at least six orders of magnitude weaker than the intensity of charge scattering. The two scattering processes are characterized by a phase shift of 90° and have a different influence on polarization: spin scattering causes a partial rotation of the polarization plane, while charge scattering causes no rotation, i.e. gives only $\sigma - \sigma'$ (or $\pi \rightarrow \pi'$) contribution. In the latter notation, σ is by definition the direction of the polarization component perpendicular to the diffraction plane and π the one parallel to it. These different properties can help distinguish between the charge and magnetic contributions in the scattered signal, in cases where charge and spin scattering occur at the same reciprocal lattice points (e.g. in ferromagnets). De Bergevin and Brunel investigated in detail the polarization dependence of the non-resonant magnetic scattering length and suggested for its expression the use of a basis with components perpendicular and parallel to the scattering plane.[5] This formalism is particularly useful for experimental applications and was later applied also in resonant x-ray diffraction investigations, as will be seen in Section 3.3.

In most electron systems, electrons form pairs of time-reversed orbits, thus the net orbital momentum is practically zero and no orbital scattering term needs to be considered for the calculation of the total elastic scattering length. However, in magnetically ordered systems this time-reversal symmetry is broken and the following term has to be added to the scattering length expression (3.3):

$$i\tau < \mathbf{Z} > \cdot\boldsymbol{\epsilon}' \times \boldsymbol{\epsilon} = -\frac{\tau}{\hbar k^2} < \sum_j e^{i\mathbf{Q}\cdot\mathbf{R}_j}(\mathbf{Q} \times \mathbf{p}_j) > \cdot(\boldsymbol{\epsilon}' \times \boldsymbol{\epsilon}) \qquad (3.8)$$

where $\mathbf{p}_j$ is the net momentum of ion j. The orbital scattering contribution to the scattering length is proportional to the vector product $\boldsymbol{\epsilon}' \times \boldsymbol{\epsilon}$, thus the $\sigma - \sigma'$ amplitude always vanishes.

3.3. Resonant X-Ray Diffraction

3.3.1. *Basic Principles and Historical Background*

While the atomic scattering factors are scalars in non-resonant x-ray diffraction, they become tensors when the x-ray energy is tuned close to an absorption edge of the investigated material. Due to the anomalous tensor components which arise, resonant x-ray diffraction is sensitive to the local electronic structure of the system. This is evidenced by the appearance of scattering at positions which are forbidden by the crystallographic space group, as a result of long-range order of either magnetic moments or orbital occupancy.

As already mentioned, non-resonant magnetic scattering is generally several orders of magnitude weaker in intensity than charge scattering. This weakness makes magnetic investigations with x-ray scattering experiments a fairly difficult task. However, by tuning the energy of the incoming x-ray beam close to an absorption edge of the investigated material, a significant resonant enhancement of the magnetic scattering cross-section can be obtained. The resulting high scattering intensities enable the magnetic x-ray study of even weakly magnetically polarized materials. The technique that has been developed to exploit this effect is called x-ray resonance exchange scattering (XRES) and is a very powerful probe, complementary to neutron diffraction, providing, among others, element- and, in some cases, ionization-state-sensitivity.

The mechanism that gives rise to the resonant enhancement involves low-order (dipole (E1) or quadrupole (E2)) electric multipole transitions from a core level into an empty state above the Fermi level. Due to the exclusion principle, which allows electrons to move only into not fully occupied orbitals, the electric transitions lead to an exchange interaction. Through this exchange the magnetic resonance appears.

The phenomenon was first experimentally observed at the K-absorption edges of ferromagnetic nickel (Ni) by Namikawa *et al.* in 1985.[6] The resonant enhancement is much stronger at the L- and M- absorption edges of metals, which correspond to electric multipole transitions from initial states with magnetic quantum number $l > 0$, i.e. from initial states that are spin-orbit split. A large resonant enhancement of the x-ray magnetic scattering cross-section was first reported at the L_{III}-edge of holmium (Ho), which is a spiral antiferromagnet, by Gibbs *et al.* in 1988.[7] The significant resonant enhancement of a factor of 50 reported in that case made clear the great potential of resonant x-ray diffraction and the new possibilities that it opened for the investigation of magnetic materials. Only one year later, Isaacs *et al.* reported a huge resonant enhancement of seven orders of magnitude at the M edges of the actinides, which resulted in a diffracted intensity of several thousand counts per second.[8]

Since then, XRES has been established as a powerful probe of magnetic order in strongly correlated electron systems. Related resonant scattering techniques have been more recently developed for the investigation of other degrees of freedom, besides magnetism, such as orbital order for instance.[9]

In all cases, a resonant enhancement of the scattering intensity is achieved through a second-order process, in which a core level electron is promoted to an intermediate excited state, which subsequently decays. In the following we refer to these techniques with the general term 'resonant x-ray diffraction' (RXD). Besides providing high scattering intensities, all resonant diffraction techniques have the additional advantage of being element-specific. Only phenomena related to the element with the absorption edge, to which the energy of the incoming x-ray beam is tuned, are observed. In cases where different valence states of the same element involve different electric transitions, the method is also ionization-state sensitive.

Resonant x-ray diffraction has been used with remarkable success for the investigation of numerous transition metal oxides. Resonant diffraction at the K- absorption edges of titanium (Ti) and manganese (Mn), for instance, has been employed for the study of $3d$ compounds, such as $YTiO_3$,[10] $LaMnO_3$,[9] $Pr_{0.6}Ca_{0.4}MnO_3$[11] and $RbMnF_3$.[12] At the K-absorption edges, the resonance is driven by electric dipole transitions from the $1s$ core level to the unoccupied $4p$ level, and gives an enhancement which is usually a factor of three to five of the scattering intensity off resonance.

Significantly larger resonance enhancements can be obtained at the L- absorption edges of transition metals, where electric dipole transitions drive the electrons directly into the partly occupied d band, which is responsible for the magnetic phenomena in these materials. The probing of the d band enables the direct observation of orbital ordering. The technique has recently been used for the study of orbitally ordered states in $3d$ ($La_{2-2x}Sr_{1+2x}Mn_2O_7$,[13] $La_{0.5}Sr_{1.5}MnO_4$,[14] $Pr_{0.6}Ca_{0.4}MnO_3$[15]), $5d$ (K_2ReCl_6[16]) and, more recently, on $4d$ electron compounds.[17-19]

Several technical difficulties are associated with working at x-ray energies close to the L-edges of $4d$ metals (around and below 3 keV). This energy regime is difficult to access both in hard x-ray and in soft x-ray synchrotron facilities, for reasons related to the operation of the insertion devices (e.g. undulators) and the monochromators (gratings), respectively. Other difficulties associated with low photon energies include the severe absorption of the x-ray beam by air, which means that the experimental setup has to be especially optimized, so that there is as little air as possible in the beam flight-path; and the small radius of the Ewald sphere, which results in a small part of the reciprocal space being accessible for investigations.

3.3.2. *Resonant Electric Dipole Scattering Length*

The total coherent elastic scattering length g for a magnetic ion, containing both non-resonant and resonant contributions, can be written as a sum of four terms:

$$g = g_0 + g' + ig'' + g_m \qquad (3.9)$$

where $g_0 \propto -Zr_e$ is the Thomson charge scattering contribution, with Z being the atomic number of the ion and r_e the classical electron radius, g_m is the non-resonant spin-dependent magnetic scattering length, and $g' + ig''$ is the contribution (both resonant and non-resonant) from dispersive and absorptive processes.[20]

For an electric 2^L-pole (EL) resonance in a magnetic ion, the resonant contribution to the coherent scattering length is:[20]

$$g_{\mathrm{EL}}(\omega) = \frac{4\pi}{|k|} W_D \sum_{M=-L}^{L} [\hat{\epsilon}'^{*} \cdot \mathbf{Y}_{\mathrm{LM}}(\hat{\mathbf{k}}') \mathbf{Y}_{LM}^{*}(\hat{\mathbf{k}}) \cdot \hat{\epsilon}] F_{LM}(\omega) \qquad (3.10)$$

In (3.10) the functions $\mathbf{Y}_{\mathrm{LM}}(\hat{\mathbf{k}})$ are vector spherical harmonics, W_D is the Debye-Waller factor (a factor which takes into consideration the reduction of the scattering intensity caused by the thermal vibrations of the ions around their equilibrium positions), $\mathbf{k}$, $\mathbf{k}'$ are the wavevectors of the incoming and scattered beams, respectively, ϵ, ϵ' are the corresponding polarization vectors, and $F_{\mathrm{LM}}(\omega)$ is a factor that determines the strength of the resonance and which depends on atomic properties of the scattering ion:

$$F_{\mathrm{LM}}(\omega) = \sum_{\alpha,n} [\frac{P_\alpha P_\alpha(\eta) \Gamma_x(\alpha M \eta; EL)}{\Gamma(\eta)}] / [x(\alpha, \eta) - i] \qquad (3.11)$$

Here $|\alpha\rangle$ is the initial ground state of the ion; $|\eta\rangle$ is the final state, where an electron has been excited to a higher level leaving a hole in the core level; P_α is the probability of the ion being in the initial state α; $P_\alpha(\eta)$ is the probability of a transition from the initial state $|\alpha\rangle$ to the final state $|\eta\rangle$, which is determined by the overlap integrals of the two states; Γ_x is the partial line width of the excited state for a pure 2^L-pole (EL) radiative decay from $|\eta\rangle$ to $|\alpha\rangle$; Γ is the total width of the excited state for all deexcitations of $|\eta\rangle$, radiative and non-radiative; and finally x is the deviation from the resonance condition in units of the total half-width: $x = \frac{E_\eta - E_\alpha - \hbar\omega}{\Gamma/2}$.

The above are valid for isotropic systems, in which the symmetry is broken by the magnetic moment. In systems where particular point group symmetries are applied, the allowed terms are altered.

In order to make clear the polarization dependence of resonant magnetic scattering and its sensitivity to individual components of the magnetic moment, it is useful to express the vector spherical harmonics in the scattering length (3.10) as a function of the polarization vectors ϵ, ϵ' of the incoming and scattered beams. For electric dipole (E1) transitions ($L = 1$, $M = \pm 1$), which usually dominate the resonant scattering cross-section, the scattering length of resonant elastic scattering can be written as follows:[21]

$$g_{E1}(\omega) = [(\hat{\epsilon}' \cdot \hat{\epsilon})F^{(0)} - i(\hat{\epsilon}' \times \hat{\epsilon}) \cdot \hat{\mu}F^{(1)} + (\hat{\epsilon}' \cdot \hat{\mu})(\hat{\epsilon} \cdot \hat{\mu})F^{(2)}] \quad (3.12)$$

where

$$F^{(0)} = \frac{3}{4k}[F_{11} + F_{1-1}] \quad (3.13)$$

$$F^{(1)} = \frac{3}{4k}[F_{11} - F_{1-1}] \quad (3.14)$$

$$F^{(2)} = \frac{3}{4k}[2F_{10} - F_{11} - F_{1-1}] \quad (3.15)$$

and $\hat{\mu}$ is a unit vector along the local magnetic moment direction.

The first term of (3.12) is the anomalous dispersion term. It is the contribution to the Bragg charge scattering and shows no dependence on the magnetic moment. The polarization dependence enters through the scalar product $\epsilon' \cdot \epsilon$ of the polarization vectors, just like in non-resonant charge scattering (3.3). The scalar product is non-zero for non-perpendicular incoming and outgoing polarization vectors. Thus no rotation of the polarization plane is caused by the anomalous dispersion term. The scattering is either $\sigma \to \sigma'$ or $\pi \to \pi'$.

The second term is the one responsible for magnetic circular dichroism in ferromagnets. It depends on the difference in magnetic resonance strength between states with magnetic quantum numbers $M = 1$ and $M = -1$. The polarization dependence enters through the vector product $\epsilon' \times \epsilon$. Thus, for a σ-polarized incoming beam, it causes a rotation of the polarization plane, i.e. leads to $\sigma \to \pi'$ scattering. For a π-polarized primary beam, both rotated ($\pi \to \sigma'$) and unrotated ($\pi \to \pi'$) scattering are allowed. As far as the dependence on the magnetic moment is concerned, this is linear, just like in non-resonant magnetic scattering. Both types of magnetic scattering produce first-order magnetic satellites, at the same reciprocal space points.

The third and last term of the total resonant scattering length is related to magnetic linear dichroism. It has a relatively complicated polarization dependence which generally leads to a partial rotation of the polarization plane. Its dependence on the magnetic moment is quadratic, producing second-harmonic magnetic satellites, which are not observed off-resonance. Magnetic x-ray linear dichroism is generally much weaker than circular dichroism, especially at hard x-ray absorption edges, thus the last term of (3.12) is also correspondingly weak compared to the other two.

Following the procedure used by de Bergevin and Brunel for the derivation of the cross-section of non-resonant elastic magnetic scattering,[5] we can similarly express the resonant scattering length with the use of 2×2 matrices in a basis, the components of which are perpendicular (σ) and parallel (π) to the scattering plane, as suggested by Hill and McMorrow.[21] Figure 3.1 shows a schematic view of a typical experimental setup for a resonant x-ray scattering experiment involving polarization analysis of the diffracted beam. If $\hat{\epsilon}_\sigma$, $\hat{\epsilon}_\pi$ are the components of the polarization vector ϵ of the incoming x-ray beam along the σ and π directions, and $\hat{\epsilon}'_\sigma$, $\hat{\epsilon}'_\pi$ the components of the polarization vector ϵ' of the diffracted beam along σ' and π', respectively, then: $\hat{\epsilon}'_\sigma \cdot \hat{\epsilon}_\sigma = 1$, $\hat{\epsilon}'_\pi \cdot \hat{\epsilon}_\pi = \cos 2\theta = \hat{k} \cdot \hat{k}'$, $\hat{\epsilon}'_\sigma \cdot \hat{\epsilon}_\pi = \hat{\epsilon}'_\pi \cdot \hat{\epsilon}_\sigma = 0$. The scalar product in the first term of (3.12) can be thus written as a diagonal matrix, the diagonal elements of which correspond to the unrotated ($\sigma \to \sigma'$, $\pi \to \pi'$) scattering processes:

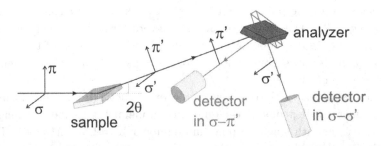

Fig. 3.1.   Schematic view of the experimental configuration in a resonant x-ray diffraction experiment.[17] The incoming beam is linearly polarized along the σ direction. The diffracted beam has, in the general case, polarization components both along σ' (perpendicular to the diffraction plane) and along π' (parallel to the d.p.). The two polarization components can be separated with the use of a proper analyzer crystal.

$$\hat{\epsilon}' \cdot \hat{\epsilon} = \begin{pmatrix} 1 & 0 \\ 0 & \hat{k}' \cdot \hat{k} \end{pmatrix} \tag{3.16}$$

As far as the second term of (3.12) is concerned, there we have: $\hat{\epsilon}'_\sigma \times \hat{\epsilon}_\sigma = 0$, $\hat{\epsilon}'_\pi \times \hat{\epsilon}_\pi = \hat{k}' \times \hat{k}$, $\hat{\epsilon}'_\sigma \times \hat{\epsilon}_\pi = \hat{k}$, and $\hat{\epsilon}'_\pi \times \hat{\epsilon}_\sigma = -\hat{k}'$. Thus the vector product in the circular magnetic dichroism term can be written as:

$$\hat{\epsilon}' \times \hat{\epsilon} = \begin{pmatrix} 0 & \hat{k} \\ -\hat{k}' & \hat{k}' \times \hat{k} \end{pmatrix} \tag{3.17}$$

The third term can be also expressed as a matrix containing products of $\hat{k}$, $\hat{k}'$ and $\hat{\mu}$. These vectors can be analyzed along the main axes $\hat{U}_1$, $\hat{U}_2$ and $\hat{U}_3$ of a coordinate system, in which the plane defined by $\hat{U}_1$, $\hat{U}_3$ is parallel to the scattering plane, and $\hat{U}_3$ is antiparallel to the scattering vector (Figure 3.2). The following expression for the total resonant electric dipole scattering length can be then found:

$$g_{E1} = F^{(0)} \begin{pmatrix} 1 & 0 \\ 0 & \cos 2\theta \end{pmatrix} - iF^{(1)} \begin{pmatrix} 0 & \mu_1 \cos\theta + \mu_3 \sin\theta \\ \mu_3 \sin\theta - \mu_1 \cos\theta & -\mu_2 \sin 2\theta \end{pmatrix}$$
$$+ F^{(2)} \begin{pmatrix} \mu_2^2 & -\mu_2(\mu_1 \sin\theta - \mu_3 \cos\theta) \\ \mu_2(\mu_1 \sin\theta + \mu_3 \cos\theta) & -\cos^2\theta(\mu_1^2 tan^2\theta + \mu_3^2) \end{pmatrix} \tag{3.18}$$

where θ is the scattering angle and μ_1, μ_2, μ_3 the components of $\hat{\mu}$ along $\hat{U}_1$, $\hat{U}_2$, $\hat{U}_3$, respectively.

Starting from Equation 3.18, one can express the resonant electric dipole scattering length as a function of the azimuthal angle ψ, namely the angle between the projection of the magnetic moment on the plane that is perpendicular to the scattering plane, and the scattering plane. By definition, $\psi = 0°$ when the magnetic moment lies in the diffraction plane. Expressing the scattering length as a function of ψ is particularly useful for resonant scattering experiments.

We assume a magnetic basis consisting of N ions with two possible spin directions, for which the notations 'spin-up': ↑ and 'spin-down': ↓ are used. According to (3.18), in the $\sigma \to \pi'$ scattering geometry, the scattering length for a 'spin-up' ion is:

$$g_{E1,\uparrow}^{\sigma \to \pi'} = -iF^{(1)}(\mu_3 \sin\theta - \mu_1 \cos\theta) + F^{(2)}\mu_2(\mu_1 \sin\theta + \mu_3 \cos\theta) \tag{3.19}$$

With α being the angle between the magnetic moment $\hat{\mu}$ and the scattering vector (Figure 3.2), the magnetic moment components can be written as

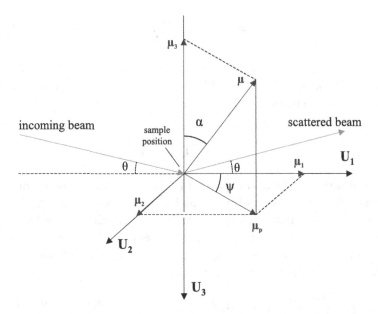

Fig. 3.2. Configuration of the scattering process, used in the calculation of the resonant electric dipole scattering length. The axis U_3 of the coordinate frame is antiparallel to the scattering vector. θ is the Bragg angle, α the angle between the magnetic moment μ and the scattering vector, and ψ the angle between the diffraction plane (defined by the incoming and scattered beams) and the projection of μ on a plane perpendicular to the diffraction plane. By definition, $\psi = 0°$ when the magnetic moment lies in the diffraction plane.

follows:

$$\mu_1 = \sin\alpha\cos\psi$$
$$\mu_2 = \sin\alpha\sin\psi$$
$$\mu_3 = -\cos\alpha \tag{3.20}$$

So (3.19) gives:

$$g_{E1,\uparrow}^{\sigma\to\pi'} = -iF^{(1)}(-\cos\alpha\sin\theta - \sin\alpha\cos\psi\cos\theta)$$
$$+ F^{(2)}\sin\alpha\sin\psi\,(\sin\alpha\cos\psi\sin\theta - \cos\alpha\cos\theta) \Rightarrow$$
$$g_{E1,\uparrow}^{\sigma\to\pi'} = iF^{(1)}\cos\alpha\sin\theta + iF^{(1)}\sin\alpha\cos\theta\cos\psi$$
$$+ F^{(2)}\sin^2\alpha\sin\theta\sin\psi\cos\psi$$
$$- F^{(2)}\sin\alpha\cos\alpha\cos\theta\sin\psi \tag{3.21}$$

Similarly for the 'spin-down' ion (opposite direction of $\hat{\mu}$) we have:

$$g_{E1,\downarrow}^{\sigma \to \pi'} = -iF^{(1)}(-\mu_3 \sin\theta + \mu_1 \cos\theta) + F^{(2)}(-\mu_2)(-\mu_1 \sin\theta - \mu_3 \cos\theta) \tag{3.22}$$

which gives:

$$\begin{aligned} g_{E1,\downarrow}^{\sigma \to \pi'} = &-iF^{(1)} \cos\alpha \sin\theta - iF^{(1)} \sin\alpha \cos\theta \cos\psi \\ &+ F^{(2)} \sin^2\alpha \sin\theta \sin\psi \cos\psi \\ &- F^{(2)} \sin\alpha \cos\alpha \cos\theta \sin\psi \end{aligned} \tag{3.23}$$

The total resonant electric dipole scattering length for the N ions of the basis in $\sigma \to \pi'$ geometry is then:

$$g_{E1}^{\sigma \to \pi'} = \sum_{j=1}^{N} g_{E1,S_j}^{\sigma \to \pi'} e^{i\mathbf{Q} \cdot \mathbf{R}_j} \tag{3.24}$$

with S_j being the spin of ion j (up or down) and $g_{E1,S_j}^{\sigma \to \pi'}$ being given by (3.21) or (3.23). The scattering intensity measured in the $\sigma \to \pi'$ polarization channel at a particular reciprocal space position $(h\,k\,l)$ is proportional to the square of the amplitude of the corresponding total scattering length:

$$I_{(hkl)}^{\sigma \to \pi'} \propto |g_{E1}^{\sigma \to \pi'}|^2 \tag{3.25}$$

For the $\sigma \to \sigma'$ scattering geometry, Equation 3.18 gives for the scattering lengths of the ions of the basis:

$$g_{E1,\uparrow}^{\sigma \to \sigma'} = g_{E1,\downarrow}^{\sigma \to \sigma'} = F^{(0)} + F^{(2)}\mu_2^2 = F^{(0)} + F^{(2)} \sin^2\alpha \sin^2\psi \tag{3.26}$$

The total scattering length is in this case:

$$g_{E1}^{\sigma \to \sigma'} = \sum_{j=1}^{N} g_{E1,S_j}^{\sigma \to \sigma'} e^{i\mathbf{Q} \cdot \mathbf{R}_j} \tag{3.27}$$

and the scattering intensity in $\sigma \to \sigma'$:

$$I_{(hkl)}^{\sigma \to \sigma'} \propto |g_{E1}^{\sigma \to \sigma'}|^2 \tag{3.28}$$

Equations 3.25 and 3.28 give the relative resonant scattering intensity in the $\sigma \to \pi'$ and $\sigma \to \sigma'$ polarization geometries as a function of the azimuthal angle ψ. The expressions, in the specific form they assume when applied on the magnetic cell of a particular material, can be used for identifying the magnetic contribution in the total scattering signal and for determining the direction of the magnetic moment in the investigated system.

3.4. Orbital Ordering in Ca_2RuO_4

3.4.1. *Introduction*

The two-dimensional ruthenate system $Ca_{2-x}Sr_xRuO_4$ is in many aspects an interesting material. Although the gradual substitution of calcium (Ca) by strontium (Sr) is isovalent, the electronic properties of the material vary significantly with the Sr content x, resulting in a rich phase diagram (Figure 3.3). The $x = 2$ end-member of the series has attracted particular attention. Sr_2RuO_4 is a metallic compound, turning below 1.5 K into an unconventional spin-triplet superconductor,[22,23] the only known layered perovskite without copper exhibiting superconducting properties. It is still an issue of debate how strongly the $4d$ electrons are correlated in this system, how exactly the electron correlations determine the material's properties, and what the role of magnetic and orbital fluctuations is in the establishment of superconductivity. On the other side of the phase diagram, Ca_2RuO_4 ($x = 0$) is a Mott insulator, exhibiting antiferromagnetic

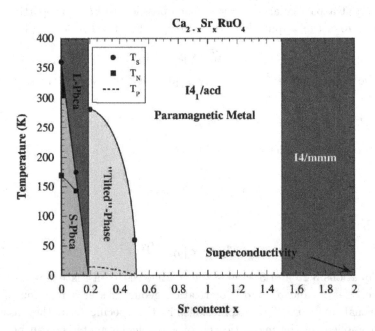

Fig. 3.3. Phase diagram of the $Ca_{2-x}Sr_xRuO_4$ system.[29] The $x = 0$ end member is an antiferromagnetic Mott insulator, turning metallic at 356 K, in a first-order phase transition, which is accompanied by substantial structural distortions. The $x = 2$ end member, on the other hand, is metallic, turning below 1.5 K into a spin-triplet superconductor.

ordering with a weak ferromagnetic component due to spin canting below 110–150 K.[24] The narrow electron bands induce strong electron correlations and the substantial spin-orbit coupling implies that the orbital degree of freedom is active and plays a crucial role in the magnetic properties of the material. Several, partly controversial theoretical predictions have been made for the orbital ordering pattern of the $4d$ orbitals in Ca_2RuO_4.[25–28] Investigating the interplay between the magnetic and orbital degrees of freedom in Ca_2RuO_4 can help understand the origin of many of the electronic properties of the $Ca_{2-x}Sr_xRuO_4$ system, including the establishment of superconductivity in Sr_2RuO_4. This may give insight into the properties of transition metal oxides in general, and in particular of high-temperature superconducting cuprates, which are isostructural to $Ca_{2-x}Sr_xRuO_4$.

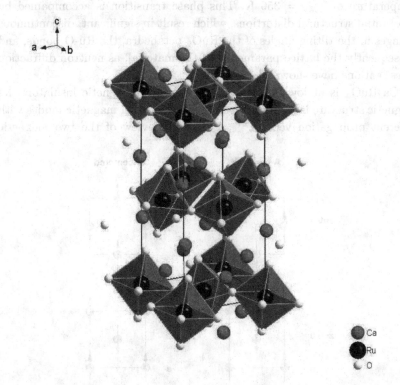

Fig. 3.4.   Crystal structure of Ca_2RuO_4 (space-group $Pbca$). The Ru ions occupy the centers of corner-sharing RuO_6 octahedra, forming RuO_2 layers that extend parallel to the ab plane. The octahedra are significantly tilted around an axis lying in the RuO_2 planes and rotated around the long crystallographic axis c. The structural distortions are strongly dependent on temperature and stoichiometry.

3.4.2. *Main Properties*

The crystal structure of Ca_2RuO_4 is orthorhombic (space-group *Pbca*), with lattice parameters a=5.4097(3) Å, b=5.4924(4) Å and c=11.9613(6) Å at room temperature.[24] It is a layered perovskite, consisting of RuO_2 layers which are made up of corner-sharing RuO_6 octahedra (Figure 3.4). Its structure is, however, strongly distorted and significant deviations from the ideal K_2NiF_4-structure (space-group *I4/mmm*) are observed. The distortion of the RuO_6 octahedra can be described as the result of a combined 'tilt plus rotation' movement, namely, a tilt of the octahedra around an axis which lies in the RuO_2 plane plus a rotation of them around the long crystallographic axis c.[24]

A first-order metal-insulator transition takes place in Ca_2RuO_4 at the temperature of T_{MI} = 356 K. This phase transition is accompanied by substantial structural distortions, which result in significant, discontinuous changes in the tilting angles of the RuO_6 octahedra, the Ru-O bonds, and subsequently the lattice parameters of the material, as neutron diffraction investigations have shown.[29]

Ca_2RuO_4 is at low temperatures an antiferromagnetic insulator. Its magnetic structure is characterized by two coexisting magnetic modes with different propagation vectors.[24] A schematic view of the two magnetic

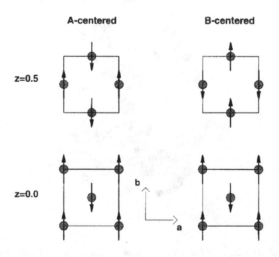

Fig. 3.5. Schematic view of the spin arrangement in the Ru^{+4} sublattice in the magnetically ordered phase of Ca_2RuO_4, as determined with neutron diffraction.[24] Two different antiferromagnetic modes coexist. The so-called A-centered mode (left) is the dominant one in stoichiometric compounds.

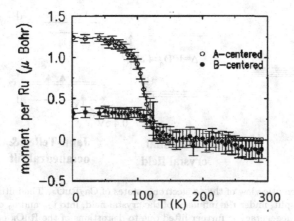

Fig. 3.6. Variation with temperature of the magnetic moment per Ru^{+4} ion for the two different magnetic modes shown in Fig. 3.5, as determined by refinement based on neutron diffraction measurements.[24] The ordered moment of the B-centered phase saturates at the temperature where the A-centered phase sets in.

modes in space-group *Pbca* is shown in Figure 3.5. In the so-called *A-centered* magnetic mode, the spin of the Ru^{+4} ion located at position (0 0 0) is aligned parallel to the one at position $(0 \frac{1}{2} \frac{1}{2})$. This magnetic arrangement has propagation vector (1 0 0) (La_2CuO_4-type) and sets in below the ordering (Néel) temperature of $T_N = 110$ K. On the contrary, in the *B-centered* magnetic mode, the Ru^{+4} spin at position (0 0 0) is antiparallel to the one at $(0 \frac{1}{2} \frac{1}{2})$ and parallel, instead, to the spin at site $(\frac{1}{2} 0 \frac{1}{2})$. This magnetic arrangement has propagation vector (0 1 0) (La_2NiO_4-type) and ordering temperature $T_N^B = 150$ K. The temperature dependence of the magnetic moment per Ru ion for the two magnetic modes, as determined by refinement based on the neutron diffraction data,[24] is shown in Figure 3.6. The ordered moment of the B-centered phase saturates at the temperature where the A-centered mode sets in. In the stoichiometric version of Ca_2RuO_4 the A-centered mode is predominant and can be considered as the main contribution to the antiferromagnetic ordering.

The formal oxidation state of ruthenium in Ca_2RuO_4 is $+4$. Every Ru^{+4} ion has four electrons in the $4d$ orbitals. The degeneracy of the electronic states of the $4d$ band is partly lifted due to the (tetragonal) crystal field effect. This is significantly stronger in Ca_2RuO_4 than in $3d$ oxides, because the larger radial extent of the $4d$ shell leads to a stronger interaction of Ru^{+4} with the surrounding ions in the lattice. The energy difference $\Delta = 10D_q$ between the three lower-lying t_{2g} (d_{xy}, d_{yz}, d_{zx}) orbitals and the two

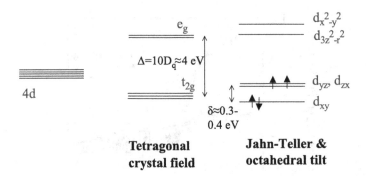

Fig. 3.7. Schematic view of the $4d$ electronic states of Ca_2RuO_4. The initially degenerate states are split, under the influence of the crystal field, into t_{2g} and e_g orbitals. The remaining t_{2g} degeneracy is further lifted due to distortions of the RuO_6 octahedra.

higher-lying e_g $(d_{x^2-y^2}, d_{3z^2-r^2})$ orbitals is estimated with band structure calculations approximately equal to 4 eV.[30] The crystal field splitting Δ is in this case larger than the Hund coupling J_H. Thus the Ru^{+4} ions are in the low-spin configuration: all four $4d$ electrons are in the t_{2g} states, leaving the e_g states completely empty. The Jahn-Teller-like distortion of the RuO_6 octahedra leads in addition to a partial splitting of the t_{2g} orbitals, with the d_{yz}/d_{zx} states now having slightly higher energy than d_{xy}. The tilting distortion of the octahedra further contributes to the lifting of the degeneracy of the t_{2g} orbitals (Figure 3.7).

3.4.3. Orbital Order

Based on the above simple picture, the d_{xy} orbital, which has the lowest energy among all t_{2g} orbitals, should be always fully occupied, as seen in Figure 3.7, i.e. the hole population in d_{xy} should be zero. The first theoretical calculations of the electronic density of the t_{2g} states, performed by Anisimov *et al.*, indeed predicted a 'ferro-orbital' (FO) ordering pattern in Ca_2RuO_4, with the d_{xy} orbitals occupied at all Ru^{+4} sites.[25] For these calculations the LDA+U method was used, that is, the usual local density approximation with the inclusion of an effective Coulomb repulsion parameter U that expresses the on-site correlations.

Experimental data, however, surprisingly put under question the homogeneous occupancy of the d_{xy} orbitals. Based on x-ray absorption spectroscopy investigations, Mizokawa *et al.* claimed that the hole population ratio of the d_{xy} and d_{yz}/d_{zx} orbitals is roughly equal to 1 : 1 at room temperature, dropping to 0.5 : 1.5 at 90 K, due to the compression of the octahedra.[31] Such a hole population would be inconsistent with the one expected from the crystal field effects and indicates that the orbital degree of freedom has to be taken into consideration as well in the determination of the electronic structure of Ca_2RuO_4. In an attempt to explain the spectroscopic data, Mizokawa *et al.* argued that the ground state may favor the occupation of complex orbitals, as a result of the strong spin-orbit coupling. In particular, it was suggested that, while at 300 K the two t_{2g} holes are located at the d_{xy} and $\frac{d_{zx}+id_{yz}}{\sqrt{2}}$ orbitals, with the spin (S) and orbital (L) angular momenta aligned along the z axis, at 90 K, where the out-of-plane Ru-O bond is shorter, the antiferromagnetic state with the two holes in the d_{yz} and $\frac{d_{xy}+id_{zx}}{\sqrt{2}}$ is stabilized instead, and the L, S momenta are aligned along the x (or y) axis.

Hotta and Dagotto took into consideration the new experimental information and proposed a theoretical model for the orbital order in Ca_2RuO_4 that was consistent with the hole populations suggested by Mizokawa *et al.*[26] Using a three-orbital Hubbard model tightly coupled to lattice distortions and by means of numerical and mean-field techniques, they suggested an 'antiferro-orbital-ordering' (AFO) pattern, that is, a pattern where *different* t_{2g} orbitals are occupied at nearest-neighbor sites.[26] This means that, for every Ru^{+4} ion with occupied d_{xy} orbital, its nearest neighbors along the x and y axes have instead their d_{zx} or d_{yz} orbitals, respectively, occupied. Thus, instead of having a uniform population of the d_{xy} orbitals, only half of the Ru^{+4} sites have these orbitals occupied. A schematic view of such a configuration is shown in Figure 3.8. It can be seen that the proposed orbital order leads to a doubling of the crystal periodicity in the *ab*-plane. If the same antiferro-orbital order is assumed along c, then the doubling of the periodicity occurs along that direction, too. Whether this doubling indeed takes place can be experimentally checked with x-ray diffraction and be used as a test of the validity of the model.

The above conclusions by Hotta and Dagotto were extracted mainly from numerical calculations applied to a small-size (2×2) plaquette cluster. Extending the study to larger lattices was not possible due to technical, computer-related limitations. For this purpose, a mean-field approximation analysis had to be employed as well. The latter can provide qualitatively correct results for the insulating ground state with static lattice distortions. In this way, up to 8×8-large clusters could be investigated. Besides the information about the orbital order, the theoretical study resulted in two more significant conclusions: firstly, that both Coulomb and phononic (electron-lattice) interactions are necessary for the stabilization of the antiferromagnetic phase of Ca_2RuO_4; and secondly, that the possibility of large magnetoresistance phenomena, reminiscent of the colossal magnetoresistance (CMR) of manganites, may exist for ruthenates, too.

A different orbitally ordered state, that actually combines the above mentioned ferro-orbital- and antiferro-orbital-ordering patterns, was proposed by Lee *et al.*[27] Based on optical spectroscopy investigations and using for their calculations the multi-orbital Hubbard model, they suggested that the FO and AFO ordering states may coexist in the ground state of Ca_2RuO_4, making it an interesting system composed of two kinds of Mott insulators. The AFO state was found to be the prevalent phase, consistent with the hole populations suggested by Mizokawa *et al.*, but the FO correlations increase and become important with decreasing temperature. The phase coexistence of these two ordered states was interpreted as the result of two competing tendencies: in the two-dimensional square network at room temperature, the AFO order is preferred, because of the gain in the kinetic energy that it provides via allowing the electron hopping between nearest neighbors, as described by Hotta *et al.*. However, at lower temperatures the distortion of the lattice favors the FO ordering state, consistent

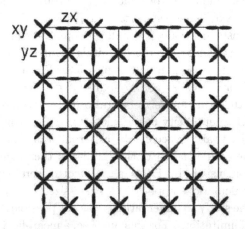

Fig. 3.8.   Schematic view of the antiferro-orbital-ordered state suggested for Ca_2RuO_4 by Hotta and Dagotto.[26] According to this model, different t_{2g} orbitals are occupied at nearest-neighbor Ru^{+4} sites, resulting in a doubling of the crystal periodicity in all three crystallographic directions. The initial lattice unit cell, corresponding to the lattice parameters a, b, c given in the text (color shaded; compare with magnetic cell in Fig. 3.5), is replaced in the orbitally ordered state by a much larger cell (thick lines).

with what Anisimov *et al.* claimed. If the energy difference between these two competing phases is small, then it is possible that they coexist in the ground insulating state of the system.

The situation changed again when Jung *et al.* put into question the orbital hole populations suggested by Mizokawa *et al.*, on which the theories of Hotta *et al.* and Lee *et al.* were based.[28] Using, as well, optical spectroscopy measurements and theoretical, first-principles LDA+U calculations, they showed that the reproduction of the x-ray absorption spectroscopy data of Mizokawa *et al.* can be achieved without necessarily assuming a low-temperature hole population ratio of 0.5 : 1.5 for the d_{xy} and d_{yz}/d_{zx} orbitals. Their study clearly indicated a predominant occupation of d_{xy} at all Ru^{+4} sites, in other words a d_{xy} ferro-orbital (FO) ordering. For the calculation of the projected density of states, a strong mutual mixture of the t$_{2g}$ orbitals due to the rotation and tilting of the RuO$_6$ octahedra, as well as due to the extended nature of the 4d states, was taken into consideration. The results show among others a substantial suppression of the d_{xy} hole population at low temperatures compared to room temperature, as well as a corresponding enhancement of the d_{yz}, d_{zx} populations, in accordance with the experimentally observed significant changes of the electronic configuration of Ca$_2$RuO$_4$ with decreasing temperature.

In the same direction, Fang *et al.* concluded, based as well on LDA+U calculations, a ferro-orbital ordering of the t$_{2g}$ orbitals with dominant d_{xy} occupation.[30] The stabilization of the d_{xy} orbital state was attributed mainly to three factors: first and most important, the energy level splitting caused by the two-dimensional crystal field in the layered structures, which had not been taken into consideration in the analysis by Hotta *et al.*; second, the splitting caused by the Jahn-Teller compression of the RuO$_6$ octahedra upon cooling; and third, the hybridization of the occupied and unoccupied orbitals, which is the origin of the superexchange. The combination of the above factors leads to an estimated total energy splitting of the d_{xy} and d_{yz}/d_{zx} orbitals of approximately 0.3–0.4 eV, which is about one tenth of the crystal field splitting Δ between the t$_{2g}$ and e$_g$ states. The suggested orbital configuration is claimed to be in good agreement both with the x-ray absorption spectroscopy (XAS) data by Mizokawa *et al.*,[31] and with optical conductivity measurements. Concerning the latter, it is argued that the spectral features which were used by Lee *et al.* to argue against the ferro-orbital order,[27] are actually not inconsistent with it, as long as one takes into consideration the orthorhombic distortion and the admixture of the d_{xy} and d_{yz}/d_{zx} states.

Definite conclusions on the orbital state in Ca_2RuO_4 were made possible with the use of resonant x-ray diffraction. As will be shown in the following, thanks to the direct probing of the orbital order, significant new information on the electronic properties of the system were made available.

3.4.4. X-Ray Investigations

The magnetic properties and orbital order in Ca_2RuO_4 were investigated with single-crystal resonant x-ray diffraction at the L_{II} and L_{III} absorption edges of ruthenium (Ru), i.e. at x-ray energies of 2.968 keV and 2.838 keV, respectively.[17] The studies revealed an orbital ordering phase transition in the paramagnetic phase of the system and provided significant information on the periodic pattern of the orbital order, despite its weak coupling to the lattice.

The A-centered magnetic arrangement of the Ru^{+4} spins, which is the dominant one in stoichiometric Ca_2RuO_4, has a characteristic propagation vector (1 0 0). Figure 3.9 shows the energy dependence of the scattered intensity of the (1 0 0) magnetic reflection around the L_{II} and L_{III} absorption edges of Ru at a sample temperature of 20 K. The measurements shown correspond to the azimuthal position at which the magnetic moment μ of the material, and thus the b direction, lies in the diffraction plane. The azimuthal angle ψ is at this position by definition equal to zero: $\psi = 0°$.

Figure 3.10 shows the energy dependence of the scattered intensity around the L_{II} edge in more detail (closed bullets). The spectrum is dominated by the large resonant enhancement of the scattering signal directly at the energy of the edge (2.968 keV). A second, weaker peak is observed in addition approximately 4 eV higher in energy (2.972 keV). The origin of this higher-energy peak will be addressed below. In the following, the lower-energy peak, at 2.968 keV, will be referred to as the 'L_{II} peak', while the one at 2.972 keV as the 'L_{II}' peak'. The energy dependence of the absorption coefficient μ is also shown in the same plot. The absorption coefficient is calculated from the fluorescence yield of the investigated sample, following Reference 32. The two peaks of the energy scan coincide with the inflection points of the absorption coefficient curve, that is, with the points where the second derivative of the function $\mu(E)$ changes sign. The lineshape of the energy scans are corrected for the absorption of the x-ray beam by the sample by multiplying the raw data with the *square* μ^2 of the absorption coefficient, as explained in Ref. 33. The branching ratio, defined as the ratio of the scattering intensities at the L_{II} and L_{III} absorption edges, is calculated approximately equal to 2.1. The result is in good agreement with the branching ratio (2.3) obtained previously with the same technique in a 5d electron system.[16] In 3d systems, on the contrary, the intensity at the L_{II} edge is smaller than at L_{III} and the branching ratio is smaller than 1 (of the order of 0.5 in manganites[15]).

Figure 3.11(a) shows the development of the lineshape of the energy scans around the L_{II} edge with increasing sample temperature. Three of these scans, corresponding to three characteristic sample temperatures (base temperature, just above the magnetic transition temperature T_N and room temperature) are selected and displayed separately in Figure 3.11(b), with the vertical (intensity) axis being now on a logarithmic scale. It is obvious that the two peaks observed in the energy scans follow quite different temperature dependences. The intensity of both is decreasing, as expected, with increasing temperature. However, while the intensity of the lower-energy (L_{II}) peak undergoes two rapid decreases, first at the magnetic ordering temperature T_N and then around 260 K, the intensity of the higher-energy (L_{II}') peak decreases smoothly and slowly with temperature without showing any anomalies. As a result, the ratio of the scattered intensity at the two energy positions changes significantly with temeparature. While at $T = 11$ K the intensity at L_{II} is a factor of 6 larger than the in-

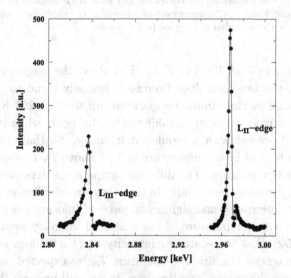

Fig. 3.9. Energy dependence of the scattered intensity at the (1 0 0) reciprocal space position around the Ru L_{II} and L_{III} absorption edges, at sample temperature $T = 20$ K and azimuth $\psi = 0°$.[17] The energy profiles are not corrected for absorption.

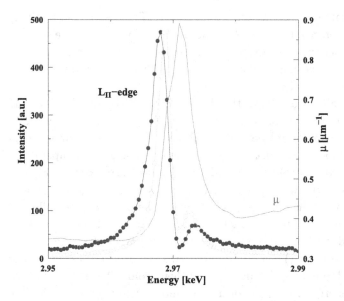

Fig. 3.10. Energy dependence of the scattered intensity of the (1 0 0) reflection around the Ru L_{II} absorption edge.[17] Besides the resonant peak at the edge, a second, weaker peak is present 4 eV higher in energy. The energy lineshape is shown both as-measured (closed bullets) and corrected for absorption (empty bullets; normalized to the uncorrected data). For the absorption correction the raw data are multiplied by the square of the absorption coefficient μ. The energy dependence of μ, as calculated from fluorescence measurements, is also included in the plot (right scale).

tensity at 2.972 keV, at $T = 114$ K, i.e. just above the magnetic transition temperature, the two intensities become practically equal to each other. Further increase of the sample temperature up to $T = 293$ K results in an almost (but not complete) vanishing of the L_{II} peak, while the L_{II}' resonance is still present with a significant intensity. So the intensity ratio changes from 6 : 1 at base temperature to 1 : 1 above T_N to approximately 1 : 8 at room temperature. The different temperature dependences of the two resonant features clearly indicate that they are of different origin.

The most interesting and significant piece of information provided by the variation with temperature of the scattered intensity around the L_{II} edge is the fact that the scattered intensity at (1 0 0) does not drop to zero at the magnetic transition temperature T_N, as expected based on the powder neutron diffraction investigations, but is still present, though very weak, even up to room temperature. This shows that an additional ordering mechanism, besides the magnetic ordering, which is characterized by the same propagation vector, but sets in at a higher temperature than magnetism, is present in Ca_2RuO_4. The temperature dependence of the integrated intensity of the the (1 0 0) reflection at the Ru L_{II} edge is shown in Figure 3.13. The integrated scattering intensity above T_N is a factor of 20 weaker than in the magnetically ordered phase. It follows a smooth, slow decrease with increasing temperature and eventually almost vanishes at approximately $T_{OO} = 260$ K in an order-parameter-like fashion. Exactly the same observations are made at position (0 1 1) of the reciprocal space, which is also magnetically allowed.

The newly discovered phase transition occuring in Ca_2RuO_4 at 260 K, far above the Néel temperature, but well below the metal-insulator transition, was attributed to orbital ordering based on three pieces of evidence.

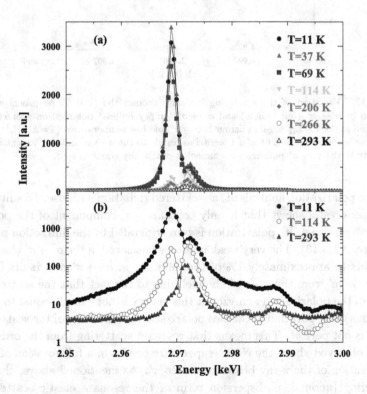

Fig. 3.11. Energy dependence of the scattered intensity of the (1 0 0) reflection around the Ru L_{II} absorption edge at azimuth $\psi = 0°$ for selected sample temperatures, on linear (a) and on logarithmic (b) intensity scale.[17] All energy scans are corrected for absorption.

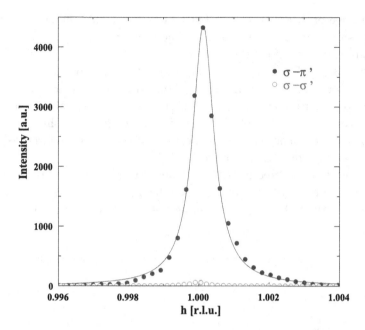

Fig. 3.12. Variation of the scattering intensity around the (1 0 0) reciprocal space position in $\sigma \to \pi'$ (full bullets) and $\sigma \to \sigma'$ (empty bullets) polarization geometries, measured at the Ru L_{II} edge, at azimuth $\psi = 0°$ and low temperatures (T=20 K).[17] The solid line is the result of a fit of a Lorentzian profile to the $\sigma \to \pi'$ data. The scattering intensity in the $\sigma \to \sigma'$ polarization channel is practically equal to zero.

Firstly, polarization analysis of the scattered radiation with use of a suitable analyzer crystal shows that it only contains a π' component of the polarization vector, i.e. its polarization is purely parallel to the diffraction plane (Figures 3.1,3.13). The very weak intensity measured in the $\sigma \to \sigma'$ channel amounts to approximately 1% of the intensity in $\sigma \to \pi'$ and rsults from the 'leakage' from the $\sigma \to \pi'$ channel, due to the fact that the scattering angle of the polarization analyzer at this energy is not exactly equal to 45°, and thus the separation of the two polarization components of the scattered beam is not perfect. This means that resonant scattering from the ordered phase observed above the Néel temperature results in a full rotation of the polarization of the x-ray beam from σ to π'. As mentioned above, charge scattering (anomalous dispersion term of the resonant elastic scattering length) does not rotate the polarization plane of the incoming beam. Thus, for σ-polarized incoming x-rays, charge scattering would give σ'-polarized diffracted radiation. The fact that such an unrotated component of the polarization vector is not experimentally observed in the diffracted beam, shows that charge scattering does not contribute to the resonant intensity in the new phase.

Secondly, complementary muon spin rotation (μSR) measurements that were carried out on powder and single-crystal Ca_2RuO_4 samples showed no ordered magnetic moment above T_N (Figure 3.14). Based on this, and given the fact that the detection limit of the technique lies well below the amplitude of the ordered moment that should be expected if the scattering intensity in the new phase were due to magnetic order, it can be concluded that the observed signal is not of magnetic origin. The phase transition at 260 K is thus a transition between two paramagnetic phases. The different origin of the ordering mechanisms below and above T_N is further underlined

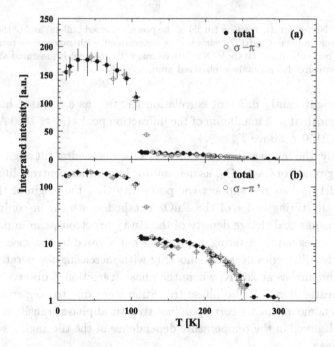

Fig. 3.13. Temperature dependence of the integrated scattering intensity at the (1 0 0) reciprocal space position at the Ru L_{II} edge, as determined from h-scans at $\psi = 0°$, both without polarization analyzer (full bullets) and in $\sigma \rightarrow \pi'$ geometry (empty bullets), presented on linear (a) and on logarithmic (b) intensity scales.[17] The two plots have been scaled to their base temperature intensities. The dependences without analyzer and in $\sigma \rightarrow \pi'$ are virtually identical to each other. Non-zero scattering intensity is observed above the Néel temperature, revealing a new phase transition at 260 K.

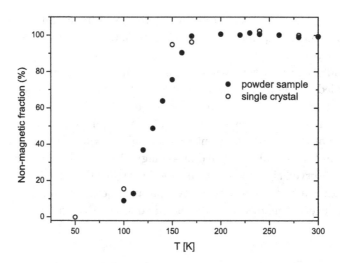

Fig. 3.14. Non-magnetic fraction (in %) of the powder (closed bullets) and of the single-crystal (open bullets) Ca_2RuO_4 samples, as determined with muon spin rotation at transverse magnetic field 100 Oe.[17] No ordered magnetic moment is measured above the Néel temperature. No anomaly is observed around 260 K.

by their significantly different correlation lengths, as calculated based on the half width at half maximum of the diffraction peaks ($\xi_a \approx 1560$ Å below T_N, $\xi_a \approx 2150$ Å above T_N).

Thirdly, the temperature dependence of the octahedral tilt angles in the insulating phase of Ca_2RuO_4, as determined by powder neutron diffraction (Figure 3.15), does not support the possibility that the origin of the new phase is the tilting order of the RuO_6 octahedra, which, in conjunction with the aspherical charge density of the Ru t_{2g} electrons, can in principle give rise to resonant scattering at positions not allowed by the space group. Indeed, the tilt angles decrease smoothly with increasing temperature and show no anomalies at 260 K, where the phase transition is observed in the x-ray investigations. If the phase transition were due to cooperative tilt order phenomena, then a corresponding structural phase transition should be also observed in the temperature dependence of the tilt angles, which is not the case.

What is more, the azimuthal angle dependences of the simulated tilt order intensities at positions (1 0 0) and (0 1 1) are different from the ones experimentally observed. The simulations were made with the use of a code previously developed by J.P. Hill for use in $YTiO_3$, appropriately modified for appication in Ca_2RuO_4. The tilt order contribution was independently

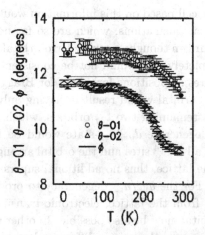

Fig. 3.15. Variation with temperature of the tilt angles θ_1 (bullets), θ_2 (triangles), and of the rotation angle ϕ of the RuO_6 octahedra in Ca_2RuO_4, as determined with neutron diffraction.[24] The definition of the angles is given in the text.

observed at reciprocal space position (1 1 0), which is not magnetically allowed. The azimuthal dependence of the scattered intensity at (1 1 0) agrees well with the tilt order simulations, which confirms the reliability of the model. The conclusion is that the origin of the L_{II} resonant signal above T_N at (1 0 0) and (0 1 1) is different from the octahedral tilt order.

Having excluded charge scattering, magnetic order and octahedral tilt order as possible origins of the resonant diffraction intensity above the Néel temperature, it is reasonable to assume that the observed intensity originates from the ordering of the Ru 4d orbitals, consistent with several theoretical studies. Since only the t_{2g} orbitals are partly occupied in the 4d band, it is these orbitals which participate in the orbital order. From the fact that the orbital order reflections are observed at the same wavevectors at which magnetic scattering occurs below T_N, we conclude that the propagation vector of the orbital order is in Ca_2RuO_4 identical to the propagation vector characterizing the low-temperature antiferromagnetic phase.

The azimuthal dependence of the integrated intensity due to orbital order is at (1 0 0) the same as the one expected for the magnetic intensity at this position, i.e. following a $\cos^2\psi$-law. This is not the case at (0 1 1), where the orbital-order intensity rather follows a $\cos^2(2\psi - \psi_o)$-law instead of the dependence corresponding to the resonant electric dipole scattering length. The precise determination of the occupancy of the t_{2g} orbitals at every site of the unit cell based on this information would demand the performance of numerical calculations, which are so far not available. Some conclusions may be drawn though: a purely ferro-orbital arrangement, with the d_{xy} orbital completely full, would not be consistent with the observation of superstructures at positions forbidden for Bragg charge scattering. A fully occupied d_{xy} orbital would result in having only one possible spin arrangement for the remaining two t_{2g} orbitals, which have the same energy. The doubly degenerate d_{yz}/d_{zx}-state would be half-occupied with two parallel spins at all Ru^{+4} sites and the orbital arrangement would have the periodicity of the lattice, thus no additional superstructure reflections would be observed. For the d_{yz}/d_{zx} degeneracy to produce an order with different periodicity from the lattice, electronic transitions from the core level into the d_{xy} orbital must be also possible. In other words, all three t_{2g} orbitals must be active and the energy difference between d_{xy} and d_{yz}/d_{zx} should be not larger than a couple of hundred meV. An antiferro-orbital order resulting in a doubling of the crystal periodicity in all three crystallographic directions would be also not consistent with our investigations, and in particular with the absence of resonant reflections at reciprocal space positions with half-integer Miller indices. Such a doubling would invoke non-zero structure factors at reciprocal space positions such as $(\frac{1}{2} \frac{1}{2} 0)$ and $(\frac{1}{2} \frac{1}{2} \frac{1}{2})$. No scattering intensity was observed at these positions, showing that such a scenario is not viable in this case. Other possibilities have to be considered instead for the interpretation of the results. The most likely of those is a predominantly ferro-orbital arrangement of the Ru $4d$ valence orbitals in the insulating phase of Ca_2RuO_4, with a phase transition associated to the onset of a staggered component of the orbital ordering pattern below 260 K.

The uniform polarization of the $4d$ orbitals was independently confirmed by Kubota *et al.* using the so-called resonant x-ray interference technique at the Ru K absorption edge.[34] This technique is based on the calculation of the interference term obtained by subtracting the fluorescence intensity spectra that correspond to two different polarization angles. In this way it is possible to observe ferro-type orbital ordering, which conventional resonant x-ray scattering (RXS) cannot achieve, since the scattering signal originating from the orbital order cannot be in this case separated from the strong coexisting charge scattering contribution. The main conclusions of the study, based mainly on the azimuthal angle dependence of the interference term, are that a ferro-type order of the Ru^{+4} t_{2g} orbitals occurs in Ca_2RuO_4 and that this order is maintained up to the metal-insulator transition temperature $T_{MI} = 357$ K. It is further argued that, since the orbital order is present even at room temperature, where the Jahn-Teller distortion is very small,[24] this distortion cannot be considered as the main origin of the orbital order. Although the onset of the staggered component at 260 K could not be observed in this case, due to the lower sensitivity of the technique, the main conclusions about the general pattern of the orbital order is in good agreement with the above mentioned RXD results.

3.5. Spin Reorientation in $Ca_3Ru_2O_7$

3.5.1. *Introduction*

The bilayer perovskite ruthenate $Ca_3Ru_2O_7$ is an interesting electron system with a particularly rich phase behavior. Its phase diagram is characterized by multiple phase transitions, reflecting its sensitivity to small perturbations, such as external magnetic fields,[35-37] uniaxial or hydrostatic pressure,[38-40] and doping.[41,42] In the Sr-substituted analog $Sr_3Ru_2O_7$ an electronic liquid crystal phase was recently discovered.[43] Since most phase transitions in $Ca_3Ru_2O_7$ have been attributed to the interplay between the magnetic and orbital degrees of freedom, resonant x-ray diffraction is an appropriate technique for the study of the electronic properties of the system. Transport, magnetic and thermal properties of $Ca_3Ru_2O_7$ are discussed in Chapter 6.

3.5.2. *Main Properties*

$Sr_3Ru_2O_7$ has an orthorhombic crystal structure described by the space-group $Bb2_1m$ with lattice parameters a=5.3677 Å, b=5.5356 Å and c=19.5219 Å at room temperature.[44] Its unit cell consists of RuO_2 bilayers formed by the equatorial planes of corner-sharing RuO_6 octahedra (Figure 3.16). The tilt-plus-rotation distortions which characterize the single-layered Ca_2RuO_4 compound (Section 3.4.1) are present in $Ca_3Ru_2O_7$ as well.

Like in Ca_2RuO_4, the formal oxidation state of ruthenium in $Ca_3Ru_2O_7$ is +4, with four electrons in the Ru^{+4} $4d$ t_{2g} orbitals, and with all e_g states completely empty. The system is thus in a low-spin state (S=1).

The magnetic and electric properties of $Ca_3Ru_2O_7$ are particularly interesting. The material is a paramagnetic metal at high temperatures. Upon cooling it orders antiferromagnetically at $T_N = 56$ K, while remaining metallic. Upon further cooling its resistivity increases abruptly at

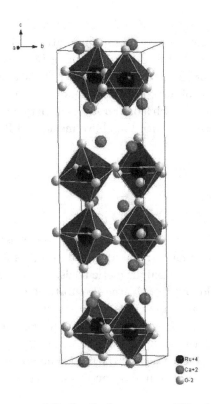

Fig. 3.16. Crystal structure of $Ca_3Ru_2O_7$ (space-group $Bb2_1m$). The Ru ions occupy the centers of corner-sharing RuO_6 octahedra, forming RuO_2 bilayers that extend parallel to the ab plane. The octahedra are tilted around an axis lying in the RuO_2 planes and rotated around the long crystallographic axis c, like in Ca_2RuO_4.

$T_{MI} = 48$ K in all three crystallographic directions (Figure 3.17). The increase in resistivity is more significant along the c-axis. While in this direction the resistivity continues increasing down to the lowest temperatures, this is not the case in the ab-plane, where it decreases again below 30 K. This indicates a quasi two dimensional metallic ground state with a large low-temperature anisotropy in conductivity. The phase transition at $T_{MI} = 48$K is considered to be a metal-insulator transition despite its unconventional character. It is a first order transition, with all structural parameters changing abruptly at T_{MI}, but the changes are much smaller than in Ca_2RuO_4 (of the order of 0.2% intead of 2%).

Susceptibility studies in $Ca_3Ru_2O_7$ indicate long-range antiferromagnetic (AFM) ordering with the magnetic moments aligned along the b axis

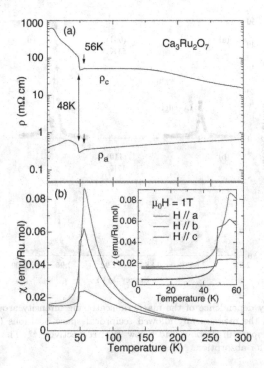

Fig. 3.17. Temperature dependence of (a) the resistivity and (b) the magnetic susceptibility in $Ca_3Ru_2O_7$.[45] The system is metallic and paramagnetic for temperatures above 56 K; metallic and antiferromagnetic for 48 K $< T <$ 56 K; and insulating and antiferromagnetic for $T <$ 48 K, exhibiting a large anisotropy in resistivity.

below T_{MI}.[45,46] Based on powder neutron diffraction results, an A-type AFM order has been proposed, i.e. ferromagnetic RuO_2 bilayers antiferromagnetically coupled along the c direction.[44] Given that this scenario was based on the observation of a single magnetic reflection, it has to be considered as only tentative. The evolution of the magnetically ordered phase as a function of temperature, pressure and applied magnetic field was further investigated by Raman scattering studies.[40] A strong coupling between the spin, charge and lattice degrees of freedom and a significant tunability of the phase behavior of the system were concluded in these studies.

3.5.3. X-Ray Investigations

The magnetic and orbital properties of $Ca_3Ru_2O_7$ were investigated with resonant x-ray diffraction at the Ru L_{II} and L_{III} absorption edges.[18] Thanks

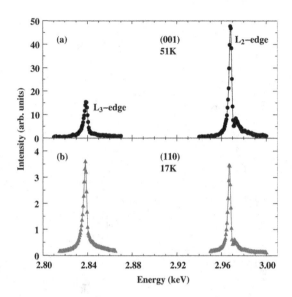

Fig. 3.18. Energy dependence of the magnetic scattering intensity around the Ru L_{II} and L_{III} edges at the magnetically allowed reciprocal space positions (a) (0 0 1) and (b) (1 1 0), at temperatures above and below T_{MI}, respectively.[18] The energy profiles are not corrected for absorption.

to the sensitivity of the technique to the magnetic moment direction, definite conclusions could be drawn about the magnetic structure of the system. The metal-insulator transition at T_{MI} is found to be accompanied by a 90° reorientation of the Ru spins within the *ab* plane. The absence of any experimental indications of either uniform of staggered orbital order indicates that the orbital polarization in $Ca_3Ru_2O_7$, if present, is much smaller than in Ca_2RuO_4.

The energy dependence of the magnetic scattering intensity around the L_{II} and L_{III} edges at the magnetic Bragg reflections (0 0 1) and (1 1 0) are shown in Figure 3.18 for two different temperatures below the magnetic ordering temperature $T_N = 56$ K. The energy profile is the same as in Ca_2RuO_4 (Section 3.4.4), exhibiting a strong resonance at each one of the edges and a second weaker peak 1-2 eV higher in energy, as a result of the electric dipole transitions into the $4d$ t_{2g} and e_g orbitals, respectively.

The azimuthal dependence of the integrated scattering intensity at (0 0 1) and (1 1 0) positions, both below and above T_{MI}, is shown in Figure 3.19. The solid curves represent the simulation results based on the formalism described in Section 3.3.2 for the resonant scattering lenth.

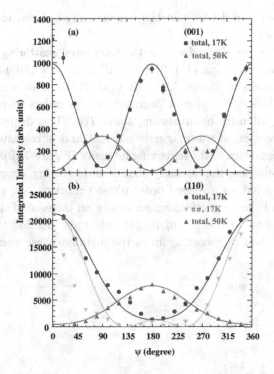

Fig. 3.19. Azimuthal dependence of the integrated scattering intensity at (a) (0 0 1) and (b) (1 1 0), at temperatures below (17 K) and above (50 K) T_{MI} at the Ru L$_{II}$ edge.[18] The solid curves represent calculations based on the electric dipole approximation and assuming the magnetic moment direction along b below T_{MI} and along a for $T_{MI} < T < T_N$.

At (0 0 1) the scattering intensity in the low-temperature insulating phase is maximum when the b axis lies in the scattering plane ($\psi = 0°$) and zero when the a axis is parallel to the scattering plane ($\psi = 90°$). Above T_{MI}, in the metallic, antiferromagnetic phase, the azimuthal dependence is shifted by 90° and the maximum intensity is observed when a is parallel to the scattering plane. At position (1 1 0) the scattering intensity at low temperatures (below T_{MI}) is maximum when the ab plane is parallel to the scattering plane and the b axis is pointing towards the incoming beam ($\psi = 0°$), while the intensity is minimum when the ab plane is parallel to the scattering plane and the a axis is pointing towards the incoming beam ($\psi = 180°$). Based on the above observations and on the theory discussed in Section 3.3.2, it is concluded that the Ru magnetic moment in $Ca_3Ru_2O_7$ is parallel to the b axis in the low-temperature insulating phase

$(T < T_{\mathrm{MI}})$, but parallel to a in the magnetically ordered metallic phase $(T_{\mathrm{MI}} < T < T_N)$.

The temperature dependence of the integrated scattering intensity at reciprocal space positions (1 1 0) ($\psi = 0°$ and $\psi = 180°$) and (0 0 1) ($\psi = 0°$) is shown in Figure 3.20. At $\psi = 0°$ the most striking feature at both positions is the abrupt decrease by two orders of magnitude of the scattering intensity upon heating above T_{MI}. This decrease is caused by the reorientation of the magnetic moment and is consistent with the azimuthal dependence of the scattering intensity (Figure 3.19). Above T_{MI}, in the metallic phase, the intensity continues decreasing, dropping to zero at T_N, as expected for a second order phase transition. At $\psi = 180°$ the reorientation of the magnetic moment causes an increase of the scattering intensity by about one order of magntitude upon heating through T_{MI}. The intensity then decreases again in the metallic phase and vanishes at

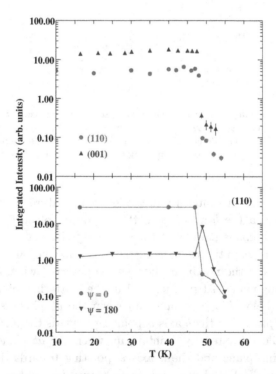

Fig. 3.20.   Temperature dependence of the integrated scattering intensity(a) at positions (1 1 0) and (0 0 1) at $\psi = 0°$, and (b) at position (1 1 0) at azimuth $\psi = 0°$ and $\psi = 180°$.[18]

T_N. From the comparison of the maximum intensities in the two phases above and below T_{MI}, the amplitude ratio of the sublattice magnetizations can be determined. The calculated value is approximately equal to $1/\sqrt{3}$.

Besides the above information about the magnetic properties of $Ca_3Ru_2O_7$, resonant x-ray diffraction can also help identify whether an orbitally ordered state is established in the system. Any kind of antiferro-orbital order, leading to a doubling of the unit cell along at least one of the three crystallographic directons, like e.g. the order suggested by Hotta and Dagotto for Ca_2RuO_4, would give rise to superstructure reflections at positions not allowed by the space group. A search for such reflections was carried out at several high-symmetry positions in reciprocal space, including $(1/2 \ 1/2 \ 0)$, $(1/2 \ 0 \ 0)$, $(0 \ 1/2 \ 0)$, $(1 \ 0 \ 0)$ and $(0 \ 1 \ 0)$.[18] No scattering intensity above the background was detected at any of these positions.

Looking for ferro-orbital order is more difficult, because a uniform orbital polarization gives rise to reflections at the same positions as the much stronger lattice Bragg reflections. To overcome this difficulty, the resonant x-ray interference technique can be used, as explained in References 47,48. The scattered intensity at a crystallographically allowed reciprocal space position is measured for two different azimuthal angles of the polarization analyzer, each of which corresponds to a small deviation from the ideal $\sigma \rightarrow \pi'$ scattering geometry (see setup in Figure 3.1). The difference between these intensities is the interference term. If a ferro-orbital order contribution is present in the total scattering intensity, its azimuthal dependence should give rise to a non-zero interference term (the Thomson scattering from the lattice is not azimuth dependent). Figure 3.21 shows the variation with energy of the interference term in $Ca_3Ru_2O_7$, as calculated for analyzer angles $\phi_A = 85°$ and $\phi_A = 95°$. No significant intereference term above background is observed either at azimuth $\psi = 0°$ or $\psi = 90°$. No indication for orbital order therefore exists.

Given the successful observation of orbital order in Ca_2RuO_4 using RXD, the absence of any indication of such an order in $Ca_3Ru_2O_7$ leads to the following conclusion: if any orbital polarization does actually exist in the bilayer compound, then it has to be much weaker than in the single layer system and lie below the sensitivity of our technique. Considering the much lower metal-insulator transition temperature in $Ca_3Ru_2O_7$ compared to Ca_2RuO_4, this conclusion does not come unexpected.

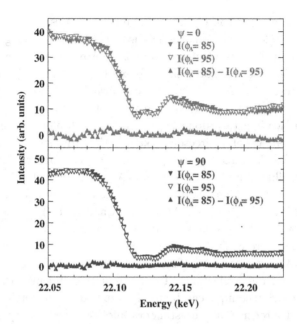

Fig. 3.21. Variation with energy of the scattering intensity of the (0 6 0) reflection around the K egde of Ru for azimuthal positions of the analyzer $\phi_A = 85°$ and $\phi_A = 95°$, as well as of their difference.[18] Data are shown for two different azimuthal positions of the sample, $\psi = 0°$ (upper panel) and $\psi = 90°$ (lower panel), for sample temperature 20 K.

3.6. Magnetic Structure Determination in RuSr$_2$GdCu$_2$O$_8$

3.6.1. *Introduction*

The hybrid ruthenocuprate system RuSr$_2$GdCu$_2$O$_8$ (Ru1212) has attracted significant scientific interest in recent years, mainly due to the intriguing coexistence of long-range magnetic order and superconductivity that it exhibits. The compound was synthesized for the first time by Bauernfeind in 1995.[49] It is a layered system, with a crystal unit cell consisting of alternating Ru-O and Cu-O planes. The Ru spins in the Ru-O planes order antiferromagnetically below the Néel temperature of $T_N = 136(2)$ K,[50] while the Cu-O planes become superconducting below a transition temperature of $T_c \approx 15 - 46$ K.[51–53] Due to the relatively high value of T_c, magnetism and superconductivity coexist within a fairly broad temperature range. This makes RuSr$_2$GdCu$_2$O$_8$ an ideal compound for investigating the interplay between the two cooperative phenomena and determining how and to what extent the corresponding order parameters interact with each other. Two major experimental difficulties have prevented the determination of the magnetic structure of the material for a number of years: 1) the difficulty to grow large single crystals of the compound, and 2) the presence of the highly neutron-absorbing element Gd in the chemical formula. The volume of the available single crystals is far too small for neutron diffraction to be applicable. In addition, the large neutron absorption coefficient of Gd dramatically reduces the scattering intensity in powder neutron diffraction studies.

3.6.2. *Main Properties*

The crystal unit cell of $RuSr_2GdCu_2O_8$ is illustrated in Figure 3.22. The Ru ions are located at the centers of corner-sharing RuO_6 octahedra, the equatorial planes of which form Ru-O layers that extend parallel to the *ab* crystallographic plane at the top and the bottom of the unit cell. The copper ions are located at the centers of the bases of corner-sharing CuO_5 square pyramids, forming Cu-O layers that are parallel to the Ru-O planes and are separated by the central gadolinium ion of the cell. The system is isostructural to the high-temperature superconductor $YBa_2Cu_3O_{6+x}$ with the yttrium, barium and chain copper ions being replaced by gadolinium, strontium and ruthenium, respectively. The crystal structure is tetragonal (space-group $P4/mmm$) with room-temperature lattice parameters $a = b = 3.836$ Å and $c = 11.563$ Å.[54] The c/a axis ratio of 3.014 at 300 K is very close to the ideal value for a triple perovskite structure. Due to the mismatch between the lengths of the in-plane Ru-O and Cu-O bonds, the RuO_6 octahedra are rotated around the crystallographic *c*-axis by about 13.8°,[54] leading to a lifting of the inversion symmetry of the four equatorial oxygen ions. A slight tilting of the octahedra that reduces the Cu-O-Ru angle to 173° is also observed.[55] These structural distortions are not significantly different at low temperatures.

The formal oxidation states of ruthenium and copper in $RuSr_2GdCu_2O_8$ are close to $+5$ and $+2$, respectively. However, the exact values are determined by the charge transfer between the Ru-O and Cu-O planes. If p is the amount of charge transferred between the planes, the oxidation states are $5 - 2p$ and $2 + p$, respectively. For $p = 0$ the oxidation state of Ru is exactly $+5$, and there are three electrons in the Ru $4d$ band. Under the influence of the crystal field, which has an O_h symmetry, the electronic states of the $4d$ band are split into t_{2g} and e_g levels. The three $4d$ electrons

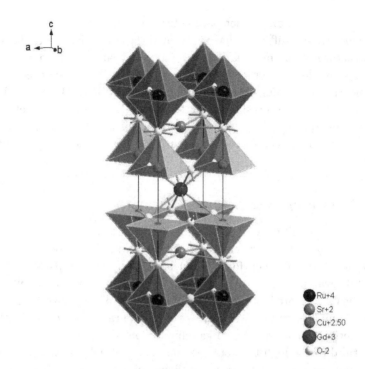

Fig. 3.22. Crystal structure of RuSr$_2$GdCu$_2$O$_8$ (space-group $P4/mmm$). The Ru ions are located at the centers of corner-sharing RuO$_6$ octahedra, the equatorial planes of which form Ru-O layers, that extend parallel to the ab crystallographic plane at the top and the bottom of the unit cell. The Cu ions are located at the centers of the bases of corner-sharing CuO$_5$ square pyramids, forming Cu-O layers, which are parallel to the Ru-O planes and are separated by the central Gd ion of the cell.

then occupy the three lower-lying t$_{2g}$ states, leading to a $t_{2g}^3 e_g^0$ configuration. Nevertheless, x-ray absorption near-edge studies clearly indicate a non-zero charge transfer between the Ru-O and Cu-O planes, with a p value around 0.2,[57] resulting in an average Ru oxidation state of +4.6. Other studies have suggested even higher values for p.[55] The latter would be, however, not consistent with the superconducting properties exhibited by RuSr$_2$GdCu$_2$O$_8$.

Long-range magnetic order is established in the Ru-O layers of RuSr$_2$GdCu$_2$O$_8$ below a critical temperature of approximately 136 K. The magnetic structure of the system remains controversial. While dc-magnetization and muon spin rotation (μ-SR) studies initially suggested a ferromagnetic type of order,[51] later neutron powder diffraction investiga-

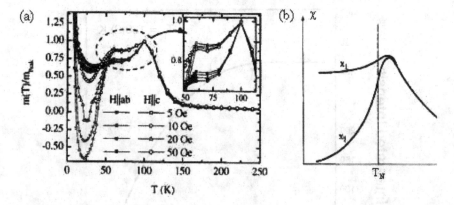

Fig. 3.23. (a) Magnetization measurements[56] on RuSr$_2$GdCu$_2$O$_8$ single crystals of the same origin as the ones used in the x-ray diffraction investigations of Ref. 19. The theoretically expected temperature dependence of the magnetic susceptibility χ of an antiferromagnet, for applied magnetic field perpendicular ($\perp$) and parallel ($\parallel$) to the magnetic moment direction, is shown in (b). The comparison with the experimental results reveals that the magnetic moment is not along c, but has instead a significant ab-component.

tions showed that the order of the Ru spins is predominantly antiferromagnetic in all three crystallographic directions (G-type antiferromagnetism), with only a small ferromagnetic component of up to 0.1 μ_B, which is enhanced in the presence of applied magnetic field.[50] The predominantly antiferromagnetic character of the spin order was supported by subsequent magnetization studies, carried out on a Eu-substituted compound.[58] Other magnetization studies concluded instead that, while the coupling between neighboring RuO$_2$ planes is antiferromagnetic, the interactions within every plane are rather ferromagnetic, i.e a 'type-I' antiferromagnetic order is established.[59] Recent nuclear magnetic resonance (NMR) measurements also support the type-I-antiferromagnetism scenario,[60] but the inconsistency of such an order with the doubling of the crystal unit cell observed in the neutron diffraction experiments has not been satisfactorily explained so far.

The temperature dependence of the magnetic order parameter of the Ru spin system as determined by neutron powder diffraction is shown in Figure 3.24. The magnetic phase transition is observed via the vanishing of the integrated scattering intensity of the $(\frac{1}{2} \frac{1}{2} \frac{1}{2})$ magnetic Bragg peak at $T_N = 136$ K. Below T_N, the magnetic intensity increases smoothly with decreasing temperature and reaches its saturation value around 60 K. The

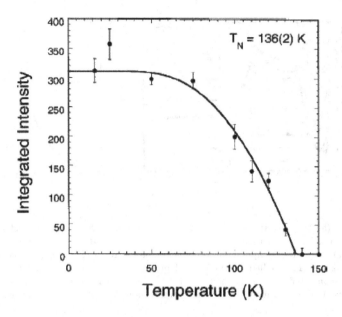

Fig. 3.24. Temperature dependence of the magnetic scattering intensity from the Ru spin system in $RuSr_2GdCu_2O_8$, as determined with powder neutron diffraction.[50] The magnetic phase transition is clearly observed at $T_N = 136$ K.

intensity remains constant thereafter, with no anomalies observed down to the lowest temperatures investigated.

The direction of the Ru magnetic moment in $RuSr_2GdCu_2O_8$ has been for a long time an issue of debate. The neutron diffraction data suggest a magnetic moment which is aligned along the crystallographic c-axis and has a magnitude of approximately 1.18 μ_B. This conclusion was drawn based on the intensity ratio of two magnetic reflections and was rather tentative. On the contrary, the above mentioned magnetization and NMR studies[59,60] strongly suggest a moment direction within the ab plane, as also indicated by the early μ-SR results.[51] More recent magnetization measurements performed for the first time on *single-crystal* $RuSr_2GdCu_2O_8$ samples could also not confirm the moment direction suggested by the neutron experiments.[56,61] The variation with temperature of the magnetization for applied magnetic field parallel and perpendicular, respectively, to the c direction is shown in Figure 3.23(a). A comparison of the data with the theoretically expected temperature dependence of the magnetic susceptibility χ of an antiferromagnet, for applied magnetic field perpendicular ($\perp$) and parallel ($\parallel$) to the magnetic moment direction (Figure 3.23(b)), reveals that the Ru spins in $RuSr_2GdCu_2O_8$ are not aligned along c, but have a significant component in the ab-plane. How large exactly this component is, cannot be determined by the magnetization data.

Independent from the Ru ions, the Gd ions also order antiferromagnetically along all three crystallographic directions. The ordering in the Gd spin system is established at the much lower temperature of $T_{N,Gd} = 2.50$ K. The magnetic moment is aligned along the c-axis, and has a magnitude of 7 μ_B. The fact that the Ru and Gd magnetic structures interact only very weakly with each other, although they correspond to identical spin configurations, is due to the position in the crystal of the Gd ions. The latter are located at the body-center of the simple tetragonal Ru lattice, symmetrically positioned with regard to the Ru ions of the nearest Ru-O planes. The antiferromagnetic alignment of the Ru spins results in an average cancellation of the magnetic interaction between the Ru and Gd ions.

While the Cu-O planes play no role in the above described magnetic ordering, they are of crucial importance when it comes to the second cooperative phenomenon that characterizes the low-temperature properties of $RuSr_2GdCu_2O_8$: superconductivity. The occurence of bulk superconductivity in the system has been confirmed with resistivity, thermopower and heat-capacity measurements, which determined a Meissner fraction of up to approximately 100% at 4.2 K.[52] The superconducting transition temperature T_c is strongly dependent on the preparation conditions of the crystal. Depending on the synthesis and annealing procedures used, T_c can have values between approximately 15 K and 46 K. The superconducting and magnetic order parameters interact only very weakly with each other.

3.6.3. X-Ray Investigations

Resonant x-ray diffraction investigations at the L_{II} and L_{III} absorption edges of Ru have revealed the magnetic structure of the electron system and the exact direction of the Ru magnetic moment, reconciling previous contradictory reports.[19] The studies were performed using tiny single crystals with sizes of the order of $100 \times 100 \times 50\ \mu m^3$.

Figure 3.25 shows the variation with energy of the magnetic scattering intensity around the Ru L_{II} edge at reciprocal space positions $(\frac{1}{2}\ \frac{1}{2}\ \frac{1}{2})$ and $(\frac{1}{2}\ \frac{1}{2}\ \frac{3}{2})$. The lineshape is the same as in the other ruthenate systems previously discussed. Polarization analysis shows that the intensity only has a $\sigma \to \pi'$ component, excluding the possibility of charge scattering

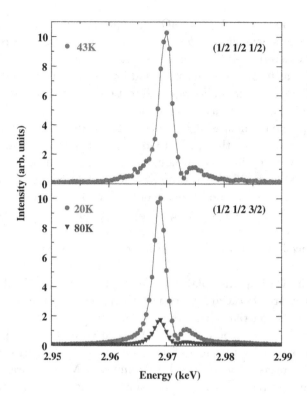

Fig. 3.25.   Variation with energy of the magnetic scattering intensity around the Ru L_{II} edge at positions $(\frac{1}{2} \ \frac{1}{2} \ \frac{1}{2})$ (upper panel) and $(\frac{1}{2} \ \frac{1}{2} \ \frac{3}{2})$ at temperatures indicated on the plot.[19]

contributing to the observed signal. The observation of magnetic intensity at half-integer positions demonstrates a doubling of the unit cell in all crystallographic directions as a result of the magnetic ordering. The results indicates a G-type antiferromagnetic ordering, consistent with the powder neutron diffraction work.

Figure 3.26 shows the temperature dependence of the integrated scattering intensity at reciprocal space positions $(\frac{1}{2} \ \frac{1}{2} \ \frac{1}{2})$ and $(\frac{1}{2} \ \frac{1}{2} \ \frac{1}{3})$. The intensity decreases smoothly with increasing temperature and vanishes at 102 K in an order-parameter-like fashion. The magnetic ordering temperature determined from the x-ray data agrees well with the one found from magnetization measurements, shown in an inset in the same figure. No anomaly in the temperature dependence is observed close to the onset of superconductivity. This is in agreement with previous studies which

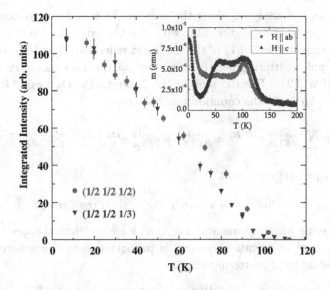

Fig. 3.26. Variation with temperature of the integrated scattering intensity of the $(\frac{1}{2}\ \frac{1}{2}\ \frac{1}{2})$ and $(\frac{1}{2}\ \frac{1}{2}\ \frac{1}{3})$ magnetic reflections, as determined from $\theta - 2\theta$-scans at azimuth $\psi = 0°$ at the Ru L_{II} edge.[19] The magnetic phase transition is clearly observed at 102 K. The Néel temperature agrees well with the one determined by magnetization measurements shown in the inset, which were carried out on single crystals at 100 Oe.

indicated no coupling between magnetic order and superconductivity in $RuSr_2GdCu_2O_8$, as a result of the spatial separation between the CuO_2 and RuO_2 layers where the two cooperative phenomena are established.

For determining the magnetic moment direction, the azimuthal dependence of the scattering intensity needs to be measured. In antiferromagnets the scattering intensity is maximum when the magnetic moment lies in the diffraction plane ($\psi = 0°$). Thus, if the magnetic moment in $RuSr_2GdCu_2O_8$ were along c, as tentatively suggested by the powder neutron diffraction studies,[50] the scattering intensity would take its maximum value at the azimuthal position where c is in the diffraction plane. The RXD investigations showed, however, that the scattering intensity is instead maximum when the c-direction is 53° off the scattering plane. This means that the magnetic moment is not along c, but has a significant ab component. The exact angle between the magnetic moment and the c-direction can be deduced from the azimuthal dependence as follows:

For the two Ru ions of the $RuSr_2GdCu_2O_8$ basis we use the notation 'spin-up': ↑ and 'spin-down': ↓ to refer to their spin directions. The mag-

netic moment is analyzed along the axes of the coordinate system introduced in Section 3.3 (Figure 3.2), with the $\hat{U}_3$ axis being antiparallel to
the scattering vector (1 1 1) of the magnetic reflection. The total resonant
electric dipole scattering length in $\sigma \to \pi'$ polarization geometry is given
by Equation 3.24. For the ions of the $RuSr_2GdCu_2O_8$ basis ($R_1 = 0$,
$R_2 = a\hat{x} + b\hat{y} + c\hat{z}$) the equation takes the form:

$$g_{E1}^{\sigma \to \pi'} = \sum_{j=1}^{2} g_{E1,S_j}^{\sigma \to \pi'} e^{i\boldsymbol{Q} \cdot \boldsymbol{R}_j} = g_{E1,\uparrow}^{\sigma \to \pi'} + g_{E1,\downarrow}^{\sigma \to \pi'} e^{i3\pi} = g_{E1,\uparrow}^{\sigma \to \pi'} - g_{E1,\downarrow}^{\sigma \to \pi'} \quad (3.29)$$

and, due to (3.21), (3.23):

$$g_{E1}^{\sigma \to \pi'} = -2iF^{(1)} \cos \alpha \sin \theta + 2iF^{(1)} \sin \alpha \cos \theta \cos \psi \quad (3.30)$$

The scattering intensity measured at the $(\frac{1}{2} \frac{1}{2} \frac{1}{2})$ reciprocal space position
in the $\sigma \to \pi'$ polarization channel is proportional to the square of the
corresponding total scattering length:

$$I_{(\frac{1}{2}\frac{1}{2}\frac{1}{2})}^{\sigma \to \pi'} \propto |g_{E1}^{\sigma \to \pi'}|^2 = |2iF^{(1)}(\sin \alpha \cos \theta \cos \psi - \cos \alpha \sin \theta)|^2 \implies$$

$$I_{(\frac{1}{2}\frac{1}{2}\frac{1}{2})}^{\sigma \to \pi'} \propto (F^{(1)})^2 (\sin^2 \alpha \cos^2 \theta \cos^2 \psi - 2 \sin \alpha \cos \alpha \sin \theta \cos \theta \cos \psi$$

$$+ \cos^2 \alpha \sin^2 \theta) \quad (3.31)$$

In the $\sigma \to \sigma'$ scattering geometry, Equation 3.27 gives:

$$g_{E1}^{\sigma \to \sigma'} = \sum_{j=1}^{2} g_{E1,S_j}^{\sigma \to \sigma'} e^{i\boldsymbol{Q} \cdot \boldsymbol{R}_j} = g_{E1,\uparrow}^{\sigma \to \sigma'} + g_{E1,\downarrow}^{\sigma \to \sigma'} e^{i3\pi} = g_{E1,\uparrow}^{\sigma \to \sigma'} - g_{E1,\downarrow}^{\sigma \to \sigma'} \quad (3.32)$$

and finally, due to (3.26):

$$g_{E1}^{\sigma \to \sigma'} = g_{E1,\uparrow}^{\sigma \to \sigma'} - g_{E1,\downarrow}^{\sigma \to \sigma'} = 0 \quad (3.33)$$

Therefore, the scattering intensity in $\sigma \to \sigma'$ is zero:

$$I_{(\frac{1}{2}\frac{1}{2}\frac{1}{2})}^{\sigma \to \sigma'} = 0 \quad (3.34)$$

Since the $\sigma \to \sigma'$ polarization channel gives no contribution to the total
scattered intensity, no polarization analysis of the diffracted signal is needed
in the experiment – the measured intensity is of purely $\sigma \to \pi'$ origin and
can be described by Equation 3.31, with a scattering angle θ of 23.37°.

Figure 3.27 shows the azimuthal dependence of the integrated scattering
intensity at the magnetic reflection $(\frac{1}{2} \frac{1}{2} \frac{1}{2})$ at temperature 45 K, well
below the magnetic ordering temperature. The experimental data points
can be fitted using the theoretical expression 3.31 (Figure 3.27). From

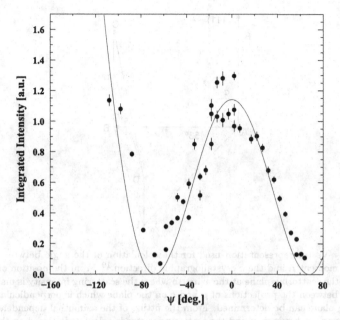

Fig. 3.27. Azimuthal dependence of the integrated scattering intensity at the $(\frac{1}{2}\,\frac{1}{2}\,\frac{1}{2})$ reciprocal space position at the Ru L_{II} edge, as determined from $\theta - 2\theta$-scans at $T = 45$ K.[19] The solid line is a fit of the data to the theoretical expression (3.31), calculated from the corresponding resonant electric dipole scattering length. From the fitting, the angle α between the magnetic moment and the scattering vector is among others determined.

the fit the values of the parameters α and $F^{(1)}$ are extracted. For $F^{(1)}$ we find: $F^{(1)} = 2.47 \pm 0.10$, while for the angle α we obtain the value: $\alpha = (0.856 \pm 0.020)$ rad $= 49.0° \pm 1.1°$.

As previously mentioned, based on the fact that the maximum of the scattered intensity occurs at the azimuthal position at which the c-direction is 53° off the scattering plane, we draw the conclusion that the angle $(\widehat{\mu_p,\,c_p})$ (Figure 3.28) between the projections of the magnetic moment μ and of the c-axis direction, respectively, on the plane which is perpendicular to the scattering vector (111) is equal to 53°. Thus: $\widehat{BAE} = (\widehat{\mu_p,\,c_p}) = 53°$. From this, we calculate the angle $\widehat{CAD} = (\widehat{\mu,\,c})$ between μ and c. We find: $\widehat{CAD} = 53.8°$.

To summarize, the direction of the magnetic moment μ is along a reciprocal space vector $(h\;k\;l)$ which forms an angle of $\alpha = 49°$ with $(1\;1\;1)$ and of 53.8° with $(0\;0\;1)$. This corresponds approximately to the $(1\;0\;2)$ reciprocal space direction: $(\widehat{(102),\,(111)}) = 45.6°$ and $(\widehat{(102),\,(001)}) = 56°$.

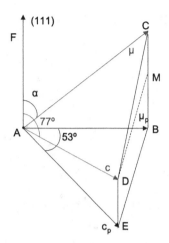

Fig. 3.28. Vector representation used for the calculation of the angle between the Ru magnetic moment μ and the c crystallographic direction.[19] From the position of c with regard to the scattering plane at the azimuth where the scattering intensity is maximum, the angle between the projections of μ and c on the plane which is perpendicular to the scattering plane can be determined. From the fitting of the azimuthal dependence (Fig. 3.27) the angle α between μ and the scattering vector is calculated. Based on these two angles, the direction of the magnetic moment can be estimated.

Thus (1 0 2) can be considered as the approximate direction of the magnetic moment.

It has to be emphasized that this result is still consistent with the previously mentioned neutron data. Indeed, the intensity ratio calculated when the magnetic moment is assumed along (1 0 2) is exactly the same with the one obtained with the moment along c, and very close to the experimentally found value of 2.49. In other words, having the magnetic moment along c is not the only possibility for the magnetic moment direction that satisfies the condition of giving an intensity ratio of approximately 2.49 for the $(\frac{1}{2} \frac{1}{2} \frac{1}{2})$, $(\frac{1}{2} \frac{1}{2} \frac{3}{2})$ peaks.

Representation analysis shows that the magnetic structure determined by RXD is not compatible with the crystallographix space-group *Pbam*. This means either that small structural distortions are present which lower the symmetry of the space-group and which have not been observed by powder neutron diffraction, or that terms in the spin Hamiltonian of order higher than the usual bilinear exchange coupling have to be taken into consderation. Such terms could arise, for example, from charge and/or orbital fluctuations in the RuO_2 layers.

3.7. Spin Orbital Mott State in Sr_2IrO_4

3.7.1. *Introduction*

In transition metal compounds which contain heavy $5d$ elements the relativistic spin-orbit coupling (SOC) is at least one order of magnitude larger than in $3d$ electron systems, and unconventional electron states may be the result. In particular, iridium oxides have recently captured significant attention in view of the analogy between their electronic structure and the one of the cuprate superconductors, and possible applications in topological quantum computation.[62,63] As in the case of $RuSr_2GdCu_2O_8$ discussed above, neutron diffraction studies of this important class of materials are difficult because of the large neutron absorption cross section of Ir. As a consequence, basic questions about the magnetic order of iridium oxides have remained unanswered until recently.

This includes the magnetic structure of Sr_2IrO_4, which is isostructural and electronically closely related to La_2CuO_4, the progenitor of a well known family of high-temperature superconductors. The weak ferromagnetism of Sr_2IrO_4 has long remained an issue of debate. Particularly puzzling has been the unusually large weak ferromagnetic moment of the layered system, which is two orders of magntitude larger than in La_2CuO_4. The phenomenon has been recently explained based both on theoretical work[64] and on resonant x-ray diffraction investigations.[65]

3.7.2. *Main Properties*

Sr_2IrO_4 is a layered perovskite with a square lattice of Ir^{+4} ions formed by corner-sharing IrO_6 octahedra. The octahedra are elongated along the c-axis and rotated about it by approximately $11°$.[66] The system is in a low-spin $5d^5$ electron configuration, with five electrons in the almost triply degenerate t_{2g} orbitals and empty e_g states. It is a Mott insulator, which undergoes a magnetic transition at 240 K, becoming a weak ferromagnet with an ordered magnetic moment of approximately 0.14 μ_B.[67] It has been sugested that it is the strong spin-orbit coupling which leads to this anomalously large magnetic moment.[64] The low-energy Hamiltonian is exclusively determined by lattice distortions, resulting in a spin canting angle which is almost as large as the rotation angle of the IrO_6 octahedra.

3.7.3. *X-Ray Investigations*

In the resonant x-ray scattering investigations presented in this chapter so far, two very important advantages of the RXD technique have been clearly demonstrated: its sensitivity to the direction of the local magnetic moment, and its ability to directly probe the valence electron orbitals. A third important advantage was used for the investigation of the magnetic properties of Sr_2IrO_4: the ability of the technique to provide information about the phase of the wave function of the valence electrons. The interference between all possible scattering paths that a core electron can follow during the electric dipole transition from the core level to the empty state above the Fermi level is reflected to the scattering intensity of the incoming photon.

Single crystals of Sr_2IrO_4 were investigated at the Ir L_{II} and L_{III} absorption edges.[65] The variation with energy of the scattering intensity around the two edges at the position of the magnetic reflection (1 0 22) for temperature 10 K is shown in Figure 3.29.

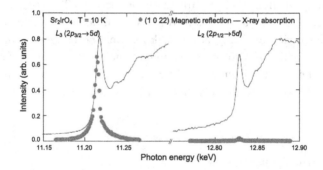

Fig. 3.29. Variation with energy of the magnetic scattering intensity around the Ir L_{II} and L_{III} absorption edges at position (1 0 22) of the reciprocal space and temperature 10 K.[65] The resonant enhancement at the L_{III} edge is two orders of magnitude larger than at L_{II}. XAS data (solid curves) are also shown for comparison.

X-ray absorption specroscopy (XAS) data are also shown in the figure (solid curves). The resonant enhancement at L_{III} is about 100 times larger than at L_{II}. The difference is due to the fact that the interference between the various scattering paths related to the virtual electric dipole transition is contsructive at L_{III} and destructive at L_{II}. Based on the ratio of the scattering amplitudes at the two edges, a ground state with total angular momentum J_{eff} very close to $1/2$ is concluded.[65] The magnetic state cannot be described as a $S = 1/2$ state, like in conventional Mott insulators. Due to the strong spin-orbit coupling, only the total effective angular momentum J_{eff} can be used to describe the state. This consists of the spin angular momentum $S = 1/2$ and the effective orbital angular momentum $L = -1$. The conclusion is in good agreement with the above mentioned theoretical studies (Ref. 64). Based on the positions of the magnetic reflections, the ordering pattern of the Ir moments can be determined. In absence of applied magnetic field, reflections $(1\ 0\ 4n+2)$ and $(0\ 1\ 4n)$ are magnetically allowed, indicating an antiferromagnetic ordering of the J moments within the IrO_2 layers. The canting of the moments results in an in-plane net moment which is antiferromagnetically ordered along the c direction. This stacking pattern is indicated by the presence of reflections at positions $(0\ 0\ L)$, with L=odd. Applying a magnetic field stronger than a critical H_c makes the $(1\ 0\ 4n+2)$ reflections vanish, while new reflections appear at $(0\ 0\ L)$, L=odd. This indicates that the net in-plane moments are aligned ferromagnetically under the influenece of the magnetic field, producing a total net moment. The reciprocal space scans showing the positions of the magnetic reflections is shown in Figure 3.30. The arrangement of the J

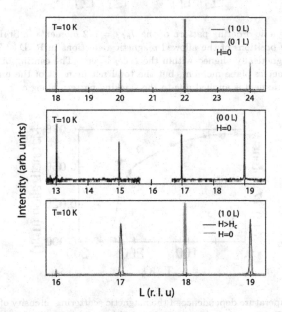

Fig. 3.30.   Reciprocal space scans along the L direction of the reciprocal space, showing several magnetic reflections.[65] From the positions of the observed reflections, the arrangement of the J moments can be determined.

moments without and with applied magnetic field is shown in Figure 3.31. Figure 3.32 shows the temperature dependence of the intergrated scattering intensity of reflection (1 0 17) for two different applied magnetic fields

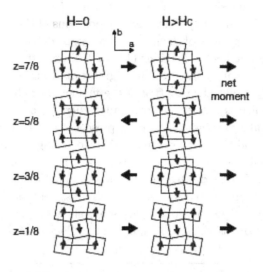

Fig. 3.31. Magnetic ordering pattern of the $J_{eff} = 1/2$ moments in Sr_2IrO_4 as determined from the positions of the allowed magnetic reflections in RXD.[65] The moments are antiferromagnetically aligned within the IrO_2 layers. The canting of the moments gives rise to a net in-plane moment, but the total net moment of the unit cell is zero when no magnetic field is applied due to the stacking pattern along c.

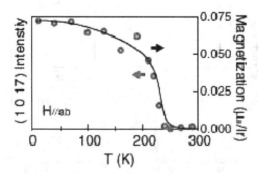

Fig. 3.32. Temperature dependence of the magnetic scattering intensity of the magnetic reflection (1 0 17) for applied magnetic field 0.3 T (red points) and 0.5 T (solid curve).[65] The magnetic phase transition shifts to higher temperatures with increasing magnetic field.

stronger than H_c. The magnetic ordering transition is clearly observed via the vanishing of the scattering intensity at 240 K.

References

1. B. Warren, *X-ray Diffraction*. (Addison-Wesley, 1969).
2. J. Als-Nielsen, *Elements of Modern X-ray Physics*. (John Wiley & Sons Ltd, 2001).
3. S. Lovesey and S. Collins, *X-Ray Scattering and Absorption by Magnetic Materials*. (Oxford University Press Inc., New York, 1996).
4. N. Ashcroft and N. Mermin, *Solid State Physics*. (Saunders College Publishing, 1976).
5. F. de Bergevin and M. Brunel, *Acta Cryst.* **A37**, 314, (1981).
6. K. Namikawa, M. Ando, T. Nakajima, and H. Kawata, *J. Phys. Soc. Jpn.* **54**, 4099, (1985).
7. D. Gibbs, D. Harshman, E. Isaacs, D. McWhan, D. Mills, and C. Vettier, *Phys. Rev. Lett.* **61**, 1241, (1988).
8. E. Isaacs, D. McWhan, C. Peters, G. Ice, D. Siddons, J. Hastings, C. Vettier, and O. Vogt, *Phys. Rev. Lett.* **62**, 1671, (1989).
9. Y. Murakami, J. Hill, D. Gibbs, M. Blume, I. Koyama, M. Tanaka, H. Kawata, T. Arima, Y. Tokura, K. Hirota, and Y. Endoh, *Phys. Rev. Lett.* **81**, 582, (1998).
10. H. Nakao, Y. Wakabayashi, T. Kiyama, Y. Murakami, M. v. Zimmermann, J. Hill, D. Gibbs, S. Ishihara, Y. Taguchi, and Y. Tokura, *Phys. Rev. B.* **66**, 184419, (2002).
11. S. Grenier, J. Hill, D. Gibbs, K. Thomas, M. v. Zimmermann, C. Nelson, V. Kiryukhin, Y. Tokura, Y. Tomioka, D. Casa, T. Gog, and C. Venkataraman, *Phys. Rev. B.* **69**, 134419, (2004).
12. A. Stunault, F. de Bergevin, D. Wermeille, C. Vettier, T. Brückel, N. Bernhoeft, G. J. McIntyre, and J. Y. Henry, *Phys. Rev. B.* **60**, 10170, (1999).
13. S. Wilkins, P. Hatton, M. Roper, D. Prabhakaran, and A. Boothroyd, *Phys. Rev. Lett.* **90**, 187201, (2003).
14. S. Dhesi, A. Mirone, C. D. Nadaï, P. Ohresser, P. Bencok, N. Brookes, P. Reutler, A. Revkolevschi, A. Tagliaferri, O. Toulemonde, and G. van der Laan, *Phys. Rev. Lett.* **92**, 056403, (2004).
15. K. Thomas, J. Hill, Y.-J. Kim, P. Abbamonte, L. Venema, A. Rusydi, Y. Tomioka, Y. Tokura, D. McMorrow, and M. van Veenendaal, *Phys. Rev. Lett.* **92**, 237204, (2004).
16. D. McMorrow, S. Nagler, K. McEwen, and S. Brown, *J. Phys.: Condens. Matter.* **15**, L59–L66, (2003).
17. I. Zegkinoglou, J. Strempfer, C. Nelson, J. Hill, J. Chakhalian, C. Bernhard, J. Lang, G. Srajer, H. Fukazawa, S. Nakatsuji, Y. Maeno, and B. Keimer, *Phys. Rev. Lett.* **95**, 136401, (2005).
18. B. Bohnenbuck, I. Zegkinoglou, J. Strempfer, C. S. ler Langeheine, C. Nelson, P. Leininger, H.-H. Wu, E. Schierle, J. Lang, G. Srajer, S. Ikeda, Y. Yoshida,

K. Iwata, S. Katano, N. Kikugawa, and B. Keimer, *Phys. Rev. B.* **77**, 224412, (2008).

19. B. Bohnenbuck, I. Zegkinoglou, J. Strempfer, C. Nelson, H.-H. Wu, C. S. ler Langeheine, M. Reehuis, E. Schierle, P. Leininger, T. Hermannsdörfer, J. Lang, G. Srajer, C. Lin, and B. Keimer, *Phys. Rev. Lett.* **102**, 037205, (2009).

20. J. Hannon, G. Trammel, M. Blume, and D. Gibbs, *Phys. Rev. Lett.* **61**, 1245, (1988).

21. J. Hill and D. McMorrow, *Acta Cryst.* **A52**, 236, (1996).

22. Y. Maeno, H. Hashimoto, K. Yoshida, S. Nishizaki, T. Fujita, J. Bednorz, and F. Lichtenberg, *Nature (London).* **372**, 532, (1994).

23. K. Ishida, H. Mukuda, Y. Kitaoka, K. Asayama, Z. Mao, Y. Mori, and Y. Maeno, *Nature (London).* **396**, 658, (1998).

24. M. Braden, G. André, S. Nakatsuji, and Y. Maeno, *Phys. Rev. B.* **58**, 847, (1998).

25. V. Anisimov, I. Nekrasov, D. Kondakov, T. Rice, and M. Sigrist, *Eur. Phys. J. B.* **25**, 191, (2002).

26. T. Hotta and E. Dagotto, *Phys. Rev. Lett.* **88**, 017201, (2002).

27. J. Lee, Y. Lee, T. Noh, S.-J. Oh, J. Yu, S. Nakatsuji, H. Fukazawa, and Y. Maeno, *Phys. Rev. Lett.* **89**, 257402, (2002).

28. J. Jung, Z. Fang, J. He, Y. Kaneko, Y. Okimoto, and Y. Tokura, *Phys. Rev. Lett.* **91**, 056403, (2003).

29. O. Friedt, M. Braden, G. André, P. Adelmann, S. Nakatsuji, and Y. Maeno, *Phys. Rev. B.* **63**, 174432, (2001).

30. Z. Fang, N. Nagaosa, and K. Terakura, *Phys. Rev. B.* **69**, 045116, (2004).

31. T. Mizokawa, L. Tjeng, G. Sawatzky, G. Ghiringhelli, O. Tjernberg, N. Brookes, H. Fukazawa, S. Nakatsuji, and Y. Maeno, *Phys. Rev. Lett.* **87**, 077202, (2001).

32. T. Brückel, D. Hupfeld, J. Strempfer, W. Caliebe, K. Mattenberger, A. Stunault, N. Bernhoeft, and G. McIntyre, *Eur. Phys. J. B.* **19**, 475, (2001).

33. N. Bernhoeft, *Acta Cryst. A.* **55**, 274, (1999).

34. M. Kubota, Y. Murakami, M. Mizumaki, H. Oshumi, N. Ikeda, S. Nakatsuji, H. Fukazawa, and Y. Maeno, *Phys. Rev. Lett.* **95**, 026401, (2005).

35. S. M. abd G. Cao and J. Crow, *Phys. Rev. B.* **67**, 094427, (2003).

36. X. Lin, Z. Zhou, V. Durairaj, P. Schlottmann, and G. Cao, *Phys. Rev. Lett.* **95**, 017203, (2005).

37. C. Nelson, H. Mo, B. Bohnenbuck, J. Strempfer, N. Kikugawa, S. Ikeda, and Y. Yoshida, *Phys. Rev. B.* **75**, 212403, (2007).

38. G. Cao, L. Balicas, Y. Xin, J. Crow, and C.S.Nelson, *Phys. Rev. B.* **67**, 184405, (2003).

39. C. Snow, S. Cooper, G. Cao, J. Crow, H. Fukazawa, S. Nakatsuji, and Y. Maeno, *Phys. Rev. Lett.* **89**, 226401, (2002).

40. J. Karpus, C. Snow, R. Gupta, H. Barath, S. Cooper, and G. Cao, *Phys. Rev. B.* **73**, 134407, (2006).

41. G. Cao, K. Abbound, S. McCall, J. Crow, and R. Guertin, *Phys. Rev. B.* **62**, 998, (2000).

42. V. Varadarajan, S. Chikara, V. Durairaj, X. Lin, G. Cao, and J. Brill, *Solid State Commun.* **141**, 402, (2007).
43. R. Borzi, S. Grigera, J. Farrell, R. Perry, S. Lister, S. Lee, D. Tennant, Y. maeno, and A. Mackenzie, *Science.* **315**, 214, (2007).
44. Y. Yoshida, S. Ikeda, H. Matsuhata, N. Shirakawa, C. Lee, and S. Katano, *Phys. Rev. B.* **72**, 054412, (2005).
45. Y. Yoshida, I. Nagai, S. Ikeda, N. Shirakawa, M. Kosaka, and N. Môri, *Phys. Rev. B.* **69**, 220411R, (2004).
46. G. Cao, L. Balicas, X. Lin, S. Chirkara, E. Elhami, V. Durairaj, J. Brill, R. Rai, and J. Crow, *Phys. Rev. B.* **69**, 014404R, (2004).
47. T. Kiyama, Y. Wakabayashi, H. Nakao, H. Ohsumi, Y. Murakami, M. Izumi, M. Kawasaki, and Y. Tokura, *J. Phys. Soc. Jpn.* **72**, 785, (2003).
48. H. Ohsumi, Y. Murakami, T. Kiyama, H. Nakao, M. Kubota, Y. Wakabayashi, Y. Konishi, M. Izumi, M. Kawasaki, and Y. Tokura, *J. Phys. Soc. Jpn.* **72**, 1006, (2003).
49. L. Bauernfeind, W. Widder, and H. Braun, *Physica C.* **254**, 151, (1995).
50. J. Lynn, B. Keimer, C. Ulrich, C. Bernhard, and J. Tallon, *Phys. Rev. B.* **61**, R14964, (2000).
51. C. Bernhard, J. Tallon, C. Niedermayer, T. Blasius, A. Golnik, E. Brücher, R. Kremer, D. Noakes, C. Stronach, and E. Ansaldo, *Phys. Rev. B.* **59**, 14099, (1999).
52. J. Tallon, J. Loram, G. Williams, and C. Bernhard, *Phys. Rev. B.* **61**, R6471, (2000).
53. E. Abel. Raman scattering in superconducting $RuSr_2GdCu_2O_8$ single crystal and thin films. Master's thesis, Max-Planck-Institute for Solid State Research / University of Stuttgart, (2001).
54. O. Chmaissem, J. Jorgensen, H. Shaked, P. Dollar, and J. Tallon, *Phys. Rev. B.* **61**, 6401, (2000).
55. A. McLaughlin, W. Zhou, J. Attfield, A. Fitch, and J. Tallon, *Phys. Rev. B.* **60**, 7512, (1999).
56. T. Nachtrab. *c-Achsen-Transporteigenschaften des intrinsischen Supraleiter-Ferromagnet-Hybrids $RuSr_2GdCu_2O_8$.* PhD thesis, Eberhard-Karls-Universität zu Tübingen, (2004).
57. R. Liu, L.-Y. Jang, H.-H. Hung, and J. Tallon, *Phys. Rev. B.* **63**, 212507, (2001).
58. G. Williams and S. Krämer, *Phys. Rev. B.* **62**, 4132, (2000).
59. A. Butera, A. Fainstein, E. Winkler, and J. Tallon, *Phys. Rev. B.* **63**, 054442, (2001).
60. Z. Han, J. Budnick, W. Hines, P. Klamut, M. Maxwell, and B. Dabrowski, *J. Mag. Mag. Mat.* **299**, 338, (2006).
61. T. Nachtrab, C. Bernhard, C. Lin, D. Koelle, and R. Kleiner, *C.R.Physique.* **7**, 68, (2006).
62. A. Shitade, H. Katsura, J. Kunes, X.-L. Qi, S.-C. Zhang, and N. Nagaosa, *Phys. Rev. Lett.* **102**, 256403, (2009).
63. J. Chaloupka, G. Jackeli, and G. Khaliullin, *Phys. Rev. Lett.* **105**, 027204, (2010).

64. G. Jackeli and G. Khaliullin, *Phys. Rev. Lett.* **102**, 017205, (2009).
65. B. Kim, H. Ohsumi, T. Komesu, S. Sakai, T. Morita, H. Takagi, and T. Arima, *Science.* **323**, 1329, (2009).
66. Q. Huang, J. L. Soubeyroux, O. Chmaissem, I. N. Sora, A. Santoro, R. J. Cava, J. J. Krajewski, and W. F. P. Jr., *J. Solid State Chem.* **112**, 355, (1994).
67. G. Cao, J. Bolivar, S. McCall, J. E. Crow, and R. P. Guertin, *Phys. Rev. B.* **57**, R11039, (1998).

Chapter 4

EXPLORING THE MAGNETOSTRUCTURAL PHASES OF THE LAYERED RUTHENATES WITH RAMAN SCATTERING

S. L. Cooper

Department of Physics and Frederick Seitz Materials Research Laboratory
University of Illinois, Urbana-Champaign, Illinois 61801, USA
E-mail: slcooper@illinois.edu

This chapter reviews the methods used and results obtained in inelastic light (Raman) scattering studies of the layered ruthenates; the particular focus of this review is on the temperature-, magnetic-field, and pressure-dependent magnetic and orbitally polarized phases of these materials. The results described here show clearly that structural changes associated with the RuO_6 octahedra—induced variously by temperature, atomic substitution, magnetic field, and pressure—play a definitive role in many of the exotic properties and phases observed in the layered ruthenates. In Raman scattering, this is evidenced, in particular, by anomalous behavior of the B_{1g} symmetry RuO_6 octahedral phonon mode in all the ruthenate materials studied. After a brief description of the theory of Raman scattering in solids to provide backround information for the reader (Section 4.2), the low temperature, high pressure, and high magnetic field Raman scattering methods used in these studies are discussed (Section 4.3). Finally, temperature-, magnetic field-, and pressure-dependent Raman scattering results are discussed for single layered $Ca_{2-x}Sr_xRuO_4$ (Section 4.4), bilayered $Ca_3Ru_2O_7$ (Section 4.5), and triple-layered $Sr_4Ru_3O_{10}$ (Section 4.6).

Contents

4.1. Introduction—Overview of the Layered Ruthenate Materials

The diverse phases and exotic phenomena exhibited by many transition metal oxides (TMO) reflect the fundamental role of the orbital degree of freedom, which interacts strongly with the spins, charges, and lattice in these materials via the Jahn-Teller, spin-orbit, and other interactions.[1,2] The strong interactions between spin-, lattice-, and orbital-degrees of freedom in TMO with perovskite or perovskite-related structures are strongly influenced by the octahedral environment of the oxygen ions (O) around the transition metal ion (M), which form the octahedral MO_6 structure illustrated in Fig. 4.1(b) (M=Ru in picture). For example, perovskite-based TMO with metal ions having a $d(e_g)$ orbital degeneracy in cubic symmetry—such as the $t_{2g}^3 e_g^1$ configurations of Mn^{3+} and Fe^{4+}, the $t_{2g}^6 e_g^1$ configuration of the low-spin state of Ni^{3+}, and the $t_{2g}^5 e_g^1$ configuration of the intermediate spin state of Co^{3+}—are susceptible to spontaneous and cooperative tetragonal or orthorhombic distortions of the MO_6 octahedra that lift the e_g degeneracy, contributing to such phenomena as orbital ordering and polaron formation.[1,2] However, perovksite-related TMO having metal ions with partially filled $d(t_{2g})$ orbitals—such as the vanadates, titanates, and ruthenates—are susceptible to cooperative rhombohedral or tetragonal distortions that lift the threefold degeneracy of the t_{2g} levels.[2]

Among TMO with partially filled $d(t_{2g})$ orbitals, the layered ruthenates $(Sr,Ca)_{n+1}Ru_nO_{3n+1}$ (n=number of connected layers of vertex sharing RuO_6 octahedra;[3,4] see Fig. 4.1(a)) are particularly noteworthy for several reasons: First, because of the extended nature of the $4d$ orbitals of Ru^{4+} ($4d^4$), the Hund's energy E_H—which favors maximizing the total spin at each Ru site—is smaller than the e_g-t_{2g} crystalline field splitting Δ_{CF} (i.e., $E_H < \Delta_{CF}$) in the layered ruthenates. Consequently, the four d-electrons of the Ru^{4+} ions in the layered ruthenates reside in the t_{2g} orbitals, while the e_g levels remain unpopulated, resulting in a $t_{2g}^4 e_g^0$ electronic configuration for the Ru^{4+} ions (see Fig. 4.1(c)). Second, the orbital angular momentum of the Ru^{4+} is not quenched for uncompressed (i.e., cubic) RuO_6 octahedral configurations, leading to a competition between c-axis compressive distortions of the RuO_6 octahedra that tend to maximize the orbital angular momentum, and c-axis expansive distortions that tend to quench the orbital angular momentum.[2] Finally, the physics of the layered ruthenates are enriched by the importance of spin-orbit coupling[2,9]—caused by the relatively large size of the Ru^{4+} ion—which leads to a rich interplay between exotic magnetic and orbital phases.

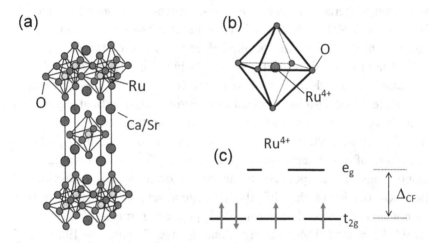

(a)

(b)

(c)

Figure 4.1. (a) Illustration of the room temperature single-layer $(Ca/Sr)_2RuO_4$ structure, consisting of planes of RuO_6 octahedra separated by layers of Ca/Sr and O. RuO_6 octahedral tilts and rotations in Ca_2RuO_4 aren't shown. (b) Illustration of RuO_6 octahedron, which is the common building block of all layered ruthenate materials. (c) Electronic structure of Ru^{4+} ion, governed by the octahedral crystal field environment of the O ions. Because the crystal field splitting Δ_{CF} in the ruthenates is larger than the Hund's energy E_H, i.e., $\Delta_{CF} > E_H$, the four d-electrons of the Ru^{4+} ions occupy the t_{2g} orbitals in the layered ruthenates.

All of these influences contribute—in ways that are not yet clearly understood—to some of the richest variety of phases found in any family of materials:

(i) Single layered (n=1) $Ca_{2-x}Sr_xRuO_4$ – $Ca_{2-x}Sr_xRuO_4$ exhibits a variety of exotic ground states with increasing substitution of Sr for Ca.[5,6] Sr_2RuO_4 (T_c=1.5 K) is a spin-triplet superconductor that is isostructural to high-T_c cuprates such as $La_{2-x}Ba_2CuO_4$ (T_c=30 K).[5,7] On the other hand, Ca_2RuO_4 is an orbital-ordered G-type antiferromagnet (AF) with a Neel ordering temperature T_N=113 K and a metal-insulator (MI) transition at T_{MI} =357 K.[8] X-ray absorption measurements show that

orbital-ordering below T_N in Ca_2RuO_4 is associated with orbital hole populations of 0.5 holes in the d_{xy} orbital and 1.5 holes in the $d_{yz/zx}$ orbital.[9] Further, neutron scattering measurements indicate that the *G*-type antiferromagnetic order in Ca_2RuO_4 is *A*-centered (La_2CuO_4 type) at ambient pressure,[10] while the application of pressure[10] (~0.5 GPa) or the substitution of Sr for Ca $(x\sim0.1)$[11] appears to stabilize *B*-centered (La_2NiO_4 type) antiferromagnetism in Ca_2RuO_4.

(ii) Bilayered (n=2) (Sr,Ca)₃Ru₂O₇ – The double-layer $(Sr,Ca)_3Ru_2O_7$ system also exhibits a rich variety of phases as a function of *x*. $Ca_3Ru_2O_7$ is an antiferromagnetic metal below T_N=56 K, exhibits a metal-insulator (MI) transition to an orbitally polarized antiferromagnetic insulating state below T_{MI}=48 K,[12,13] and has a rich spectrum of magneto-structural and -conducting phases with applied magnetic field.[13,14,15,16] On the other hand, $Sr_3Ru_2O_7$ is a strongly enhanced paramagnetic metal near a ferromagnetic instability,[17] but has a magnetic-field-tuned quantum phase transition to an electronically anisotropic phase,[18] and exhibits induced ferromagnetism upon application of hydrostatic pressure[19] or magnetic field.[20,21] Density functional calculations indicate that induced ferromagnetism in $Sr_3Ru_2O_7$ results from rotations of the RuO_6 octahedra, which lead to an orthorhombically distorted unit cell.[22]

(iii) Triple-layered (n=3) Sr₄Ru₃O₁₀ – $Sr_4Ru_3O_{10}$ is a structurally distorted (antiferromagnetically) canted ferromagnet with a Curie temperature of T_C=105 K,[23] a metamagnetic transition at H_c=2 T for *H∥ab*-plane,[23] strong spin-lattice coupling,[24] and a large structural sensitivity to applied pressure and magnetic field.[24]

In this chapter, we review inelastic light (Raman) scattering measurements of the layered ruthenates, focusing particularly on the temperature-, magnetic-field, and pressure-dependent magnetic and orbitally polarized phases of these materials. For reasons described in more detail in the next section, Raman scattering measurements are particularly well suited to studying structural and magnetic transitions in materials as functions of temperature, magnetic field, and pressure. Thus, Raman scattering provides an ideal means of exploring the influence of structure on the numerous exotic temperature-, magnetic-field, and pressure-dependent phases observed in the layered ruthenates. Indeed, the results described here show clearly that structural changes associated with the RuO_6 octahedra—induced variously by temperature, atomic substitution, magnetic field, and pressure—play a definitive role in many of the exotic properties and phases observed in the layered ruthenates.

The outline of the rest of this review is as follows: **Section 4.2** provides sufficient details on the theory of Raman scattering in solids for the reader to understand the results described in the balance of the review. **Section 4.3** gives detailed descriptions of the Raman scattering experimental methods—and in particular, of the high pressure and high magnetic field methods—used to obtain the Raman results described. Finally, temperature-, magnetic field-, and pressure-dependent Raman scattering results are discussed for single layered $Ca_{2-x}Sr_xRuO_4$ in **Section 4.4**, for bilayered $Ca_3Ru_2O_7$ in **Section 4.5**, and for triple-layered $Sr_4Ru_3O_{10}$ in **Section 4.6**.

X-ray scattering studies of these materials are presented in Chapter 3; transport, magnetic and thermal properties are discussed in Chapters 5 and 6.

4.2. Raman Scattering as a Probe of Correlated Materials

Inelastic light (Raman) scattering[25,26] has proven to be a valuable technique for studying a wide range of magnetic and strongly correlated materials—including high T_c superconductors,[27-31] ferroelectric and magnetic oxides,[32,33,34] and rare earth and actinide materials.[35] The efficacy of inelastic light scattering for studying strongly correlated materials is rooted in two important benefits of this technique: First, inelastic light scattering can provide important symmetry, energy, and lifetime information regarding a remarkable range of excitations, including one- and two-phonon excitations,[36,37] one- and two-magnon excitations,[26] crystal-electric-field excitations,[35] spin-flip excitations,[38] and superconducting gap excitations.[27-31] Consequently, this technique provides an opportunity to obtain detailed microscopic information regarding all the important excitations in strongly correlated materials. Second, because inelastic light scattering employs visible lasers as excitation sources—the collimated light from which can be tightly focused and easily directed to accommodate complex experimental arrangements—this technique lends itself to spectroscopic studies of phase transitions in high-magnetic fields, e.g., using superconducting solenoid magnet systems, and at high pressures, e.g., using diamond anvil cells. The following provides some of the basic details of the theory of Raman scattering so the reader can better appreciate the results presented later in this chapter.

4.2.1. *General Raman Scattering Details*

In the classical description of the Raman scattering process, one considers an electric field vector $E_I(r,t)$—having polarization ε_I, wave vector k_I, and angular frequency ω_I—that is incident on a material. The incident field induces a polarization $P(r,t)$ in the material,

$$P_i(r,t) = \varepsilon_o \sum_j \chi_{ij} E_{jI}(r,t), \tag{4.1}$$

where χ_{ij} is a second-rank tensor component describing the linear electric susceptibility of the material, and ε_o is the permittivity of free space. Excitations in a material cause fluctuations in the induced polarization P—which are reflected in the susceptibility tensor χ—resulting in a scattered electric field $E_S(r,t)$. The scattered light will in general be unpolarized and have a continuum of scattered frequencies and scattered wave vectors; however, the specific design of the Raman scattering apparatus (see Fig. 4.3, Section 4.3) allows one to detect scattered light having a specific angular frequency ω_S, a particular wave vector k_S, and a specific polarization ε_S.

The inelastic light-scattering cross-section resulting from induced polarization fluctuations includes two contributions: (i) the Stokes component, in which the incident light field $E_I(r,t)$ creates an excitation of frequency, ω, and wave vector, q, causing a reduction in the frequency of the scattered field $E_S(r,t)$ given by $\omega_S = \omega_I - \omega$; and (ii) the anti-Stokes component, in which the scattered field gains energy from thermal excitations in the sample, $\omega_S = \omega_I + \omega$. In scattering processes having time-reversal symmetry (i.e., non-magnetic excitations) and no resonant enhancements (i.e., enhancements associated with a matching of the incident or scattered light frequencies with real electronic transitions in the material), the principle of detailed balance relates the Stokes and anti-Stokes intensities,[25,39]

$$I_{AS}(\omega) = I_S(\omega) \left(\frac{\omega_I + \omega}{\omega_I - \omega} \right)^2 e^{-\hbar\omega/k_B T},$$ (4.2)

where ω_I is the incident light frequency, ω is the frequency of the excitation, and T is the temperature. The relationship between the Stokes' and anti-Stokes' Raman intensities therefore provides a method for determining the "actual" temperature of the sample in the scattering region, e.g., allowing one to account for such important effects as laser heating of the sample.

In an ideal crystal with full translational symmetry, kinematical constraints imposed by wave vector conservation dictate the following restriction on the Raman scattering process, $q = k_I - k_S$, where k_I and k_S are the wave vectors of the incident and scattered photons, respectively, and q is the wave vector of the excitation. In typical light scattering experiments—with incident and scattered photons in the visible frequency range ($\omega_{I,S} \approx 10^{14\text{-}15}$ Hz) and excitation frequencies in the far- to near-infrared range ($\omega \approx 10^{11\text{-}13}$ Hz) (which covers the frequency range in which optical phonon, magnon, and intraband electronic excitations are typically observed in ruthenates)—the conditions $\omega \ll \omega_I$ and $|k_I| \approx |k_S|$ apply. Consequently, the excitation wave vectors typically probed in light-scattering experiments have magnitudes in the range, $0 < |q| < 3 \times 10^{-3}$ Å^{-1},[25] which are generally several orders of magnitude smaller than the size of the Brillouin zone boundary, $|k_{ZB}| \approx 2\pi/a \approx 1$ Å^{-1}, where a is the lattice parameter of the crystal. Thus, Raman scattering measurements generally probe excitations only very near the Brillouin zone center, i.e., at $|q| \approx 0$. There are several exceptions to this constraint on the wavevector q in light scattering measurements: First, in light scattering from two-particle (e.g., two-phonon and two-magnon) excitations, excitations of equal and opposite momenta, q and $-q$, throughout the Brillouin zone can be excited by the incident light; second, in light scattering from amorphous and disordered materials, the kinematic constraint is relaxed due to the absence of wave-vector conservation.[40,41]

Finally, note that the energy scale typically used when plotting Raman scattering results is the wavenumber, "cm^{-1}", which has the following conversions to other energy units:

$$0.03 \text{ THz} = 1 \text{ cm}^{-1} \ (\nu[THz] = c\bar{\nu}\left[cm^{-1}\right] = 0.03\bar{\nu}\left[cm^{-1}\right]);$$

$$1 \text{ meV} = 8.06 \text{ cm}^{-1} \ (E[meV] = hc\bar{\nu}\left[cm^{-1}\right] = 0.124\bar{\nu}\left[cm^{-1}\right]);$$

$$1.44 \text{ K} = 1 \text{ cm}^{-1} \ (T[K] = \frac{hc}{k_B}\bar{\nu}\left[cm^{-1}\right] = 1.44\bar{\nu}\left[cm^{-1}\right]).$$

4.2.2. *Raman Scattering Cross Section*

The measured quantity in a typical Raman scattering measurement is related to the photon differential cross-section per unit scattered frequency range, $d^2\sigma/d\Omega d\omega_s$, which reflects the fraction of photons inelastically scattered into the differential solid angle $d\Omega$ with a scattered frequency in the range ω_s to $\omega_s+d\omega_s$. There are numerous detailed microscopic (quantum mechanical) derivations of the Raman scattering cross section $d^2\sigma/d\Omega d\omega_s$ in the literature.[25,26,29] In this review, we emphasize the classical derivation of the Raman scattering cross section to provide the reader a more intuitive understanding of the various important contributions to the Raman scattering intensity.

In the classical derivation, the differential scattering cross section for light polarized in the direction ε_S is related to the power spectrum of fluctuations in the polarization P,[25,26,39]

$$\frac{d^2\sigma}{d\Omega d\omega_s} = \frac{\omega_I\omega_s^3 V}{(4\pi\varepsilon_o)^2 c^4}\frac{n_S}{n_I}\frac{1}{|E_I|^2}\left\langle \varepsilon_s \cdot P_s^*\left(k_s,\omega_s\right)\varepsilon_s \cdot P_s\left(k_s,\omega_s\right)\right\rangle, \quad (4.3)$$

where V is the volume of the sample that contributes to scattering ("scattering volume"), n_I and n_S are the indices of refraction of the incident and scattered light *inside the sample*, respectively, and $\langle\cdots\rangle$ represents the power spectrum of polarization fluctuations induced by phononic, magnetic, or electronic excitations in the material. In the following, we specifically discuss the two most important excitations involved in Raman scattering studies of the layered ruthenates (Sections 4.4, 4.5, and 4.6): phonons and magnons.

4.2.2.1. *Phonon Raman Scattering*

Phonon Raman scattering can provide a substantial amount of information regarding the structure, lattice-stiffness, and even electronic and orbital properties of a material. This is particularly true in Raman studies of layered ruthenates, which show that B_{1g}-symmetry (referenced to tetragonal D_{4h} or orthorhombic D_{2h} space groups) phonons in the energy range 380–415 cm^{-1}—which involve distortions of the RuO$_6$ octahedra that are intimately tied to many properties of the layered ruthenates[9-10,13-15,42,43]—exhibit anomalous behavior as functions of temperature, magnetic field, and/or pressure.[14,15,24,44-49] These anomalous phonon behaviors will be discussed in greater detail in Sections 4.4, 4.5, and 4.6 of this review. Below, we provide some basic background regarding phonon Raman scattering.

For non-magnetic excitations such as phonons, the relationship between the induced polarization and the incident electric field is given by Eq. 4.1; classically, a phonon excitation in the material is represented by a dynamical variable, $X(r,t)$, which is associated with the deviation of atomic position from equilibrium caused by the lattice excitation. The effect of phonon excitations on the induced polarization in the material is obtained by first expanding χ in powers of the dynamical variable, $X(r,t)$, around the static susceptibility, χ_o:

$$\chi_{ij} = \left(\chi_{ij}\right)_o + \sum_k \left(\frac{\partial \chi_{ij}}{\partial X_k}\right)_o X_k + \frac{1}{2}\sum_{k,m}\left(\frac{\partial^2 \chi_{ij}}{\partial X_k \partial X_m}\right)_o X_k X_m + \cdots . \quad (4.4)$$

Ignoring the first term in this expansion—which is associated with elastic scattering—and considering only the first-order inelastic contribution to the power spectrum for polarization fluctuations in Eq. 4.3, the differential scattering cross section can be written:

$$\frac{d^2\sigma}{d\Omega d\omega_s} = \frac{\omega_I \omega_s^3 V}{(4\pi)^2 c^4}\frac{n_s}{n_I}\left|\varepsilon_s \cdot \frac{d\chi(\omega)}{dX}\cdot \varepsilon_I\right|^2 \left\langle X(q,\omega)X^*(q,\omega)\right\rangle . \quad (4.5)$$

There are several noteworthy features in the expression for the Raman scattering cross section given in Eq. 4.5: First, this scattering cross section depends on the susceptibility derivative, $\chi'=d\chi/dX$, which indicates that the Raman intensity of an excitation provides a measure of how strongly that excitation (i.e., the phonon in this case) modulates the electronic susceptibility χ.

Second, Eq. 4.5 shows that by varying the incident and scattered polarizations, ε_I and ε_S, one can couple to different components of the susceptibility derivative tensor, $\chi'=d\chi/dX$. This is a powerful feature of the Raman scattering technique, which enables one to identify the symmetries of excitations studied, by varying the polarizations of the incident and scattered light relative to the material's crystalline axes. The nomenclature often used to indicate the geometry of a particular Raman scattering experiment is $k_I(\varepsilon_I,\varepsilon_S)k_S$, where k_I and k_S are the wave vectors of the incident and scattered photons, respectively, and ε_I and ε_S are the polarizations of the incident and scattered photons, respectively. For example, $z(x,y)\bar{z}$, indicates a "true-backscattering" and "depolarized" scattering geometry in which the incident and scattered wave vectors are directed along the z- and -z-axes of the crystal, respectively, and the incident and scattered photons are polarized along the x- and y-axes, respectively.

Finally, Eq. 4.5 shows that the Raman scattering cross section for Stokes scattering is related to the correlation function associated with the dynamical variable, $<X(q,\omega)X^*(q,\omega)>$. This correlation function is related to the imaginary part of the Raman scattering response function, $Im\chi(q,\omega)$, by the fluctuation-dissipation theorem:[25]

$$\langle X(q,\omega)X^*(q,\omega)\rangle = S(q,\omega) = \frac{\hbar}{\pi}\big[n(\omega)+1\big]Im\,\chi(q,\omega), \quad (4.6)$$

where $[n(\omega)+1]$ is the Bose-Einstein thermal factor and $n(\omega) = [\exp(\hbar\omega/k_BT)-1]^{-1}$.

In the absence of disorder, Raman scattering only couples to a subset of the $q=0$ phonons, the so-called "Raman-allowed" modes. The number and symmetry of $q=0$ normal modes for a particular structure can be determined using group theoretical methods, such as the "correlation method" for identifying the symmetry of $q=0$ phonons in a crystal.[50] Note that "Raman-allowed" phonons have symmetry properties that transform like the polarizability tensor α, the tensor character of Raman-allowed modes reflects the two-photon Raman scattering process and the tensor character of $|\varepsilon_S \cdot \chi' \cdot \varepsilon_I|$ in the Raman scattering cross section in Eq. 4.5.

4.2.2.2. *Magnon Raman Scattering*

Raman scattering is also an excellent technique for studying magnetic excitations—such as the one- and two-magnon excitations observed in Ca_2RuO_4 (Section 4.4.2) and $Ca_3Ru_2O_7$ (Section 4.5.2.1)—and the magnetic phases of materials.[26,38] A classical description of light scattering from magnetic excitations is obtained by noting that the electric susceptibility in Eq. 4.4 is spin-dependent in magnetic materials, and hence can be expanded in powers of the spin operator S:[26,51]

$$\chi_{ij}(r) = \left(\chi_{ij}\right)_o + \sum_k K_{ijk}(r)S_k^r + \sum_{k,m} G_{ijkm}(r)S_k^r S_m^r$$
$$+ \sum_\delta \sum_{k,m} H_{ijkm}(r,\delta)S_k^r S_m^{r+\delta} + \cdots \quad , \tag{4.7}$$

where the first term is the susceptibility in the absence of magnetic excitations, S^r is the spin operator at site r, and the magneto-optical coefficients K and G are tensors related to magnetic circular birefringence and magnetic linear birefringence, respectively; the second and third terms in Eq. 4.7 are primarily responsible for one-magnon scattering. The magneto-optical coefficient H is a tensor involving spin operators at different sites, and so the fourth term in Eq. 4.7 is primarily associated with two-magnon scattering.[25,26]

One-magnon scattering – Microscopically, in one-magnon light-scattering, the electric field of the incident photon can couple to the spin via the spin-orbit interaction, $\lambda L \cdot S$.[26,52] For example, consider the simple case of a magnetic ion having a ground state with $L=0$, as shown in Fig. 4.2: the spin-orbit interaction splits the excited $L=1$ state into different spin components, resulting in finite transition probabilities for the pairs of electric dipole transitions shown with arrows in Fig. 4.2. The net result of these transitions on the orbital and spin states are, respectively, $\Delta L=0$ and $\Delta S= -1$, which corresponds to the excitation of a single magnon. A

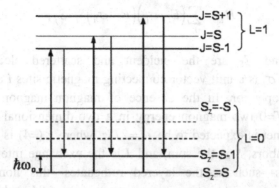

Figure 4.2. Illustration of the one-magnon light scattering mechanism, after Fleury and Loudon:[52] The spin-orbit interaction splits the excited $L=1$ state into different spin components, resulting in finite transition probabilities for the pairs of transitions shown. The net result of the transitions on the orbital and spin states is $\Delta L=0$ and $\Delta S=-1$, which corresponds to the excitation of a single magnon.

more detailed discussion of the mechanism for magnon light scattering is beyond the scope of this introductory review, but a good description is provided in Cottam and Lockwood.[26]

Two-magnon scattering – Two-magnon Raman scattering involves the excitation by an incident photon of two magnons having opposing momenta, q and $-q$.[52,53] Although, magnon excitations throughout the Brillouin zone can, in principle, contribute to the two-magnon Raman scattering intensity, two-magnon Raman scattering is dominated by short-wavelength spin excitations near the Brillouin zone boundary, where the magnon density of states is large.

The two-magnon scattering process occurs in three steps: (i) an incident photon excites a magnetic ion to an excited state via an electric dipole transition; (ii) a double spin flip occurs between the excited state spin and the spin on a nearest neighbor magnetic ion due to direct spin exchange $(S_i \cdot S_j)$; and (iii) another electric dipole transition results in the emission of a scattered photon and the transition of the magnetic system back to its original electronic configuration, but with spins flipped on nearest neighbor magnetic sites.[31] This two-magnon scattering process is described by the Hamiltonian,[54]

$$H_R = \sum_{ij} \left(E_I \cdot \sigma_{ij} \right) \left(E_S \cdot \sigma_{ij} \right) S_i \cdot S_j , \tag{4.8}$$

where E_I and E_S are the incident and scattered electric fields, respectively, σ_{ij} is a unit vector connecting magnetic sites i and j, and S is the spin operator. In the absence of magnon–magnon interaction effects, the $T=0$ two-magnon energy in a two-dimensional Heisenberg antiferromagnet is expected to be $\omega_b=2JSz$, where z (=4) is the number nearest neighbors, S is the spin, and J is the exchange interaction. For $S=1$ systems—such as the layered ruthenates—this non-interacting picture predicts a two-magnon energy of $\omega_b=8J$, which simply corresponds to the energy associated with breaking 4 exchange bonds at each of the two sites on a square lattice; however, consideration of magnon-magnon interaction effects renormalizes the two-magnon energy to $\omega_b=6.7J$ for $S=1$ systems.[55,56] Notably, for spin interactions involving only nearest-neighbor spins, spin-1 antiferromagnets with D_{4h} symmetry are predicted by Eq. 4.8 to have a two-magnon Raman scattering intensity only in the B_{1g} ((E_i,E_s)=(x,y)) scattering geometry,[55] which is indeed the symmetry observed for two-magnon scattering in Ca_2RuO_4 (see Section 4.4.2).[46,47]

4.3. Experimental Details

4.3.1. Raman Scattering System

A schematic of the Raman scattering system used in these experiments is shown in Fig. 4.3. The major components of the typical Raman system include a *laser* as the excitation source, a *spectrometer* to spectrally disperse the scattered light collected in the Raman scattering experiment, a *detector* to detect the scattered light after it has been dispersed by the spectrometer, and various optical components to direct, filter, and control the polarization of the incident laser beam or the collected scattered light. Below, we briefly describe these different components of the Raman scattering system: (i) *Laser* – A laser is the primary excitation source used in most modern Raman scattering applications because of its high power and collimated beam, which can be easily directed to and tightly

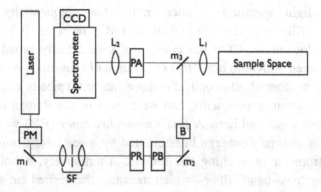

Figure 4.3. Schematic illustration of a typical Raman scattering system, showing the prism monochromator (PM), spatial filter (SF), polarization rotator (PR), polarizing beamsplitter (PB), Berek compensator (B), polarization analyzer (PA), lenses (L), mirrors (m), and CCD camera.

focused on the sample, even in complicated experimental arrangements. The numerous laser light sources available for spectroscopy applications are discussed in Asher and Bormett.[57] (ii) *Spectrometer* – The function of the spectrometer is to separate the relatively weak light that is inelastically scattered by the sample, from the unwanted—but significantly more intense (by ~10^{11}-10^{14} orders of magnitude)—specularly reflected incident light that is elastically scattered by the sample. The Raman scattering experiments described here employ a triple-grating ("three-stage") spectrometer, which enables the detection of inelastically scattered light with energy shifts lower than 8 cm^{-1} (~1 meV). For more details on common spectrometer systems in Raman experiments, see Asher and Bormett,[57] Kuzmany,[58] and Cooper *et al.*[59] (iii) *Detector* – A charge-coupled-device (CCD) detector was used to detect the scattered light in these Raman scattering measurements. A CCD is an analog solid-state image sensor consisting of a two-dimensional array of pixels typically consisting of doped silicon.[60] The dimensions of each pixel—typically on the order of 20×20 μm^2—define the smallest linear dispersion that can be detected, and therefore provide a limit on the spectrometer resolution. A CCD detector provides several benefits in comparison to more conventional photomulitiplier tube detectors in Raman scattering measurements, including the ability to record the entire scattered light spectrum at once, rather than sequentially; greater quantum efficiency (70-90%) than a PMT detector (<30%); and relatively low noise. CCD detectors also have a fairly broad spectral range between 300-1000 nm.[60] (iv) *Other optical elements* – As shown in Fig. 4.3, a number of other optical components are typically employed in Raman scattering experiments, and were used in the Raman scattering experiments described here. A *prism monochromator (PM)* was used to prevent the plasma discharge lines emitted by a gas laser from entering the spectrometer or exciting the sample. Alternatively, a holographic "notch" or "pass-band" filter—which transmits the desired emission line of the laser, but rejects light away from the fundamental emission line—can be used for this purpose.[57] *Polarizers and polarization rotators (PR)* were used to generate and orient the polarization of the incident light and couple to particular components of the Raman susceptibility tensor, as discussed in Section 2.2. *Polarization beamsplitters (PB)* were also used in conjunction with the polarization rotators to create a clean polarized beam in the incident plane. A *beam expander and spatial filter (SF)* was used between the prism monochromator and polarization rotator to help achieve a uniform Gaussian beam that could be tightly focused on the sample. A $\lambda/4$-plate or *Berek polarization compensator (B)* was used in the "incident" side of the optical chain to convert linear polarized light to circularly polarized light in the high magnetic field experiments. The use of circularly polarized light is important in Raman measurements involving high magnetic fields, since linearly polarized light experiences a Faraday rotation in the sample that can complicate the interpretation of magnetic-field-dependent Raman intensity data. The final component of the "input optics" of the Raman system is a *focusing lens (L1)*, which is generally chosen to be a convex lens that can focus light onto the sample. When using a commercial camera lens as the focusing lens, a laser spot size of ~10–50 μm on the sample can be obtained. However, by using a microscope objective as the focusing lens, laser spot sizes ≤ 1 μm can be obtained for the study of small samples. A *collection lens* was used to collect the scattered light from the sample and direct a collimated beam toward the spectrometer. This lens is generally a camera lens or high quality achromatic doublet, chosen to minimize spherical and chromatic aberrations. A *spectrometer focusing lens (L2)* focused the light collected from the sample onto the spectrometer entrance slit. A *polarization analyzer (PA)* was used to select the polarization of the scattered beam. By controlling the polarization states of both the incident and scattered light, specific components of the Raman susceptibility tensor could be selected and excitation symmetries could be identified, as discussed in Section 4.2.2.1. The polarization analyzer consisted of a $\lambda/4$ retarder plate and linear polarizer to select a particular polarization direction, followed by a $\lambda/2$ retarder plate to rotate the selected polarization to maximize the transmission efficiency of the spectrometer gratings.

4.3.2. *High Magnetic Field Measurements*

To make the high magnetic field, low temperature Raman measurements described here, a ~10 Tesla open bore superconducting solenoid magnet was used; a separate optical helium flow-through cryostat—into which the sample was placed for temperature control—was inserted into the open bore of the magnet (see Fig. 4.4(a)). This experimental arrangement not only separates the magnet and sample cryogenics—which is particularly useful for high temperature, high field measurements—but also provides greater flexibility for placing optics close to the sample space, allowing f-numbers as low as $f/1.4$ to be attained. A limitation of this configuration is that it provides optical access in the Faraday geometry ($k\|H$) (see Fig. 4.4(a)), but not the Voigt geometry ($k\bot H$). However, a Voigt geometry was readily obtained in our experiments by mounting the sample "sideways" in the magnet and inserting a 45° mirror near the sample, as shown in Fig. 4.4(b).[14,15]

4.3.3. *High Pressure Measurements*

High pressure Raman scattering has proven to be a very effective method for studying phononic, electronic, and magnetic excitations in complex materials, as will be discussed below for the specific case of the layered ruthenates. In the following, we discuss some of the essential features of the anvil cell used in these experiments, schematically illustrated in Fig. 4.5, which is the most common pressure transmitting device used for Raman scattering.[61]

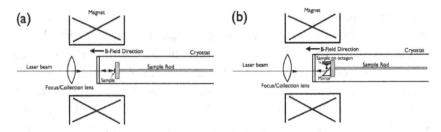

Figure 4.4. Sample orientation in (a) Faraday ($k\|H$) and (b) Voigt ($k\perp H$) geometries, as described in Section 4.3.2.

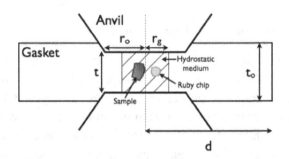

Figure 4.5. Illustration of anvil cell (AC) described in Section 4.3.3.

An anvil cell consists primarily of a pair of *anvils* that exert pressure on the sample (indirectly through some *pressure transmitting medium*, generally) and serve as the window through which the incident and scattered photons travel to and from the sample. The three most common anvil materials used in optical experiments are diamond,[62] sapphire,[63] and moissanite, i.e., *6H*-SiC.[64,65] The hardness of these anvil materials varies; the highest reported pressures have been obtained with diamond anvils (~2000 kbar=200 GPa), but pressures exceeding 300 kbar have been attained with moissanite anvils,[64] and pressures close to 150 kbar have been achieved using sapphire anvils.[63] Moissanite anvils were used in the experiments described here.

An anvil cell also contains a metal *gasket*, as shown in Fig. 4.5. The metal gasket in an anvil cell not only contains the fluid that serves as the medium for transmitting quasi-hydrostatic pressure to the sample, but also provides important massive support to the edges of the anvil, allowing higher pressures to be generated. Eremets[62] and Snow[66] list common gasket materials. BeCu or phosphor bronze were generally used as gasket materials in the low temperature, high pressure experiments described here, as these materials have high thermal conductivities, become harder and more plastic at low temperatures, and have low electrical conductivities. Additional details on gasket design can be found in Dunstan[67] and Cooper *et al.*[59]

In high pressure optical experiments, a *hydrostatic medium* is also generally used between the anvils and surrounding the sample to uniformly distribute the pressure generated by the anvils around the sample. Jayaraman[68] lists a number of common quasi-hydrostatic pressure media used in high pressure Raman experiments. It is important to note that no pressure medium provides completely hydrostatic pressure transfer from the anvils to the sample, and all pressure transmitting media undergo liquid-to-solid phase transitions at low temperatures and high pressure. Helium liquid probably provides the most hydrostatic pressure transfer at low temperatures,[69] and several studies have shown that argon liquid also provides quasi-hydrostatic (nonhydrostaticity ~1% of pressure applied) pressure transfer to the sample at low temperatures for $P<100$ kbar.[65,66,70,71] Argon liquid was used as the pressure-transmitting medium in the experiments described here.

4.4. Raman Scattering Studies of Single-Layer $(Ca,Sr)_2RuO_4$

4.4.1. *Overview*

The single-layered ruthenates, $Ca_{2-x}Sr_xRuO_4$, have drawn much experimental and theoretical attention because they are isostructural to the high-T_c cuprate $La_{2-x}Ba_2CuO_4$ (T_c=30 K) and they have strongly correlated magnetic, electronic, phononic, and orbital degrees of freedom, which result in a rich phase diagram and exotic phenomena.[5,6,8] $Ca_{2-x}Sr_xRuO_4$ (Fig. 4.6) exhibits various ground states with increasing substitution of Sr for Ca. For example, Sr_2RuO_4 is a spin-triplet superconductor (T_c=1.5 K),[5] while Ca_2RuO_4 is an orbital-ordered G-type antiferromagnet (AF) with a Neel ordering temperature T_N=113 K and a metal-insulator (MI) transition at $T_{MI} = 357$ K.[8] X-ray absorption measurements show that orbital-ordering below T_N in Ca_2RuO_4 is associated with orbital hole populations of 0.5 holes in the d_{xy} orbital and 1.5 holes in the $d_{yz/zx}$ orbital.[9] Neutron scattering measurements indicate that the *G*-type antiferromagnetic order in Ca_2RuO_4 is *A*-centered (La_2CuO_4 type) at $P = 0$,[10] although pressure (~0.5 GPa)[10] or Sr-substitution $(x\sim0.1)$[11] appears to stabilize *B*-centered (La_2NiO_4 type) antiferromagnetism with a higher Neel temperature in Ca_2RuO_4.

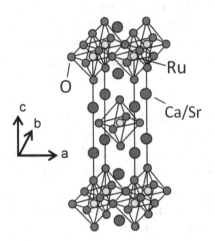

Figure 4.6. Illustration of the room temperature $(Ca/Sr)_2RuO_4$ structure. RuO_6 octahedral tilts and rotations in Ca_2RuO_4 aren't shown.

4.4.1.1. *Temperature-Dependent Effects in Ca_2RuO_4*

Orbital ordering below T_{MI} in Ca_2RuO_4 is associated with a change from a high temperature *L-Pbca* phase—characterized by a long *c*-axis lattice parameter (expanded RuO_6 octahedra in *c*-axis direction) and metallic behavior—and a low temperature *S-Pbca* phase—characterized by a short c-axis lattice parameter (compressed RuO_6 octahedra in *c*-axis direction), insulating behavior, and orbital ordering due to a lowering of the d_{xy} orbital relative to the $d_{yz/zx}$ orbitals.[9,11,72] In the antiferromagnetic insulating phase of Ca_2RuO_4, the *hole* distribution in the Ru t_{2g} levels has been measured to be 0.5 and 1.0 for the *xy*, *yz/zx* hole populations, respectively, while ~0.5 hole is transferred from the Ru t_{2g} orbitals to the O $2p$ orbitals due to hybridization.[73] With increasing temperature above T_N, there is an expansion of the RuO_6 octahedra along the *c*-axis direction,[11,42] which destabilizes d_{xy} orbital ordering by increasing the d_{xy} hole population and suppressing the $d_{yz/zx}$ orbital population,[74] until a homogeneous hole population ($n_{xy}:n_{yz/zx}=1:1$) is realized near $T=300K$.[9] The MI transition is also driven by the continued expansion of the RuO$_6$ octahedra along the c-axis direction with increasing temperature through T_{MI}; consequently, the insulating and metallic states are characterized by short (S-$Pbca$) and long (L-$Pbca$) c-axis lattice parameters, respectively.[11,42]

4.4.1.2. *Doping Dependence of (Ca,Sr)$_2$RuO$_4$*

Sr$_2$RuO$_4$ has the ideal tetragonal K$_2$NiF$_4$ structure (*I4/mmm*) (see Fig. 4.6), and this structure is maintained through the superconductor phases of Ca$_{2-x}$Sr$_x$RuO$_4$, i.e., for $1.5<x<2.0$.[11] However, for Sr concentrations in the range $0.5<x<1.5$, there is a simple rotation of the RuO$_6$ octahedra around the c-axis, resulting in an *I4$_1$/acd* structural phase and the development of a non-superconducting, paramagnetic metal phase. At even lower Sr concentrations, $0.2<x<0.5$, there is a combined rotation of the RuO$_6$ octahedra around the c axis and a tilt of the octahedra around an axis lying in the RuO$_2$ planes.[11,75] This phase regime is metallic at low temperatures, but exhibits a large magnetic susceptibility and a large magnetic anisotropy between the a- and b-axis, suggesting a connection between magnetism and the tilt of the RuO$_6$ octahedra.[11,75] Further, magnetoresistance measurements reveal a substantial change in the Fermi surface of Sr$_2$RuO$_4$ upon Ca substitution,[76] indicating that the octahedral rotations and tilts induced with Ca substitution have a profound impact on the electronic band structure of Ca$_{2-x}$Sr$_x$RuO$_4$. At Sr concentrations lower than $x=0.2$, there is a first-order transition to a *Pbca* phase; as discussed previously, two such phases have been identified as a function of temperature, a higher temperature *L-Pbca* phase characterized by a longer c-axis lattice parameter and metallic behavior, and a lower temperature *S-Pbca* phase characterized by a short c-axis lattice parameter and insulating behavior.[11] Thus, unlike the doping driven high-T_c cuprates, the MI transition in Ca$_{2-x}$Sr$_x$RuO$_4$ is bandwidth driven: the increase of Ca content causes a compression of the RuO$_6$ octahedra, which reduces the in-plane Ru-O separation and lowers the d_{xy} orbital relative to the $d_{yz/zx}$ orbitals,[11,77] causing orbital ordering and a decrease in the $4d$ bandwidth W relative to the large effective Coulomb energy U.[75]

4.4.2. *Phonon and Magnon Scattering in Ca$_2$RuO$_4$*

As discussed above, single layer Sr$_2$RuO$_4$ is tetragonal ($I4/mmm$-D_{4h}) with a Ru–O–Ru angle in the ab-plane of ~180° (see Fig. 4.6).[49,78,79] A factor group analysis indicates that that there should be four Raman-active modes associated with this structure, $2A_{1g} + 2E_g$; the two A_{1g} modes involve stretching vibrations of Sr and the apical oxygens along the c-axis, while the two doubly degenerate E_g modes involve Sr and O vibrations in the ab-plane.[49] The Ru and planar oxygen atoms are at centers of inversion symmetry, and hence are not Raman active.

By contrast, the replacement of Ca for Sr in Ca$_2$RuO$_4$ strongly distorts the RuO$_6$ octahedra, causing a rotation of the octahedra around the c-axis and a tilt of the octahedra around an axis in the RuO$_2$ plane.[11,42] As a result, Ca$_2$RuO$_4$ has an orthorhombic crystal structure (space group $Pbca$-D_{2h}[15]) with four formula units per unit cell. A factor-group analysis predicts a total of 81 Γ-point phonons in Ca$_2$RuO$_4$, of which 36 [9(A_g+B_{1g}+B_{2g}+B_{3g})] are Raman-active modes involving the Ca, in-plane oxygen, and apical oxygen ions, 33 [11(B_{1u}+B_{2u}+B_{3u})] are infrared-active modes, and 12 (12A_u) are silent modes. The Ru ions are located at a center of inversion symmetry in Ca$_2$RuO$_4$, and do not participate in any Raman-active phonon modes.

As shown in Fig. 4.7, polarized Raman spectra of Ca$_2$RuO$_4$ in a back-scattering geometry (i.e., with the propagation vector $k\|c$-axis) exhibit the following phonon modes: [46,47] 9 B_{1g} symmetry modes in the scattering geometry $(E_i,E_s)=(x,y)$, 9 A_g symmetry modes in the scattering geometry $(E_i,E_s)=(x,x)$, and 9B_{1g}+9A_g symmetry modes in the scattering geometry $(E_i,E_s)=(x',x')$. Two distinct temperature regimes can be identified in the Raman spectrum:

(a) $T_N<T<T_{MI}$ – With increasing temperature toward T_{MI}, the optical phonons exhibit a substantial decrease in energy (softening) (Fig. 4.8), a significant broadening, and an increased asymmetry (see Fig. 4.7).[47] The dramatic temperature-dependent changes in the phonon spectrum of

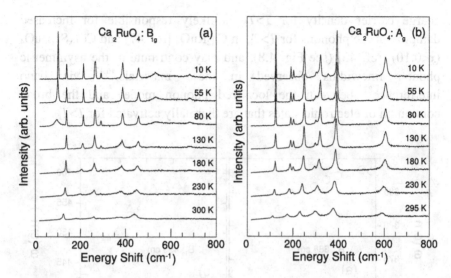

Figure 4.7. Temperature-dependences of (a) B_{1g} $((E_i Es)=(x, y))$ and (b) A_g $((E_i E_s)=(x, x))$ phonon spectra for Ca_2RuO_4.[ref. 47]

Ca_2RuO_4—particularly for $T < T_{MI}$—likely derives from a combination of factors:

(i) There is a shortening of the out-of-plane Ru-O(2) bondlengths—and a corresponding elongation of the in-plane Ru-O(1) bondlengths—below T_{MI}.[11,42] This MI transition is associated with a change from a high temperature *L-Pbca* phase, characterized by a long c-axis lattice parameter and metallic behavior, and a low temperature *S-Pbca* phase, characterized by a short c-axis lattice parameter, insulating behavior, and orbital ordering associated with a lowering of the d_{xy} orbital relative to the $d_{xy/yz}$ orbitals.[11,72,80] The gradual crossover from *L-Pbca* to *S-Pbca* phases with decreasing temperature between $T_N < T < T_{MI}$ is at least partly responsible for the dramatic increase in phonon mode frequencies with decreasing temperature observed in Ca_2RuO_4 (see Fig. 4.8), which don't stabilize until the Neel temperature (T_N) is reached.

(ii) The expansion of the RuO_6 octahedra in the c-axis direction leads to an increase in the number of effective charge carriers—and a decrease in the optical charge gap—for $T > T_N$, ultimately closing the charge gap and disrupting d_{xy} orbital ordering for $T > T_{MI}$.[74] This increase in the charge carrier density for $T>T_N$ is likely responsible for increased damping of the phonons for $T>T_N$ in Ca_2RuO_4 [ref. 47] and $Ca_{2-x}Sr_xRuO_4$ ($x<0.10$) [ref. 46] (see Fig. 4.8), and may contribute to the asymmetric phonon linewidths observed in these materials,[46,47] via Fano interference[81] between the localized phonon modes and the broad continuum of electronic states that are thermally activated for $T>T_N$.

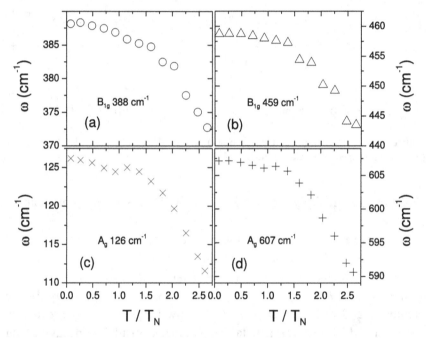

Figure 4.8. Ca_2RuO_4 phonon frequency ω vs. normalized temperature T/T_N (T_N=113 K) for B_{1g} symmetry modes at (a) 388 cm^{-1}, and (b) 459 cm^{-1}, and of A_g symmetry modes at (c) 126 cm^{-1} and (d) 607 cm^{-1}.[ref. 47]

Notably, the MI transition in $Ca_{2-x}Sr_xRuO_4$ ($0<x<0.2$) is first-order, as evidenced by hysteresis in the Raman spectra measured at T=175 K.[46] Hysteresis in the MI transition of $Ca_{2-x}Sr_xRuO_4$ ($0<x<0.2$) has also been observed in transport and neutron diffractions measurements, based upon which estimates of T_{MI}=155 K upon cooling and T_{MI}=220 K upon warming have been made for $Ca_{1.91}Sr_{0.09}RuO_4$.[11,42]

(b) $T<T_N$ – At temperatures below the Neel temperature (T_N=113 K), a Raman-active mode near 102 cm^{-1} appears only in the B_{1_g} scattering geometry $((E_i,E_s)=(x, y))$ (see Fig. 4.7(a)). This mode is clearly associated with the onset of AF ordering in the Neel state, and is likely two-magnon scattering (ΔS=0) (see Section 4.2.2.2) associated with the photon-induced superexchange of pairs of spins on nearest-neighbor Ru $4d^4$-orbital sites via the intervening oxygen $2p$ orbital.[46,47] This interpretation is supported by: (i) the temperature-dependent frequency and linewidth of this mode below T_N (see Fig. 4.9), which are consistent with those observed in other antiferromagnets such as FeF$_2$ and K$_2$MnF$_4$;[26] and (ii) the B_{1g} symmetry of this excitation, which is consistent with that expected for the Heisenberg Hamiltonian in an antiferromagnet with tetragonal (D_{4h}) symmetry. We note, however, that the possibility that this excitation is a single-magnon excitation ($\Delta S=\pm 1$) (see Section 4.2.2.2) can't be completely ruled out, although this interpretation is less likely because the 102 cm^{-1} mode in Ca$_2$RuO$_4$ exhibits no field-induced Zeeman splitting of the two-fold magnon

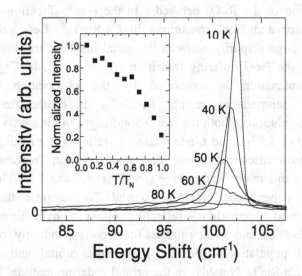

Figure 4.9. Temperature dependence of the two-magnon scattering response of Ca$_2$RuO$_4$. The inset shows the normalized integrated intensity $I(T)/I(T=10 \text{ K})$ vs. normalized temperature T/T_N (T_N=113 K) for the two-magnon scattering response of Ca$_2$RuO$_4$. [ref. 46]

degeneracy (qualitatively corresponding to the spins precessing in opposite senses on the two magnetic sublattices of Ca_2RuO_4) for fields oriented in either the in-plane or c-axis directions of Ca_2RuO_4.[47]

Assuming the two-magnon interpretation of the 102 cm^{-1} mode in Ca_2RuO_4 is correct, the energy of this mode can be used to estimate the in-plane exchange coupling $J_{||}$ between nearest-neighbor Ru-$4d^4$ sites:[46,47] for a layered $S=1$ AF insulator, energy cost of a two-magnon excitation is roughly $\omega_0 = 6.7 J_{||}$,[47,55] where $J_{||}$ is the in-plane spin-spin exchange coupling between nearest-neighbor Ru-$4d^4$ sites. For Ca_2RuO_4, the measured 102 cm^{-1} magnon mode energy gives an estimate of $J_{||}=15.2$ cm^{-1}. This value is consistent with the following rough estimate of the exchange constant from the Neel temperature (assuming $J_{||} \approx J_{\perp}$ and using a molecular-field approximation[82]): $|J_{||}/k_B|=3T_N/2zS(S+1) \sim 14$ cm^{-1}, where z (=6) is the number of nearest-neighbor Ru spins, S (=1) is the magnitude of the Ru spin, and T_N (=113 K) is the Néel temperature for Ca_2RuO_4.

The above results clearly demonstrate the strong connection between structural changes—specifically those associated with the compression and expansion of the RuO_6 octahedra in the c-axis direction—and the metal-insulator and Neel transitions in Ca_2RuO_4.[83] Less well under-stood is the large disparity between the metal-insulator transition, $T_{MI} \sim$ 350 K, and the Neel ordering transition, $T_N \sim$ 110 K, in Ca_2RuO_4. A possible explanation is motivated by the observation that the intermediate temperature regime $T_N<T<T_{MI}$ is characterized by a continuous evolution of both the Ru-O bondlength in the c-axis direction and the ratio of S-$Pbca$ and L-bca phases, as reflected for example in the rapid variation discussed above in optical phonon frequencies and linewidths in this temperature range (see Figs. 4.7 and 4.8). This strong temperature dependence between T_N and T_{MI} suggests that strong orbital/structural fluctuations—reflecting either a dynamic or spatial mixture of L-$Pbca$ and S-$Pbca$ phases (and, correspondingly, of d_{xy} and $d_{yz/zx}$ orbital populations)—prevent long-range orbital and magnetic ordering in Ca_2RuO_4. Notably, in the orbital ordering material $KCuF_3$, a similar discrepancy between orbital-ordering ($T_{OO}\sim$800 K) and Neel ordering ($T_N\sim$40 K) temperatures has also been associated with strong

structural/orbital fluctuations present in the intermediate temperature regime, $T_N < T < T_{OO}$.[84]

4.4.3. *Franck-Condon Effects in the Orbital-Ordered Phase of* Ca_2RuO_4

Another interesting prediction in the orbital ordered state of complex oxides are enhanced multi-phonon effects due to Franck-Condon effects.[85,86] In conventional phonon Raman scattering, the electronic excited states do not modify the atomic positions, and the Raman intensity associated with multi-phonon scattering (i.e., the creation of multiple phonons of a particular mode) is expected to fall off very rapidly with increasing phonon number n, i.e., according to $(\gamma_{ep})^{2n}$, where γ_{ep} is the electron-phonon coupling parameter having a typical value $\gamma_{ep} \leq 0.1$.[86,87] However, Allen and Perebeinos have proposed that photon-induced excitations of the orbital ordered state of transition metal oxides—for example, an excited orbital state generated by the dipole interaction in the first step of the Raman scattering process—can be self-trapped by oxygen rearrangements in the MO_6 octahedra (M=transition metal).[85,86] The resulting Frenkel exciton, or "orbiton," state exists in a superposition of multiphonon states; consequently, the virtual "orbiton" excitation can decay back to the ground state orbital configuration in any number of one-phonon or multi-phonon vibrational states in the final step of the Raman process. The resulting Raman intensity of a particular vibration state generated by this Franck-Condon mechanism is expected to decrease much less weakly with increasing phonon number, n, than in conventional phonon Raman scattering, i.e., according to $(\gamma_{ep})^n$, where γ_{ep} is the electron-phonon coupling parameter.[87,88]

Although the Franck-Condon multi-phonon mechanism described above was originally proposed for orbital ordering in $LaMnO_3$,[85] Fig. 4.10 shows that there is strong evidence for anomalously intense multiphonon scattering in the orbital ordered state of Ca_2RuO_4. Specifically, at T=10 K, the 388 cm^{-1} B_{1g} phonon of Ca_2RuO_4—which involves out-of-phase ("Jahn-Teller"-like) Ru-O vibrations of the RuO_6 octahedra, similar to the 416 cm^{-1} B_{1g} mode in $Ca_3Ru_2O_7$ (see Section 4.4.2)—is followed by a succession of phonon modes that appear at regular intervals of ~73 cm⁻¹ from the 388 cm⁻¹ phonon frequency; these higher-order phonon peaks decrease only slightly in intensity compared with the 388 cm⁻¹ phonon intensity, consistent with the Franck-Condon process described by Allen and Perebeinos.[85,86] These results strongly suggest that the regularly spaced modes shown in Fig. 4.10 are multi-phonon excitations involving the excitation of a 388 cm⁻¹ B_{1g} phonon + n 73cm⁻¹ excitations. The nature of the 73 cm⁻¹ excitation has not yet been identified, but it is likely to be an infrared-active or "silent" phonon mode in Ca_2RuO_4. Interestingly, the 3rd mode in the multi-phonon series shown in Fig. 4.10, expected near 607 cm⁻¹, is missing in the B_{1g} spectrum, but the strong breathing mode (A_g symmetry) vibration of the RuO_6 octahedra near 607 cm⁻¹ is clearly observed in the $(E_i,E_s)=(x, x)$ spectrum. The reason for the missing order in the B_{1g} multiphonon spectrum is not yet clear, but it may reflect interference between B_{1g} multiphonon and A_g breathing mode scattering near 607 cm⁻¹ in Ca_2RuO_4.

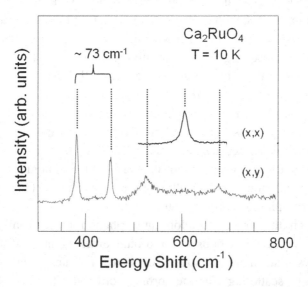

Figure 4.10. Phonon spectrum at T=10 K for Ca_2RuO_4 in both B_{1g} ((E_i,E_s)=(x,y)) and A_g ((E_i,E_s)=(x,x)) scattering geometries. Dashed lines represent a ~73 cm⁻¹ energy separation between multiphonon peaks in Ca_2RuO_4.

4.4.4. *Doping Dependence of Magnon and Phonon Scattering in* $(Ca,Sr)_2RuO_4$

As discussed previously, Sr_2RuO_4 has the ideal tetragonal K_2NiF_4 structure (*I4/mmm*) (see Fig. 4.6), and this structure is maintained through the superconductor phases of $Ca_{2-x}Sr_xRuO_4$, i.e., for $1.5<x<2.0$.[11] However, there are significant structural changes associated with the RuO_6 octahedra for Sr concentrations $x<1.5$ that have substantial impact on the electronic and magnetic properties of $Ca_{2-x}Sr_xRuO_4$ (see Section 4.2.2.2): For example, for $0.5<x<1.5$, there is a simple rotation of the RuO_6 octahedra around the c-axis—resulting in an $I4_1/acd$ structural phase and the development of a non-superconducting, paramagnetic metal phase—while for $0.2<x<0.5$, there is a combined rotation of the RuO_6 octahedra around the c-axis and a tilt of the octahedra around an axis in the ab-plane.[11,75] Magnetoresistance measurements reveal a substantial change in the Fermi surface of Sr_2RuO_4 upon Ca substitution,[76] indicating that the octahedral rotations and tilts induced with Ca substitution have a profound impact on the electronic band structure of $Ca_{2-x}Sr_xRuO_4$. As discussed in Section 4.2.2.2, at small Sr concentrations, $x<0.2$, there is a first-order transition to different *Pbca* phases, including a higher temperature *L-Pbca* phase, characterized by a longer c-axis lattice parameter and metallic behavior, and a lower temperature *S-Pbca* phase, characterized by a short c-axis lattice parameter and insulating behavior.[11]

Not surprisingly, these strongly x-dependent structural changes in $Ca_{2-x}Sr_xRuO_4$ are evident in the phonon spectrum: Fig. 4.11(a) shows that the optical phonon spectrum of Ca_2RuO_4 exhibits narrow and symmetric phonon line shapes at 10 K; however, many of the phonons develop broadened, asymmetric lineshapes with increasing Sr substitution, indicative of an increased interaction between the optical phonons and the electronic states. This interaction can be quantified by fitting the observed $Ca_{2-x}Sr_xRuO_4$ Raman-active phonon intensities, $I(\omega)$, to a Fano profile, $I(\omega)=I_o(q+\varepsilon)^2/(1+\varepsilon^2)$, where $\varepsilon=(\omega-\omega_o)/\Gamma$, ω_o is the phonon frequency, Γ is the effective phonon linewidth, and q is the asymmetry parameter related to the electron-phonon coupling strength V and the imaginary part of the electronic susceptibility ρ according to $1/q\sim V\rho$.[46,81]

Rho *et al.* find that Sr substitution results in a significant increase in the electron-phonon coupling constant and phonon linewidth associated with the 388 and 460 cm^{-1} B_{1g} modes in $Ca_{2-x}Sr_xRuO_4$.[46] The importance of electron-phonon coupling in $Ca_{2-x}Sr_xRuO_4$ is consistent with Hartree-Fock calculations, which show that the electron-lattice interaction is important in determining the phase diagram of $Ca_{2-x}Sr_xRuO_4$, particularly the *G*-type antiferromagnetic phase for low values of *x*.[89]

Figure 4.11(a) shows that two-magnon scattering near 100 cm^{-1} is also observed throughout the antiferromagnetic insulating doping regime (0<*x*<0.2) of $Ca_{2-x}Sr_xRuO_4$. Interestingly, there is very little doping dependence of the two-magnon energy or linewidth, suggesting that the local antiferromagnetic exchange coupling $J_\parallel$ is relatively unaffected by Sr substitution in the doping range (0<*x*<0.2). Indeed, using the fact that the two-magnon peak energy for *S*=1 antiferromagnetic insulators is given by $\omega=6.7J_\parallel$,[46,55] estimates of the in-plane exchange energies between nearest-neighbor Ru-4d^4 sites are $J_\parallel$=15.22, 15.37, and 15.97 cm^{-1} for the *x*=0, 0.06, and 0.09 samples, respectively.[46] The insensitivity of the exchange coupling $J_\parallel$ to Sr substitution in $Ca_{2-x}Sr_xRuO_4$ for (x<0.10) is consistent with the absence of a suppression of the Neel temperature in this substitution range (T_N=113 K, 150 K, and 141 K for *x*=0, 0.06, and 0.09, respectively),[75] and is consistent with pressure-dependent Raman studies of Ca_2RuO_4—discussed below—which show that the two-magnon energy is relatively insensitive to pressure up to the pressure-induced collapse of the orbital-ordering phase.[45] Both of these results support the conclusion that antiferromagnetism and orbital order in $Ca_{2-x}Sr_xRuO_4$ is not suppressed with *x* and pressure by reducing the antiferromagnetic exchange coupling, but rather by reducing the volume fraction of distorted *S-Pbca* phase regions that support orbital polarization and antiferromagnetic order.[45,46]

On the other hand, Fig. 4.11(b) shows that the two-magnon responses in $Ca_{2-x}Sr_xRuO_4$ exhibit substantially different temperature-dependences in the *x*=0, 0.06, and 0.09 samples. In particular, in comparison to the *x*=0 sample, the x=0.06 and 0.09 Sr substituted samples exhibit a much more dramatic renormalization of the two-magnon energy and linewidth with increasing temperature, suggesting that long-range magnetic order is less robust against thermal fluctuations in the Sr-substituted samples than in pure Ca_2RuO_4. This is presumably due to the disorder introduced in the magnetic lattice when Sr is introduced and/or to the increase in electronic states with Sr substitution, which interact more strongly with the magnons in the doped material.[46]

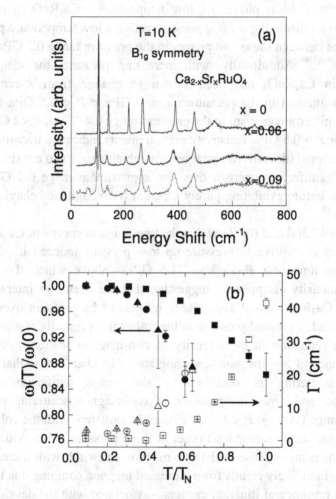

Figure 4.11. (a) B_{1g} symmetry $((E_i,E_s)=(x,y))$ Raman spectra of $Ca_{2-x}Sr_xRuO_4$ for $x=0$, $x=0.06$, and $x=0.09$. Spectra have been offset for clarity. (b) Summary of the normalized two-magnon energy $\omega(T)/\omega(T=0)$ vs. normalized temperature T/T_N in $Ca_{2-x}Sr_xRuO_4$ for $x=0$ (filled squares), $x=0.06$ (filled circles), and $x=0.09$ (filled triangles), and the associated two-magnon linewidth (FWHM) vs. normalized temperature for $x=0$ (open squares), $x=0.06$ (open circles), and $x=0.09$ (open triangles).[ref. 46]

4.4.5. *Pressure Dependence of Magnon Scattering in Ca₂RuO₄*

At room temperature and low pressures, the structural properties of Ca_2RuO_4 are similar to those observed as a function of Sr substitution (x):[10] at $P=0.5$ GPa, there is a discontinuous transition from the *S-Pbca* phase to the *L-Pbca* phase. At low temperatures, Ca_2RuO_4 partially reverts into an insulating *S-Pbca* phase, creating a low temperature phase coexistence between these two phases in the pressure range 0.3 GPa < P < 2 GPa.[10,45,90] Additionally, with increasing pressure, the magnetic ordering in Ca_2RuO_4 has been shown to change from *A*-centered antiferromagnetism in the pressure range 0 GPa < P < 0.5 GPa to *B*-centered antiferromagnetism in the pressure range 0.4 GPa < P <2 GPa.[10] Finally, for $P > 0.5$ GPa, resistivity measurements indicate a transition to two-dimensional itinerant ferromagnetism, which appears to coexist with *B*-centered antiferromagnetism over the approximate range 0.5 GPa < P < 2 GPa, before exhibiting purely itinerant ferromagnetic behavior for $P > 2$ GPa.[90]

Figures 4.12(a) and (b) show that the two-magnon energy of Ca_2RuO_4 is relatively insensitive to pressure up to a pressure-induced insulator-metal transitions near P~15 kbar (1.5 GPa)—above which the two-magnon intensity disappears—suggesting that the exchange interaction energy in Ca_2RuO_4 is not appreciably influenced by pressure over this range. Instead, increased pressure in the ruthenates primarily results in a systematic reduction in the intensity of two-magnon scattering, and a disappearance of two-magnon scattering above 15 kbar.[45] Note that two-magnon scattering is sensitive to short-range antiferromagnetic correlations, and hence the presence of two-magnon scattering in the pressure range 0 GPa < P < 1.5 GPa is not inconsistent with the collapse of long-range antiferromagnetic order at these pressure. Also evident in Fig. 4.12(b) is an increase in the two-magnon linewidth with increasing pressure, which likely results from increased magnon damping due to the increasing presence of itinerant carriers—associated with the developing *L-Pbca* phase regions—with increased pressure in Ca_2RuO_4.[45]

The pressure-dependence of two-magnon Raman scattering in Ca_2RuO_4 suggests that—like Sr substitution—increased pressure suppresses antiferromagnetic correlations in Ca_2RuO_4 not by suppressing

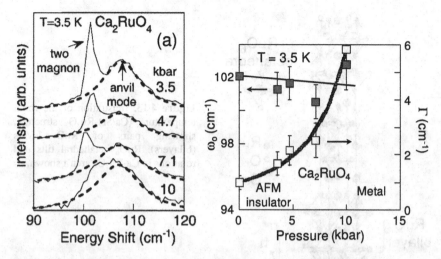

Figure 4.12. (a) Two-magnon scattering response in Ca_2RuO_4 as a function of pressure. Dashed line indicates a SiC anvil phonon. The spectra have been offset for clarity. (b) Summary of the two-magnon energy (filled squares) and linewidth (open squares) as a function of pressure in Ca_2RuO_4.[ref. 45]

antiferromagnetic exchange, but by suppressing the RuO_6 octahedral distortions associated with the *S-Pbca* phase, which help stabilize antiferromagnetism in the ruthenates. Note that this is quite different from the pressure dependence of conventional Mott insulators such as La_2CuO_4 [ref. 91] and NiO,[92] in which the two-magnon energy increases systematically with increasing pressure, reflecting an increase in the *d*-electron hopping matrix element t with increasing pressure—and a corresponding increase in the exchange constant J—according to the simple Hubbard model prediction, $J=t^2/U$, where U is the Coulomb interaction.

4.5. Raman Scattering in $Ca_3Ru_2O_7$

4.5.1. *Overview*

Bilayered $Sr_3Ru_2O_7$ and $Ca_3Ru_2O_7$ consist of pairs of closely coupled RuO planes separated from adjacent bilayers along the c axis by layers of Ca and O ions (see Fig. 4.13).[3,4] While $Sr_3Ru_2O_7$ has a body-centered

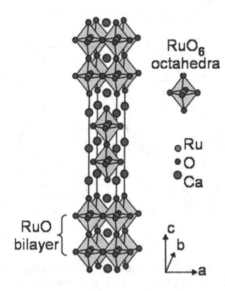

Figure 4.13. Illustration of the room temperature $(Ca/Sr)_3Ru_2O_7$ structure, showing pairs of RuO layers (bilayers). RuO_6 octahedral tilts and rotations in $Ca_3Ru_2O_7$ aren't shown.

tetragonal crystal structure ($I4/mmm$-D_{4h} symmetry) with an ideal Ru-O-Ru bond angle of ~180°, $Ca_3Ru_2O_7$ has an orthorhombic ($A2_1ma$-C_{2v}) structure[48] with lattice parameters a=5.3720(6) Å, b=5.5305(6) Å, and c=19.572(2) Å. The RuO_6 octahedra in $Ca_3Ru_2O_7$ have a large tilt angle that projects primarily onto the ac plane (a Ru-O-Ru bond angle projection of ~153.22°), but only slightly onto the bc plane (a Ru-O-Ru bond angle projection of ~172.0°).[93,94] Like Ca_2RuO_4, $Ca_3Ru_2O_7$ exhibits a reduction of the c-axis lattice parameter—and a corresponding compression of the RuO_6 octahedra—in the c-axis direction as a function of decreasing temperature, leading to a transition to a metallic antiferromagnetic state below T_N=56 K, and a metal-insulator (MI) transition to an antiferromagnetic insulating state below T_{MI}=48 K.[13-15] The low temperature compression of the RuO_6 octahedra is expected to lower the d_{xy} orbitals relative to the d_{zx} and d_{yz} orbitals, likely resulting in a polarization of the orbital configuration on the Ru^{4+} sites.[13-15,43,95]

4.5.2. Magnon and Phonon Scattering in $Ca_3Ru_2O_7$

As discussed above, $Ca_3Ru_2O_7$ has an orthorhombic crystal structure ($A2_1ma$-C_{2v}) structure. A factor-group analysis predicts a total of 69 Γ-

point Raman-active phonons ($18A_1+17A_2+16B_1+18B_2$) for this structure.[48] Figure 4.14 shows the Raman scattering spectrum of $Ca_3Ru_2O_7$ in both the antiferromagnetic (T=11.5 K, H=0 T, and P=0 kbar) and metallic (T=58 K, H=0 T, and P=0 kbar) phases. A complete symmetry analysis of the Raman-active phonons in $Ca_3Ru_2O_7$ and $Sr_3Ru_2O_7$ is presented by Iliev *et al.*[48] In comparison to the rather simple Raman spectrum of $Sr_3Ru_2O_7$,[48] numerous optical phonons are observed in $Ca_3Ru_2O_7$ due to the lower structural symmetry caused by the RuO_6 octahedral rotations and tilts in the latter compound.

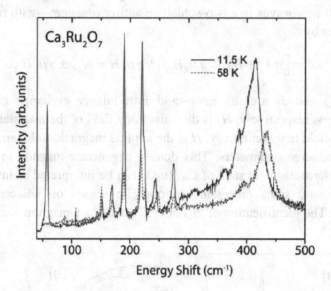

Figure 4.14. Raman spectrum of $Ca_3Ru_2O_7$ at 11.5 K (solid line) and 58 K (dashed line) in the $(E_i,E_s)=(x,y)$ scattering geometry.[ref. 15]

Two Raman spectroscopic features in the B_{1g}-symmetry $(E_i,E_s)=(x,y)$ scattering geometry of $Ca_3Ru_2O_7$ (Fig. 4.14) have particularly interesting behaviors as functions of temperature, pressure, and magnetic field: (i) A low frequency mode near 56 cm^{-1}, which develops below T_N and is associated with spin wave (magnon) excitations of the antiferromagnetic ground state, and (ii) a broad, asymmetric phonon mode near 416 cm^{-1} (at 5 K), which exhibits a substantial decrease in energy ("softening") with decreasing temperature through the metal-insulator transition temperature, T_{MI}, as shown in Fig. 4.14.[14,15,44,48] As discussed in the following, a substantial amount of information regarding the magnetic and structural/orbital phases can be gleaned from careful studies of these excitations.

4.5.2.1. *Magnetic Phases in Ca₃Ru₂O₇: Magnon Scattering*

Figures 4.14 and 4.15(a) show that a narrow peak near 56 cm⁻¹ evolves in the Raman scattering spectrum of $Ca_3Ru_2O_7$ below T_N=56 K. This mode can be identified as one-magnon scattering associated with scattering from q=0 spin waves in a two-sublattice antiferromagnet—with frequencies given by:[26]

$$\omega^{\pm} = \left[\alpha J_{\perp} + \beta J_{\parallel} + g\mu_B H_A \right] \pm g\mu_B H = \Delta_{AF} \pm g\mu_B H , \qquad (4.9)$$

where $J_{\perp}$ and $J_{\parallel}$ are the inter- and intra-bilayer exchange coupling parameters, respectively, H_A is the anisotropy field of the material, Δ_{AF} is the zero-field magnon energy, H is the applied magnetic field, and α and β are exchange parameters. This doubly degenerate magnon in the A-type antiferromagnetic state of $Ca_3Ru_2O_7$ can be interpreted as involving ferromagnetic spins precessing in opposite senses on adjacent RuO bilayers. The identification of this 56 cm⁻¹ mode as a magnon is aided by

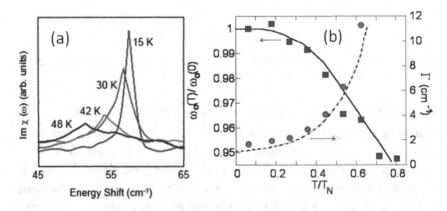

Figure 4.15. (a) Temperature-dependence of the magnon in $Ca_3Ru_2O_7$ (T_N=56K). (b) Magnon normalized frequency, $\omega_o(T)/\omega_o(0)$ (squares), and linewidth, Γ (circles), as a function of normalized temperature, T/T_N, in $Ca_3Ru_2O_7$.[ref. 15]

the following characteristics:[15] (i) with increasing temperature below T_N, the magnon mode frequency ω_o decreases (squares, Fig. 4.15(b))—and the linewidth systematically increases (circles, Fig. 4.15(b))—reflecting an increase in magnon-magnon interactions with increasing temperature, as expected theoretically and observed in previous studies of one-magnon excitations in antiferromagnets;[26] (ii) the magnon is observed in the "crossed" (or "depolarized") $(= z(x, y)\bar{z})$ scattering geometry, which is consistent with theoretical expectations that magnon scattering involves off-diagonal terms in the Raman scattering tensors;[26,52] and (iii) the Zeeman splitting, $\Delta E = g\mu_B H$ ($g=2$), of this mode in the presence of a magnetic field ($H\|ab$-plane), as shown in Fig. 4.16(a), is consistent with a lifting of the degeneracy of the single magnon mode frequency (see Eq. 4.9), involving ferromagnetic spins that precess in opposite senses on adjacent RuO bilayers.

Substantial microscopic information can be obtained regarding the magnetic phases and magnetic exchange parameters in $Ca_3Ru_2O_7$ from the field-dependence of the magnon scattering in this material. For example, the large zero-field energy of the magnon, i.e., $\omega^{\pm}(H=0)=\Delta_{AF}=$ 57 cm^{-1} (~7 meV), reveals a very large magnon gap in this materials, presumably caused by the large exchange and anisotropy fields in $Ca_3Ru_2O_7$. The large anisotropy field in this material presumably reflects the large c-axis compression of the RuO_6 octahedra in the antiferro-magnetic insulating phase of $Ca_3Ru_2O_7$.[13] Additionally, Fig. 4.16(b) shows that the 56 cm^{-1} magnon mode energy in $Ca_3Ru_2O_7$ exhibits no field dependence for $H\|c$-axis (diamonds in Fig. 4.16(b)). However, the degenerate magnon energies (Eq. 4.9) exhibit a Zeeman splitting for $H\|ab$-plane—with a maximum splitting of $\Delta = 2g\mu_B H$ for $H\|[110]$—and an angular dependence given by $\omega^{\pm}(\theta) = \Delta_{AF} \pm g\mu_B H \cos\theta$ (triangles in Fig. 4.16(c)). In agreement with neutron scattering measurements,[94] the observation of a magnon splitting only for $H\|ab$-plane shows that the Ru-spins are aligned in the plane; further, the observation that the magnon splitting is maximum for $H\|[110]$ shows that the Ru spins in $Ca_3Ru_2O_7$ are oriented at 45° to the [100] direction below T_N.

Figures 4.16(a) and (b) also show that the split magnon modes collapse into a single magnon mode for $H>H_c=5.9$ T, reflecting a metamagnetic transition from A-type antiferromagnetic alignment of

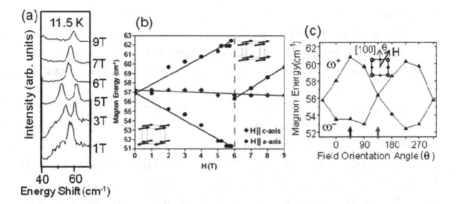

Figure 4.16. (a) Magnetic field dependence of the magnon in $Ca_3Ru_2O_7$ for $H\|[110]$ and at T=11.5 K. (b) Summary of the magnon energy in $Ca_3Ru_2O_7$ for $H\|[001]$ (diamonds) and $H\|[110]$ (circles). (c) Dependence of the magnon energy in $Ca_3Ru_2O_7$ on the in-plane magnetic field orientation with H=4 T, with θ=0 corresponding to $H\|[100]$.[ref. 15]

adjacent RuO bilayers (i.e., ferromagnetic alignment within the RuO bilayers, antiferromagnetic alignment between RuO bilayers) to a (forced) ferromagnetic alignment of the bilayers. This transition is expected to occur at a critical field value H_c=$2H_E$-H_A—where H_E and H_A are the exchange and anisotropy field, respectively—above which the magnon energy is given by:[26]

$$\omega^{FM} \cong \left[g\mu_B H_E + \alpha J_\perp + g\mu_B H_A \right] + g\mu_B H , \qquad (4.10)$$

where H is the applied field. From the critical field, H_c, one can estimate an interbilayer exchange energy—roughly, the exchange energy gained by the system when adjacent RuO bilayers switch between AF and FM alignments—of $\alpha J_\perp$=$g\mu_B H_c$~5.5 cm^{-1} (~0.7 meV). Additionally, from the measured zero-field AF magnon energy and the relationship between the critical, exchange, and anisotropy fields, $H_c(T=0)$=$2H_E$-H_A, one can estimate H_E=20 T and H_A=35 T in $Ca_3Ru_2O_7$ at 11.5 K; these values are somewhat larger than those estimated from the low-T magnetic susceptibility, H_E~14.2 T H_A~22.4 T.[93]

Notably, the metamagnetic transition at H_c=5.9 T in $Ca_3Ru_2O_7$ is associated with a small change in the c-axis conductivity,[16] but there are no structural changes that are concomitant with this transition.[13-15] This suggests that the change in *c*-axis conductivity accompanying this metamagnetic transition is associated with a "spin-valve" effect—i.e., a decrease in resistance due to a change from an antiparallel-to-parallel alignment of neighboring ferromagnetic layers—and is not attributable to field-induced structural changes.[13-15]

4.5.2.2. *Structural/Orbital Phases in $Ca_3Ru_2O_7$: Octahedral Phonon Scattering*

The 416 cm^{-1} phonon in $Ca_3Ru_2O_7$ also exhibits an anomalous temperature dependence through the metal insulator transition (see Figs. 4.14 and 4.17). According to lattice dynamical calculations by Iliev *et al.*,[48] the 416 cm^{-1} phonon mode of $Ca_3Ru_2O_7$ involves mostly in-plane vibrations of the oxygen atoms in the RuO plane, as well as *c*-axis vibrations of the apical oxygen atoms. Consequently, this mode should couple strongly to

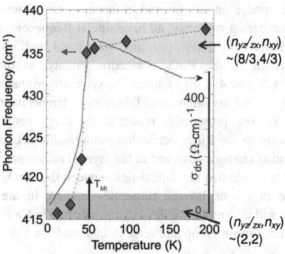

Figure 4.17. Temperature dependence of the ~416 cm^{-1} RuO$_6$ octahedral phonon frequency (diamonds) through the MI transition of $Ca_3Ru_2O_7$. The solid curve compares the dc conductivity of $Ca_3Ru_2O_7$. Light shaded region at the top indicates the frequencies at which $Ca_3Ru_2O_7$ is in the "high conducting" phase, where the *d*-orbital occupancy of the Ru^{4+} sites is presumed to be roughly $(n_{yz/zx}, n_{xy})$~(8/3,4/3) (degenerate *d*-levels). The darker shaded region on bottom indicates the frequencies at which $Ca_3Ru_2O_7$ is in the "low conducting" phase, where the *d*-orbital occupancy of the Ru^{4+} sites is presumed to be roughly $(n_{yz/zx}, n_{xy})$~(2,2).[ref. 14,15]

the c-axis compressions of the RuO_6 octahedra, which are responsible for the abrupt changes in d-orbital population that accompany the MI transition in $Ca_3Ru_2O_7$.[13-15,43,95] This strong coupling is consistent with the dramatic softening of this RuO mode frequency through the metal-insulator transition of $Ca_3Ru_2O_7$ (diamonds in Fig. 4.17), coincident with the abrupt decrease in the dc conductivity through T_{MI} (solid line in Fig. 4.17).

Note that the 416 cm^{-1} mode has A_2 symmetry when referenced to the true orthorhombic $A2_1ma$-C_{2v} space group of $Ca_3Ru_2O_7$.[48] However, in the literature, the 416 cm^{-1} mode in $Ca_3Ru_2O_7$ is sometimes referenced to the D_{2h} orthorhombic space group, with respect to which this phonon has B_{1g} symmetry. This B_{1g} symmetry identification emphasizes the strong similarities between the 416 cm^{-1} phonon mode of $Ca_3Ru_2O_7$ and the ~380 cm^{-1} B_{1g}-symmetry phonons in $Sr_3Ru_2O_7$,[48] $Ca_{2-x}Sr_xRuO_4$ (see Sections 4.4.2, 4.4.3),[46,47] and $Sr_4Ru_3O_{10}$ (see Section 4.6.2).[24,49] In particular, like the 416 cm^{-1} phonon mode of $Ca_3Ru_2O_7$, the ~380 cm^{-1} B_{1g}-symmetry modes in all layered ruthenates are observed in the $(E_i, E_s) = (x, y)$ scattering geometry, all have similar frequencies (380 – 415 cm^{-1}), asymmetric Fano lineshapes, and vibrational motions involving similar distortions of the RuO_6 octahedra. Finally, as discussed in Sections 4.4, 4.5, and 4.6, all of these B_{1g}-symmetry ruthenate phonon modes exhibit anomalous behavior as functions of temperature, magnetic field, and/or pressure, presumably reflecting the strong coupling of this vibrational mode to the RuO_6 octahedral compressions/expansions and orbital-population changes observed in the layered ruthenates.[9,11,13-15,42,43] Thus, to further emphasize the similarities between this RuO_6 octahedral phonon mode in all the layered ruthenates studied, in the following discussion we will also reference the 416 cm^{-1} phonon in $Ca_3Ru_2O_7$ to the D_{2h} orthorhombic space group and refer to this mode as a B_{1g}-symmetry mode.

4.5.3. *Temperature-Dependent Structural/Orbital Phases in* $Ca_3Ru_2O_7$

The strong sensitivity of the ~416 cm^{-1} B_{1g} phonon mode of $Ca_3Ru_2O_7$ to the RuO_6 octahedral compressions/expansions and d-orbital population makes this phonon mode useful as a sensitive probe of changes in RuO_6 structure and Ru^{4+} orbital population as functions of temperature, magnetic field,[14,15] and pressure.[15,45] The temperature dependence of this B_{1g} RuO phonon mode (Fig. 4.17) reveals two distinct orbital and conducting phase regimes, discussed in the following.

4.5.3.1. *High-Conducting, Orbital-Degenerate Regime*

In the high conducting phase regime ($T>T_{MI}=48$ K for $H=0$ and $P=0$) of $Ca_3Ru_2O_7$, the B_{1g} octahedral RuO_6 phonon mode has a peak energy of ~433 cm^{-1}. In this phase, the RuO_6 octahedra are less compressed in the c-axis direction.[13] Assuming that the three orbitals in the t_{2g} subshell are approximately degenerate in this phase regime, with the 4 Ru d electrons populating the t_{2g} states approximately equally, the high temperature phase ($T>T_{MI}$) of $Ca_3Ru_2O_7$ has an approximate orbital occupancy of $(n_{yz/zx},n_{xy})=(8/3,4/3)$.[96]

4.5.3.2. *Low-Conducting, Orbital-Polarized Regime*

In the insulating phase regime ($T<T_{MI}=48$ K for $H=0$ and $P=0$) of $Ca_3Ru_2O_7$, the B_{1g} octahedral RuO_6 phonon mode frequency decreases abruptly to less than ~418 cm^{-1}. Through the metal-insulator transition, the RuO_6 octahedra exhibit an abrupt compression,[13,42] lowering the energy of the d_{xy} orbital relative to the energies of the $d_{yz/zx}$ orbitals. For a fully polarized orbital state, and in the absence of hybridization, this distortion would result in an orbital occupancy of $(n_{yz/zx},n_{xy})=(2,2)$ in the low temperature antiferromagnetic state of $Ca_3Ru_2O_7$. However, the actual orbital occupancy in the antiferromagnetic insulating phase of bilayer $Ca_3Ru_2O_7$ has not yet been definitively established. Using resonant x-ray diffraction, Bohnenbuck *et al.* find no evidence for orbital-order within the experimental sensitivity of their measurement; these authors attribute the weak orbital-order parameter in $Ca_3Ru_2O_7$ to charge or orbital quantum fluctuations.[95] Recent calculations indeed show that the competing effects of orbital exchange and crystal field create a competition between antiferro-orbital (AFO) and ferro-orbital (FO) phases that cause these orbital phases to be fragile in the absence of

an applied field.[43] This competition is expected to make the experimental observation of orbital ordering in $Ca_3Ru_2O_7$ difficult at $H=0$;[43] however, as discussed below, this competition also leads to a strong field- and pressure-dependence of the structural and orbital phases in $Ca_3Ru_2O_7$ that is ultimately responsible for this material's rich field- and pressure-tuned phase diagrams.

4.5.4. *Field-Dependent Structural/Orbital Phases in $Ca_3Ru_2O_7$*

The strong phase competition in $Ca_3Ru_2O_7$ created by the interplay between crystal-field, orbital-exchange, and spin-orbital interactions leads to a large sensitivity of the structural, conductive, and orbital phases in this material to both the magnetic field strength and direction.[13-15,16,43] The properties of $Ca_3Ru_2O_7$ are fairly insensitive to applied field for $H \| c$-axis,[16] at least for field strengths typically accessible in the laboratory. However, two distinct structural/conductive HT phase diagrams can be observed for a magnetic field applied in the ab-plane of $Ca_3Ru_2O_7$.

4.5.4.1. *Structural/Orbital Phases in $Ca_3Ru_2O_7$ for $H \bot$Magnetic Easy-Axis*

Figure 4.18 shows representative field-dependent Raman spectra of $Ca_3Ru_2O_7$ with the applied magnetic field oriented perpendicular to the magnetic easy axis direction. While there is little change in the B_{1g} phonon frequency at low temperatures (11.5 K), at intermediate temperatures (41 K) the B_{1g} phonon frequency increases abruptly for $H>6$ T, as summarized in Fig. 4.19(b), reflecting a rapid expansion of the RuO_6 octahedra in this field range.[14,15] This field-induced octahedral expansion is associated with a loss of magnon scattering intensity (Figs. 4.18(c) and 4.19(a)) and an abrupt increase in the c-axis conductivity (Fig. 4.19(c)). Qualitatively similar behavior has been reported in field-dependent x-ray diffraction measurements of the $Ca_3Ru_2O_7$ structure by Nelson *et al.*,[13] further supporting the key role that field-induced changes in the octahedral structure of $Ca_3Ru_2O_7$ play in governing the dramatic magnetoconductive behavior in $Ca_3Ru_2O_7$.[13-15]

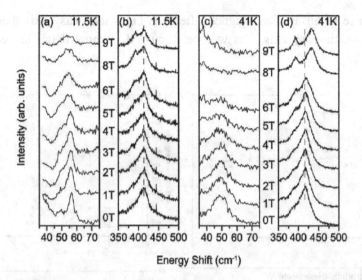

Figure 4.18. Field-dependence ($H\perp$magnetic easy axis) of $Ca_3Ru_2O_7$ for: (a) the magnon mode at T=11.5 K; (b) the B_{1g} RuO octahedral phonon mode at T=11.5 K; (c) the magnon mode at T=41 K, and (d) the B_{1g} phonon mode at T=41 K.[ref. 15]

Notably, the very large field-induced shift of the B_{1g} phonon frequency in $Ca_3Ru_2O_7$, $d\omega/dB\sim4.7$ cm^{-1}/T at 41 K (see Fig. 4.19(b)), reflects a large spin-phonon coupling in the layered ruthenates, as will be discussed further in Section 4.6.2.

Taken together, the results in Figs. 4.18 and 4.19—summarized in the phase diagram in Fig. 4.19(d)—suggest that for $H\perp$easy-axis direction, the magnetic field induces a transition from a low field antiferromagnetic, orbitally polarized regime—in which the applied magnetic field is insufficient to reorient the moments from the easy-axis direction—to a high field orbital degenerate metallic regime—in which the applied field is sufficiently strong to reorient the moments along the hard-axis direction, resulting in an expansion of the RuO_6 octahedra in the c-axis direction due to strong spin-orbital coupling. This description is supported by calculations of Forte *et al.*[43] and x-ray measurements of Nelson *et al.*,[13] both of which indicate (i) that there is an intimate connection between field-induced structural changes in $Ca_3Ru_2O_7$ and the dramatic magnetoconductive properties of this material; and (ii) the magnetostructural phases of $Ca_3Ru_2O_7$ are more robust ("stiff") in response to an applied magnetic fields $H\perp$magnetic easy axis than for $H\|$magnetic easy axis,[43] as will be discussed in more detail in Section 4.5.4.2.

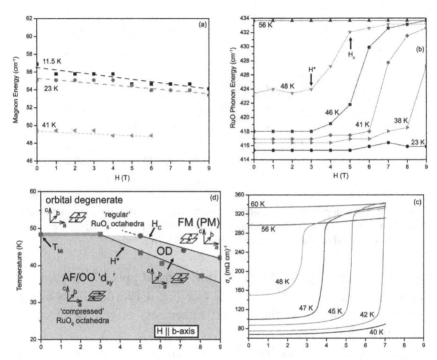

Figure 4.19. Summaries of field- ($H\perp$magnetic easy axis) and temperature-dependences in $Ca_3Ru_2O_7$ for: (a) the magnon energy; (b) the B_{1g} RuO phonon energy; and (c) the c-axis dc conductivity. (d) The H-T phase diagram for $Ca_3Ru_2O_7$ with ($H\perp$magnetic easy axis), deduced from the data in (a)–(c). Squares denote a transition boundary between the antiferromagnetic orbital-ordered (AF/OO) phase (blue region) and the orbital disordered (OD) phase (orange region), while the filled circles represent transitions between the OD and forced ferromagnetic orbital-degenerate phase (white region).[ref. 15]

4.5.4.2. *Structural/Orbital Phases in $Ca_3Ru_2O_7$ for $H\|$Magnetic Easy-Axis*

Figure 4.20 shows representative field-dependent Raman spectra of $Ca_3Ru_2O_7$ with the field-oriented parallel to the magnetic easy axis. At low temperatures, $T\sim30$ K, the Ru-O B_{1g} phonon has a field-independent frequency near 416 cm^{-1} (Fig. 4.20(b))—and the magnon scattering response persists at all fields (Fig. 4.20(a))—indicating that $Ca_3Ru_2O_7$ remains in the antiferromagnetic insulating phase (with compressed RuO_6 octrahedra) in this field and temperature range. By contrast, at higher temperatures, 30 $K<T<T_N$, the Ru-O B_{1g} phonon frequency increases rapidly at intermediate fields, then subsequently decreases with increasing field back to its zero-field value (Fig. 4.20(d)). Likewise, the magnon (Fig. 4.20(c)) disappears at intermediate fields, but weakly reappears at high magnetic field strengths. The latter field-induced changes in the magnon and B_{1g} phonon energy reflect an evolution in $Ca_3Ru_2O_7$ from an antiferromagnetic insulating phase at low fields, to a metallic, relaxed RuO_6 octahedral phase at intermediate fields, then back to an antiferromagnetic insulating phase with compressed octahedra at high fields.[14,15] Figure 4.21(c) shows that the field-dependent c-axis conductivity of $Ca_3Ru_2O_7$ again correlates very well with the field-dependent phonon frequency (Fig. 4.21(b)), indicating that the dramatic magnetoconductivity in $Ca_3Ru_2O_7$ is directly connected with field-induced structural changes associated with the RuO_6 octahedra.[13-15]

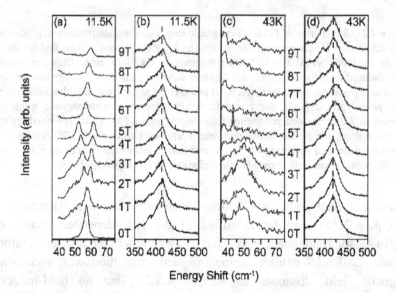

Figure 4.20. Field-dependence (H||magnetic easy axis) of $Ca_3Ru_2O_7$ for: (a) the magnon mode at T=11.5 K; (b) the B_{1g} RuO octahedral phonon mode at T=11.5 K; (c) the magnon mode at T=43 K, and (d) the B_{1g} phonon mode at T=43 K.[ref. 15]

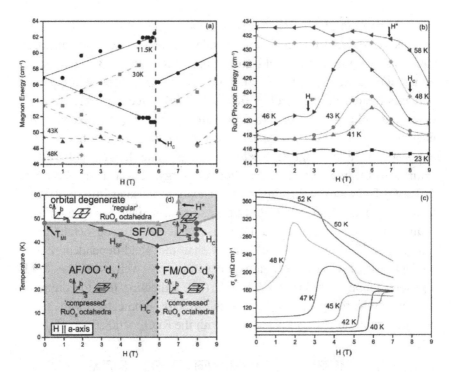

Figure 4.21. Summaries of field- (*H*∥magnetic easy axis) and temperature-dependences in $Ca_3Ru_2O_7$ for: (a) the magnon energy; (b) the B_{1g} RuO phonon energy; and (c) the *c*-axis dc conductivity. (d) The *H-T* phase diagram for $Ca_3Ru_2O_7$ with (*H*∥magnetic easy axis), deduced from the data in (a)–(c). Squares denote a transition boundary between the antiferromagnetic orbital-ordered (AF/*OO*) phase (blue region) and the orbital disordered (OD) phase (pink region), the green diamonds represent a metamagnetic transition between an AF orbital ordered regime (blue region) and a FM orbital polarized regime (yellow region), the purple circles represent a transition from the SF/OD regime to a FM/OO regime, and the orange triangles represent a field-induced transition from the metallic orbital degenerate regime (white region) to the OD regime.[ref. 15]

From these results, two distinct temperature regimes—summarized in Fig. 4.21(d)—can be identified in the field-dependent-phases of $Ca_3Ru_2O_7$ for *H*∥magnetic easy-axis: (i) *T*<40 K – In the low temperature regime, $Ca_3Ru_2O_7$ exhibits a metamagnetic transition with increasing magnetic field, discussed in Section 4.5.2.1, but no field-induced structural changes or orbital rearrangements. The absence of field-induced structural and conductivity changes in this low temperature regime is presumably caused by the highly compressed RuO_6 octahedral structure at these temperatures, which result in large anisotropy fields that prevent field-induced changes in the Ru^{4+} moments for the $H<10$ T. (ii) $T>40K$ – In the higher temperature regime, the complex and interesting field-tuned behavior of the orbital population for the H||magnetic easy axis is likely associated with the appearance of a "spin flop" (SF) phase at intermediate temperatures and fields in $Ca_3Ru_2O_7$, in which regime it is energetically favorable for the Ru^{4+} moments of $Ca_3Ru_2O_7$ to become oriented perpendicular to the applied field direction (i.e., perpendicular to the magnetic easy axis) at intermediate fields.[26] The development of a spin-flop phase in this higher temperature regime likely reflects the gradual expansion in the c-direction of the RuO_6 octahedra with increasing temperature in $Ca_3Ru_2O_7$, which reduces the anisotropy field relative to the exchange field at intermediate temperatures.

Interestingly, the strong connection between field-induced structural and magnetoconductivity, illustrated in Figs. 4.21(b) and (c), suggests that field-induced "flopping" of some fraction of the Ru^{4+} moments from the magnetic easy axis to the magnetic hard axis directions in $Ca_3Ru_2O_7$ gives rise a field-dependent reorganization of orbital populations on the affected Ru-sites. This interpretation is also supported by x-ray diffraction measurements[13] and theoretical calculations.[43] The broad intermediate field range over which the RuO B_{1g} phonon mode frequency is observed to rapidly change for 41 K < T < 47 K (Fig. 4.21(b)) also suggests that the most dramatic field-dependent conductivity observed for H||magnetic easy axis (Fig. 4.21(c)) is associated with a magnetically and orbitally disordered phase regime (Fig. 4.21(d))—i.e., a regime consisting of a field-dependent random mixture of hard- and easy-axis-oriented Ru^{4+} moments and orbital populations on different Ru sites.[14,15] Finally, the reduction in the B_{1g} phonon frequency at high fields in Fig. 4.21(b) indicates that above the intermediate field regime ($H>8$ T), the Ru^{4+} moments gradually become oriented along the applied field for H||magnetic easy axis in $Ca_3Ru_2O_7$. This field-induced alignment of the Ru^{4+} moments results in a high magnetic field "reentrance" into an antiferromagnetic insulating phase with compressed RuO_6 octahedra and orbital polarization.[14,15]

The resulting *HT* phase diagram of $Ca_3Ru_2O_7$ (Fig. 4.21(d)) for *H*∥magnetic easy axis indicates that the magnetostructural phases of $Ca_3Ru_2O_7$ are less robust ("softer") in response to an applied magnetic field along the magnetic easy axis, *H*∥magnetic easy axis, in agreement with theoretical calculations.[43] The distinct differences between the *HT* phase diagrams $Ca_3Ru_2O_7$ for *H*∥magnetic easy axis (Fig. 4.21(d)) and *H*∥magnetic hard axis (Fig. 4.19(d)) also reveal a strong difference between the magnetoconductance in $Ca_3Ru_2O_7$ and more conventional "colossal magnetoresistance" behavior observed in the manganese perovskites.[1] First, in contrast to the manganites, in which a sufficiently strong field in any direction is expected to generate a high-field ferromagnetic (FM) metal phase, the high-field FM phase of $Ca_3Ru_2O_7$ is *not* energetically favorable for electron itinerary with *H*∥magnetic easy axis. Rather, an antiferromagnetic insulating, orbitally polarized (compressed RuO_6 octahedra) phase is apparently stabilized at high magnetic fields for *H*∥magnetic easy axis in $Ca_3Ru_2O_7$ (see Fig. 4.21(d)). This difference reflects the importance in $Ca_3Ru_2O_7$ of strong spin-lattice coupling, which connects the compression of the RuO_6 octahedra—and the corresponding polarization of the Ru^{4+} *d*-orbitals—to the alignment of the Ru^{4+} moments along the magnetic easy axis of $Ca_3Ru_2O_7$.[14,15,43]

4.5.5. *Pressure-Dependent Structural/Orbital Phases in $Ca_3Ru_2O_7$*

The structural properties of the RuO_6 octahedra in the layered ruthenates—and therefore the *d*-orbital populations on the Ru^{4+} ions—are strongly modified with applied pressure.[10,15,45] In particular, the metal-insulator transition in $Ca_3Ru_2O_7$ is associated with a compression of the RuO_6 octahedra,[13] which should lower the energy of the d_{xy} orbital relative to the $d_{yz/zx}$ orbital energy, leading to orbital polarization below T_{MI}=48 K. Hydrostatic pressure is expected to reduce the octahedral distortion by favoring the lower volume, or less *c*-axis compressed, configuration of the unit-cell.[10] Figure 4.22(a) shows the effects of pressure on the magnon Raman spectrum of $Ca_3Ru_2O_7$.[15,45] At low pressures, *P*<50 kbar, there is a slight decrease in the magnon energy with increasing pressure (Fig. 4.22(b)), which is attributable to a gradual reduction of the anisotropy field H_A as the *c*-axis compressive distortions of the octahedral configuration are reduced with pressure.[15] Most dramatically, there is an abrupt disappearance of the ~56 cm^{-1} one-magnon excitation in $Ca_3Ru_2O_7$ above a critical pressure P^*~55 kbar, suggesting a collapse of the AF insulating state above this pressure.[45] As discussed above for Ca_2RuO_4 (see Section 4.4.5), the pressure dependence of the magnon energy in the ruthenates is quite different than the pressure-dependent magnon behavior observed in conventional anti-ferromagnetic (AF) Mott insulators such as La_2CuO_4 [ref. 91] and NiO,[92] in which antiferromagnetism is stabilized, and the Neel temperature increased, with increasing pressure.

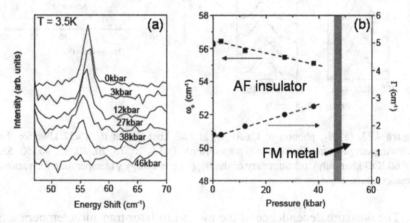

Figure 4.22. (a) Magnon spectra for $Ca_3Ru_2O_7$ at various pressures at T=3.5 K. (b) Summary of the magnon energy (squares) and linewidth (circles) as a function of pressure.[refs. 15,45]

The anomalous pressure dependence of the magnon in $Ca_3Ru_2O_7$ suggests that pressure disrupts antiferromagnetism in this system by thwarting the octahedral distortions that result in orbital polarization. This interpretation is supported by examining the pressure-dependence of the RuO octahedral B_{1g} Raman mode frequency in $Ca_3Ru_2O_7$, which—as discussed above—is sensitive to the RuO_6 octahedral structure. Figure 4.23 shows representative Raman spectra of the B_{1g} phonon mode at varying temperatures for different fixed pressures. The transition between antiferromagnetic insulating and metallic regimes is clearly identified by the abrupt shift in the phonon energy at the metal-insulator transition T_{MI} (see Section 4.5.2.2). Figure 4.23(b) shows that the metal-insulator transition temperature in $Ca_3Ru_2O_7$ decreases systematically with increasing pressure.

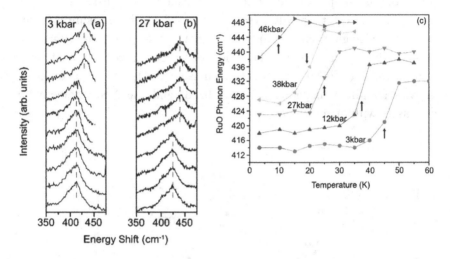

Figure 4.23. (a) B_{1g} phonon of $Ca_3Ru_2O_7$ at $P=3$ kbar and at (b) $P=27$ kbar for the following temperatures (bottom to top spectrum): T=3.5, 10, 15, 20, 30, 35, 40, 50, 55, and 60 K (3 kbar only). (c) Summary of the B_{1g} phonon energy vs temperature for various pressures.[refs. 15,45]

The pressure dependence of the metal-insulator transition temperature (T_{MI}) in $Ca_3Ru_2O_7$ is summarized in Fig. 4.24, which shows that there is an approximately linear decrease of T_{MI} with increasing pressure, with a rate $T_{MI}/P=-0.85$ K/ kbar, and a $T=0$ critical point between AF insulating and FM metal phases at $P^*\sim55$ kbar.[15,45] This critical pressure is close to the pressure at which magnon scattering disappears in $Ca_3Ru_2O_7$ (see Fig. 4.22), suggesting that there is a pressure-tuned quantum ($T\sim0$) phase transition near 55 kbar in $Ca_3Ru_2O_7$ from an orbitally polarized, antiferromagnetic insulating phase to an orbitally degenerate, metallic phase.[15,45] However, recent pressure-dependent measurements of the c-axis[97] and ab-plane[98] resistivities of $Ca_3Ru_2O_7$ provide evidence that—while both the structural transition T_{MI} and Neel temperature T_N are systematically suppressed with increasing pressure—the structural transition temperature is suppressed to a zero value at a somewhat lower pressure than the magnetic transition temperature. These transport results suggest the evolution of a new pressure-induced ground state in $Ca_3Ru_2O_7$ in the pressure range 30 kbar$<P<$60 kbar, characterized by metallic behavior and by ferromagnetic and/or short range antiferromagnetic order.[97,98]

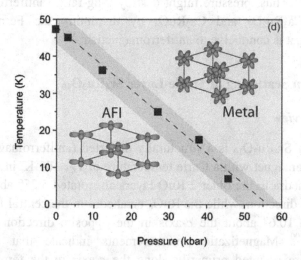

Figure 4.24. Pressure-temperature phase diagram showing the phase boundary between orbital-ordered insulating and orbital-degenerate metallic regimes, as well as the T=0 critical point near P^*=55 kbar.[refs. 15,45]

Although the details of the PT phase diagram of $Ca_3Ru_2O_7$ are still being worked out, it is clear from the pressure dependences of the magnon and B_{1g} phonon in $Ca_3Ru_2O_7$ that antiferromagnetism in $Ca_3Ru_2O_7$ is stabilized by the same octahedral compressions that give rise to orbital-polarization in this system. A qualitative explanation for this connection has been offered by Goodenough,[2] who has pointed out that in transition metal perovskites with t_{2g}-orbital occupancy, there is a tendency for the octahedra to become compressed along the c-axis at low temperatures so that the material can maximize the orbital angular momentum in the low temperature magnetically ordered phase. At higher temperatures, by contrast, increasing thermal fluctuations makes it energetically favorable for the material to reduce the c-axis octahedral compressions and quench the orbital angular momentum.[2] A more specific mechanism by which pressure might disrupt antiferromagnetism in $Ca_3Ru_2O_7$ is suggested by the observation in angle-resolved photoemission studies of a substantial amount of Fermi surface nesting in $Ca_3Ru_2O_7$;[99] electronic structure calculations have shown that such nesting is conducive to antiferromagnetic instabilities in the layered ruthenates.[22] Thus, pressure might destroy long-range antiferromagnetic order in $Ca_3Ru_2O_7$ and Ca_2RuO_4 by disrupting the Fermi-nesting condition that is conducive to antiferromagnetism.[15]

4.6. Raman Scattering in Triple-Layer $Sr_4Ru_3O_{10}$

4.6.1. *Overview*

Triple-layer $Sr_4Ru_3O_{10}$ is a structurally distorted (antiferromagnetically) canted ferromagnet with a Curie temperature of $T_C=105$ K, in which the RuO_6 octahedra in the outer 2 RuO layers are rotated 5.25° about the c-axis in one direction, while the RuO_6 octahedra in the central RuO layer are rotated 10.6° about the c-axis in the opposite direction (see Fig. 4.25(c)).[23,100] Magnetization measurements indicate that the Ru^{4+} moments are oriented primarily along the c-axis in the ferromagnetic phase, as shown in Fig. 4.25(c)).[100] Additionally, there is a strong sensitivity of the B_{1g} phonon frequency in $Sr_4Ru_3O_{10}$ to applied magnetic fields[24] and the onset of ferromagnetic order[24,49]—as discussed below—providing strong evidence that the magnetic moments in $Sr_4Ru_3O_{10}$ are localized on the Ru^{4+} ions and that RuO_6 octahedral rotations are closely connected with magnetism in this material. As discussed previously, the latter point is also supported by density functional calculations of bilayer $Sr_3Ru_2O_7$, which indicate that RuO_6 octahedral rotations are closely connected with magnetism in the layered ruthenates.[22]

4.6.2. *T-Dependent Phonon Spectrum and Spin-Phonon Coupling in $Sr_4Ru_3O_{10}$*

Figure 4.25(d) shows that the temperature dependence of the B_{1g} phonon frequency—which is associated with internal vibrations of the RuO_6 octahedra[49]—exhibits a distinct change in slope, $d\omega/dT$, below T_C.[24,49]

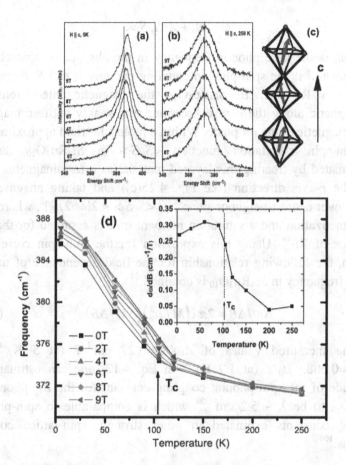

Figure 4.25. Raman spectra of the B_{1g} phonon mode in $Sr_4Ru_3O_{10}$ as a function of magnetic field with $H \parallel c$ axis for (a) $T = 5$ K and (b) $T = 250$ K. (c) Picture of the orientation of the Ru moments in the FM phase in the three layers of octahedral RuO_6. (d) Temperature dependence of the phonon frequency (ω) at different magnetic fields. The inset shows the slope $d\omega/dB$ as a function of temperature.[ref. 24]

This anomalous frequency dependence is indicative of a strong spin-phonon coupling between the B_{1g} phonon mode and the c-axis ordered Ru-moments. From the field dependence of the B_{1g} phonon frequency, a rough estimate of the magnetoelastic coupling between the RuO phonon and the Ru spins in $Sr_4Ru_3O_{10}$ can be obtained: the contribution of spin-spin correlations $\langle S_i \cdot S_j \rangle$ to the phonon frequency ω can be approximated as[101,102]

$$\omega = \omega_0 + \lambda < S_i \cdot S_j >, \qquad\qquad (4.11)$$

where ω_0 is the bare phonon frequency in the absence of spin-phonon interactions, λ is the spin-phonon coupling parameter, and S_i is the spin on the ith Ru site. In $Sr_4Ru_3O_{10}$, the magnetic interactions are ferromagnetic along the c-axis direction, and weakly antiferromagnetic or paramagnetic in the ab-plane. Within a molecular field approximation, the spin-spin correlation function $<S_i \cdot S_j>$ in $Sr_4Ru_3O_{10}$ can be approximated by treating this layered material as a ferromagnetic chain along the c-axis direction (see Fig. 4.25(c)) and taking an ensemble average over nearest-neighbor sites:[24,103] $<S_i \cdot S_j> \approx 2[M/2\mu_B]^2$, where M is the magnetization and a saturation moment of 2 is assumed for the $S=1$ spin state of Ru^{4+}. Using this expression for the spin-spin correlation function, the following relationship for the field-dependence of the B_{1g} phonon frequency in $Sr_4Ru_3O_{10}$ is obtained:[24]

$$\Delta\omega/\Delta B = 2\pi\lambda \left(M/\mu_B^2 \right) \left(\Delta M/\Delta B \right) \qquad\qquad (4.12)$$

Using measured values of $\Delta\omega/\Delta B$=0.27 cm^{-1}/T (at 5 K)[24] and $\Delta M/\Delta B$=0.0083 μ_B/T (at 1.7 K)[100] in Eq. 4.12, one can estimate the magnitude of the spin-phonon coupling constant for the B_{1g} phonon in $Sr_4Ru_3O_{10}$ to be $\lambda \sim 5.2$ cm^{-1},[24] which is comparable to spin-phonon coupling constants estimated in other strongly spin-lattice coupled materials.[104]

4.6.3. *Field-Dependent Phonon Spectrum of Sr₄Ru₃O₁₀*

4.6.3.1. *Field-Dependent Phonon Spectrum, H∥c-axis*

For a magnetic field applied along the c axis direction of $Sr_4Ru_3O_{10}$, the frequency of the B_{1g} phonon increases with increasing field (see Figs. 4.25(a) and (d)), with a maximum frequency shift of $\sim$3 cm^{-1} at T$\sim$ 50 K. This field-induced phonon frequency shift below T_C for $H\|c$-axis is likely associated with a field-dependent rotation of the RuO$_6$ octahedra— and a reduced canting of the Ru^{4+} moments—which causes the octahedra to expand along the c axis and contract in the ab-plane.[24] Thus, field-induced octahedral rotations appear to be responsible for a weakly field-dependent c-axis magnetization observed in $Sr_4Ru_3O_{10}$.[23,24,100]

4.6.3.2. *Field-Dependent Phonon Spectrum, $H \parallel ab$-plane*

As summarized in Fig. 4.26, very different magnetoelastic effects associated with the B_{1g} phonon are observed upon applying a magnetic

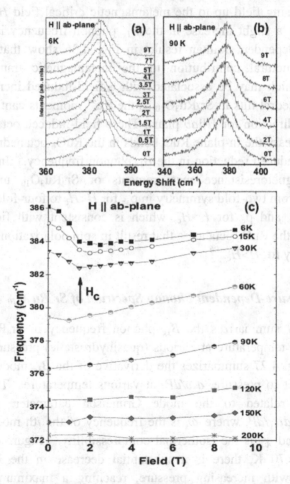

Figure 4.26. Raman spectra of the B_{1g} phonon mode in $Sr_4Ru_3O_{10}$ as a function of magnetic field with $H \parallel ab$-plane for (a) $T = 6$ K and (b) $T = 90$ K. (c) Field dependence of the B_{1g} phonon frequency for different temperatures.[ref. 24]

field in the *ab*-plane direction of $Sr_4Ru_3O_{10}$. Magnetization measurements with $H\|ab$ plane revealed a metamagnetic transition at $H_c \sim 2$ T for $T<50$ K in $Sr_4Ru_3O_{10}$;[23] the specific nature of this transition was not identified, but transport measurements showed abrupt resistive jumps near this field-induced transition, indicative of some form of switching behavior.[105] Figure 4.26 shows that there is little change in the B_{1g} phonon frequency with increasing field for temperatures 50 K$<T<T_C$. However, for $T<50$ K, the B_{1g} phonon frequency decreases significantly with increasing field up to the metamagnetic critical field $H_c=2$ T. For $H>H_c$, only a slight increase in the B_{1g} phonon frequency is observed. The field-dependent Raman results in Fig. 4.26 show that there is a distinct structural contribution to the metamagnetic transition near $H_c=2$ T, which may be associated with a field-induced increase in the RuO_6 octahedral tilts in $Sr_4Ru_3O_{10}$ as the Ru^{4+} moments cant away from the *c*-axis direction for $H\|ab$ plane. These field-induced octahedral tilts should increase the in-plane RuO bonds in the RuO_6 octahedra, resulting in a field-induced reduction in the B_{1g} phonon frequency.[24] Interestingly, recent magnetoresistance measurements of $Sr_4Ru_3O_{10}$ uncovered a transition from two-fold symmetry in ρ_{ab} for $H<H_c$ to four-fold symmetry in both ρ_{ab} and ρ_c for $H>H_c$, which is consistent with field-induced changes in the RuO_6 octahedra that result in spin polarization of the $d_{xz/yz}$ orbitals only for $H>H_c$.[106]

4.6.4. *Pressure-Dependent Phonon Spectrum of $Sr_4Ru_3O_{10}$*

Figure 4.27 summarizes the B_{1g} phonon frequency in $Sr_4Ru_3O_{10}$ as a function of temperature at various (quasihydrostatic) pressures, and the inset of Fig. 4.27 summarizes the derivative of the B_{1g} mode frequency with respect to pressure, $d\omega/dP$, at various temperatures. The quantity $d\omega/dP$ is related to the mode Gruneisen parameter, defined as $\gamma=(1/\omega_i\chi_T)d\omega_i/dP$, where ω_i is the frequency of the *i*th mode, P is the pressure, and χ_T is the isothermal compressibility.[24] Figure 4.27 shows that for $T<70$ K, there is a substantial decrease in the B_{1g} phonon frequency with increasing pressure, reaching a maximum value of $d\omega/dP=-0.32$ cm^{-1}/kbar at $T=5$ K. This pressure-induced "softening" of the B_{1g} phonon in $Sr_4Ru_3O_{10}$ is likely associated with the pressure-induced buckling of the RuO_6 octahedra on adjacent RuO layers, which forces the Ru^{4+} moments to cant toward the *ab* plane with increasing pressure, i.e., in opposition to the tendency for *c* axis FM ordering at *P*=0. Interestingly, Fig. 4.27 shows that for *P*≥24 kbar, the B_{1g} phonon frequency has a roughly linear temperature dependence, with no anomalous change in slope through T_C, suggesting that the pressure-induced buckling of the RuO_6 octahedra on adjacent RuO layers at these pressures suppresses *c*-axis ferromagnetism for isobaric temperature sweeps, presumably by suppressing the RuO_6 rotations that accompany *c* axis FM ordering in $Sr_4Ru_3O_{10}$.[24] Pressure-dependent magnetization measurements of $Sr_4Ru_3O_{10}$ are needed to confirm this interpretation of the data.

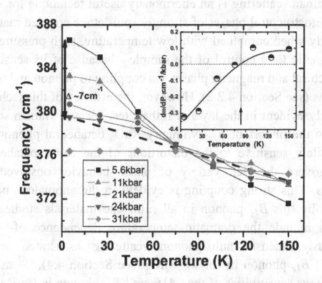

Figure 4.27. Temperature dependence of the B_{1g} phonon mode in $Sr_4Ru_3O_{10}$ for different pressures. The inset shows the pressure derivative of the B_{1g} phonon frequency, $d\omega/dP$, as a function of temperature; the vertical line indicates the temperature (~75 K) below which the derivative has a negative value.[ref. 24]

4.7. Summary

As we have hoped to make clear in this review, the layered ruthenate family, $(Sr,Ca)_{n+1}Ru_nO_{3n+1}$, exhibits a unique variety of phases and properties that result—in ways still not fully understood—from a complex interplay of factors, particularly the extended $4d$ orbitals of the Ru^{4+} ions, the complex Ruddlesden-Popper structure, and strong spin-orbital coupling in these materials. In particular, the experimental[10,11,13-15,24,42,.45-49] and theoretical[22,43,80] results described in this review show clearly that structural changes associated with the RuO_6 octahedra—induced variously as functions of temperature, atomic-substitution, magnetic-field, and pressure—play an enormous role in generating the dramatic highly "tunable" phenomena and diverse phase diagrams observed in the layered ruthenates.

In this review, we have focused on the special perspective that Raman scattering provides on the layered ruthenates. As discussed above, Raman scattering is an enormously useful technique for studying the magnetostructural phases of strongly spin-lattice coupled materials—particularly when combined with low temperature, high pressure, and/or high magnetic field control of the sample—because of its sensitivity to both structural and magnetic phases via coupling to phonon and magnetic excitations (see Section 4.2.2). However, the efficacy of this technique is particularly evident in the layered ruthenates, because Raman scattering couples to the important B_{1g}-symmetry RuO_6 octahedral phonon, which is exquisitely sensitive to the distortions of the RuO_6 octahedra that largely govern the rich variety of phase behavior observed in the ruthenates. This strong coupling is evident in the anomalous behaviors exhibited by this B_{1g} phonon in all ruthenate materials studied. These behaviors include the dramatic temperature dependence of—and the anomalously intense multi-phonon scattering associated with—the ~380 cm^{-1} B_{1g} phonon in $Ca_{2-x}Sr_xRuO_4$ (see Section 4.4),[46,47] as well as the enormous sensitivities of the ~416 cm^{-1} B_{1g} phonon in $Ca_3Ru_2O_7$ (see Section 4.5) and the ~380 cm^{-1} B_{1g} phonon in $Sr_4Ru_3O_{10}$ (see Section 4.6) to the temperature-, magnetic-field, and pressure-dependent phases of these materials.

Yet, while there is now a fairly clear picture of how RuO_6 octahedral distortions influence much of the rich phase behavior observed in the layered ruthenates, a number of important unanswered questions remain. For example, in single layer $Ca_{2-x}Sr_xRuO_4$, what role does the orbital degree of freedom and strong spin-orbit coupling—which are so important in the antiferromagnetic-insulating phase of Ca_2RuO_4—play in the spin-triplet pairing superconductor phase in Sr_2RuO_4? Both experimental[107] and theoretical[108] results suggest the importance of spin-orbit coupling to unconventional superconductivity in Sr_2RuO_4, but this connection needs to be explored more fully. Pressure-dependent studies of Ca_2RuO_4 (see Section 4.4.5) using experimental probes that are sensitive to the orbital degree of freedom might provide insight into the relationship between the ground states of Ca_2RuO_4 and Sr_2RuO_4. Similarly, in double layer $(Ca,Sr)_3Ru_2O_7$, it is important to determine the extent to which the orbital degree of freedom—which is evidently important in $Ca_3Ru_2O_7$ (see Section 4.5)—plays in the field-induced anisotropic low temperature phase observed in $Sr_3Ru_2O_7$.[18] Recent calculations suggest a possible role of orbital-ordering in the anisotropic low temperature state in $Sr_3Ru_2O_7$,[109] although mechanisms for generating this state via electronic correlations alone have also been proposed.[110] Again, pressure-tuned studies that allow a more careful exploration of the evolution of the orbital degree of freedom between the distorted insulating and undistorted metallic phases might elucidate the role of the orbital degree of freedom in the various novel low temperature phases of $Sr_3Ru_2O_7$.

Acknowledgments

This material is based on work supported by the U.S. Department of Energy, Division of Materials Sciences, under Award No. DE-FG02-07ER46453, through the Frederick Seitz Materials Research Laboratory at the University of Illinois at Urbana-Champaign, and by the National Science Foundation under Grant NSF DMR 08-56321. I would like to thank all the students, postdoctoral research associates, and collaborators who've made the work described in this review possible, particularly Gang Cao, Rajeev Gupta, John Karpus, Hsiang-Lin Liu, Yoshi Maeno, Satoru Nakatsuji, Heesuk Rho, and Clark Snow. I would also like to acknowledge fruitful discussions on issues involving orbital physics with collaborators Peter Abbamonte, Eduardo Fradkin, Paul Goldbart, Siddhartha Lal, and Minjung Kim, and I would like to thank Minjung Kim for her critical reading of this manuscript.

References

1. Y. Tokura and N. Nagaosa, *Science* **288**, 462 (2000).
2. J. B. Goodenough, *Rep. Prog. Phys.* **67**, 1915 (2004).
3. H. K. Muller-Buschbaum and J. Wilkens, *Z. Anorg. Allg. Chem.* **591**, 161 (1990).
4. S. N. Ruddlesden and P. Popper, *Acta Crystallogr.* **11**, 54 (1958).
5. Y. Maeno, H. Hashimoto, K. Yoshida, S. Nishizaka, T. Fujita, J.G. Bednorz, and F. Lichtenberg, *Nature (London)* **372**, 532 (1994).
6. T. Hotta and E. Dagotto, *Phys. Rev. Lett.* **88**, 017201 (2002).
7. Y. Liu, *New J. Phys.* **12**, 075001 (2010).
8. C.S. Alexander, G. Cao, V. Dobrosavljevic, S. McCall, J.E. Crow, E. Lochner, and R.P. Guertin, *Phys. Rev. B* **60**, R8422 (1999).
9. T. Mizokawa, L.H. Tjeng, G.A. Sawatzky, G. Ghiringhelli, O. Tjernberg, N.B. Brookes, H. Fukazawa, S. Nakatsuji, and Y. Maeno, *Phys. Rev. Lett.* **87**, 077202 (2001).
10. P. Steffens, O. Friedt, P. Alireza, W.G. Marshall, W. Schmidt, F. Nakamura, S. Nakatsuji, Y. Maeno, R. Lengsdorf, M.M. Abd-Elmeguid, and M. Braden, *Phys. Rev. B* **72**, 094104 (2005).
11. O. Friedt, M. Braden, G. André, P. Adelmann, S. Nakatsuji, and Y. Maeno, *Phys. Rev. B* **63**, 174432 (2001).
12. G. Cao, L. Balicas, X. N. Lin, S. Chikara, E. Elhami, V. Duairaj, J. W. Brill, R. C. Rai, and J. E. Crow, *Phys. Rev. B* **69**, 014404 (2004).
13. C. S. Nelson, H. Mo, B. Bohnenbuck, J. Strempfer, N. Kikugawa, S. I. Ikeda and Y. Yoshida, *Phys. Rev. B* **75**, 212403 (2007).
14. J.F. Karpus, R. Gupta, H. Barath, S.L. Cooper, and G. Cao, *Phys. Rev. Lett.* **93**, 167205 (2004).
15. J.F. Karpus, C.S. Snow, R. Gupta, H. Barath, S.L. Cooper, and G. Cao, *Phys. Rev. B* **73**, 134407 (2006).
16. X. N. Lin, Z. X. Zhou, V. Durairaj, P. Schlottmann, and G. Cao, *Phys. Rev. Lett.* **95**, 017203 (2005).
17. R. S. Perry, K. Kitagawa, S.A. Grigera, R.A. Borzi, A.P. Mackenzie, K. Ishida, and Y. Maeno, *Phys. Rev. Lett.* **92**, 166602 (2004).
18. R. A. Borzi, S. A. Grigera, J. Farrell, R. S. Perry, S. J. S. Lister, S. L. Lee, D. A. Tennant, Y. Maeno, and A. P. Mackenzie, *Science* **315**, 214 (2007).
19. S. I. Ikeda, Y. Maeno, S. Nakatsuji, M. Kosaka, and Y. Uwatoko, *Phys. Rev. B* **62**, R6089 (2000).
20. R. S. Perry, L.M. Galvin, S.A. Grigera, L Capogna, A.J. Schofield, A.P. Mackenzie, M. Chiao, S.R. Julian, S.I. Ikeda, S. Nakatsuji, Y. Maeno, and C. Pfleiderer, *Phys. Rev. Lett.* **86**, 2661 (2001).
21. S. A. Grigera, R.S. Perry, A.J. Schofield, M. Chiao, S.R. Julian, G.G. Lonzarich, S.I. Ikeda, Y. Maeno, A.J. Millis, and A.P. Mackenzie, *Science* **294**, 329 (2001).
22. D. J. Singh and I. I. Mazin, *Phys. Rev. B* **63**, 165101 (2001); I. I. Mazin and D. J. Singh, *Phys. Rev. Lett.* **82**, 4324 (1999).
23. M. K. Crawford, R.L. Harlow, W. Marshall, Z. Li, G. Cao, R.L. Lindstrom, Q. Huang, and J.W. Lynn, *Phys. Rev. B* **65**, 214412 (2002).

24. R. Gupta, M. Kim, H. Barath, S. L. Cooper, and G. Cao, *Phys. Rev. Lett.* **96**, 067004 (2006).
25. W. Hayes and R. Loudon, *Scattering of Light by Crystals* (New York, Wiley, 1978).
26. M.G. Cottam and D.J. Lockwood, *Light Scattering in Magnetic Solids* (Wiley, New York, 1986).
27. C. Thomsen and M. Cardona, *Physical properties of the high T_c superconductors I.* ed. by D.M. Ginsberg (Singapore, World Scientific, 1989).
28. C. Thomsen, *Topics in Applied Physics 68*, ed. by M. Cardona and G. Guntherodt (Berlin, Springer-Verlag, 1991).
29. T.P. Devereaux and R. Hackl, *Rev. Mod. Phys.* **79**, 175 (2007).
30. S.L. Cooper, *Handbook on the Physics and Chemistry of Rare Earths 31*, ed. by K.A. Gschneidner and M.B. Maple, (Amsterdam: Elsevier Science, 2001).
31. M. Cardona, *Raman scattering in materials science*, ed. by W.H. Weber and R. Merlin (Berlin, Springer, 2000).
32. J.F. Scott, *Rev. Mod. Phys.* **46**, 83 (1974).
33. S.L. Cooper, *Structure and Bonding* **98**, 161 (2001).
34. V.B. Podobedov and A. Weber, *Raman scattering in materials science*, ed. by W.H. Weber and R. Merlin (Berlin, Springer, 2000).
35. E. Zirngiebl and G. Guntherodt, *Topics in Applied Physics 68*, ed. by M. Cardona and G. Guntherodt (Berlin, Springer-Verlag, 1991).
36. A. Pinczuk and E. Burstein, *Topics in Applied Physics vol. 8,* ed. by M. Cardona (Berlin, Springer-Verlag, 1975).
37. M.V. Klein, *Light scattering in solids III,* ed. by M. Cardona and G. Guntherodt (Berlin, Springer-Verlag, 1982).
38. A.K. Ramdas and S. Rodriguez, *Topics in Applied Physics 68*, ed. by M. Cardona and G. Guntherodt (Berlin, Springer-Verlag, 1991).
39. M. Cardona, *Light scattering in solids II*, ed. by M. Cardona and G. Guntherodt (Berlin, Springer-Verlag, 1982).
40. M.H. Brodsky, *Topics in Applied Physics vol. 8,* ed. by M. Cardona (Berlin, Springer-Verlag, 1975).
41. M.V. Klein, J.A. Holy, and W.S. Williams, *Phys. Rev. B* **17**, 1546 (1978).
42. M. Braden, G. André, S. Nakatsuji, and Y. Maeno, *Phys. Rev. B* **58**, 847 (1998).
43. F. Forte, M. Cuoco, and C. Noce, *Phys. Rev. B* **82**, 155104 (2010).
44. H. L. Liu, S. Yoon, S. L. Cooper, G. Cao, and J. E. Crow, *Phys. Rev. B* **60**, R6980 (1999).
45. C.S. Snow, S.L. Cooper, G. Cao, J.E. Crow, S. Nakatsuji, and Y. Maeno, *Phys. Rev. Lett.* **98**, 226401 (2002).
46. H. Rho, S.L. Cooper, S. Nakatsuji, H. Fukazawa, and Y. Maeno, *Phys. Rev. B* **68**, 100404(R) (2003).
47. H. Rho, S.L. Cooper, S. Nakatsuji, H. Fukazawa, and Y. Maeno, *Phys. Rev. B* **71**, 245121 (2005).
48. M. N. Iliev, S. Jandl, V. N. Popov, A. P. Litvinchuk, J. Cmaidalka, R. L. Meng, and J. Meen, *Phys. Rev. B* **71**, 214305 (2005).

49. M.N. Iliev, V.N. Popov, A.P. Litvinchuk, M.V. Abrashev, J. Backstrom, Y.Y. Sun, R.L. Meng, and C.W. Chu, *Physica B* **358**, 138 (2005).
50. W.G. Fateley, F.R. Dollish, N.T. McDevitt, and F.F. Bentley, *Infrared and Raman selection rules for molecular and lattice vibrations: the correlation method* (New York, Wiley-Interscience, 1972).
51. T. Moriya, *J. Phys. Soc. Japan* **23**, 490 (1967).
52. P.A. Fleury and R. Loudon, *Phys. Rev.* **166**, 514 (1968).
53. P.A. Fleury and H.J. Guggenheim, *Phys. Rev. Lett.* **24**, 1346 (1970).
54. J.B. Parkinson, *J. Phys. C* **2**, 2012 (1969).
55. S. Sugai, M. Sato, T. Kobayashi, J. Akimitsu, T. Ito, H. Takagi, S. Uchida, S. Hosoya, T. Kajitani, and T. Fukuda, *Phys. Rev. B* **42**, R1045 (1990).
56. Y. Endoh, K. Yamada, K. Kakurai, M. Matsuda, K. Nakajima, R.J. Birgeneau, M.A. Kastner, B. Keimer, G. Shirane, and T.R. Thurston, *Physica B* **174**, 330 (1991).
57. S.A. Asher and R. Bormett, *Raman scattering in materials science*, ed. by W.H. Weber and R. Merlin (Berlin, Springer, 2000).
58. H. Kuzmany, *Solid state spectroscopy: an introduction* (Berlin, Springer, 1998).
59. S. L. Cooper, P. Abbamonte, N. Mason, M. Kim, H. Barath, C. Chialvo, J.P. Reed, Y.I. Joe, X. Chen, and D. Casa, *Raman scattering as a tool for studying complex materials*, in *Optical techniques for materials characterization* (Taylor & Francis), in press (2011).
60. J.R. Janesick, SPIE Press Monograph vol. PM83 (2001).
61. D.J. Dunstan and W. Scherrer, *Rev. Sci. Instrum.* **59**, 3789 (1988).
62. M.I. Eremets, *High pressure experimental methods* (Oxford, Oxford University Press, 1996).
63. A.M. Patselov, K.M. Demchuk, and A.A. Starostin, *Pri. Tek. Eksp.* **6**, 159 (1989).
64. J. Xu and H. Mao, *Science* **290**,783 (2000).
65. C.S. Snow, J.F. Karpus, S.L. Cooper, T.E. Kidd, and T.-C Chiang, *Phys. Rev. Lett.* **91**, 136402 (2003).
66. C.S. Snow, Probing the dynamics of pressure- and magnetic-field-tuned transitions in strongly correlated electron systems: Raman scattering studies. PhD Thesis, University of Illinois at Urbana-Champaign, (2003).
67. D.J. Dunstan and I.L. Spain, *J. Phys. E: Sci. Instrum.* **22**, 913 (1989).
68. A. Jayaraman, *Rev. Mod. Phys.* **55**, 65 (1983).
69. J.H. Burnett, H.M. Cheong, and W. Paul, W, *Rev. Sci. Instrum.* **61**, 3904 (1990).
70. S. S. Rosenblum and R. Merlin, *Phys. Rev. B* **59**, 6317 (1999).
71. U. Venkateswaran, M. Chandrasekhar, H.R. Chandrasekhar, B.A. Vojak, F.A. Chambers, and J.M. Meese, *Phys. Rev. B* **33**, 8416 (1986).
72. I. Zegkinoglou, J. Strempfer, C.S. Nelson, J.P. Hill, J. Chakhalian, C. Bernhard, J.C. Lang, G. Srajer, H. Fukazawa, S. Nakatsuji, Y. Maeno, and B. Keimer, *Phys. Rev. Lett.* **95**, 136401 (2005).
73. T. Mizokawa, L. H. Tjeng, H.-J. Lin, C. T. Chen, S. Schuppler, S. Nakatsuji, H. Fukazawa, and Y. Maeno, *Phys. Rev. B* **69**, 132410 (2004).
74. J.H. Jung, Z. Fang, J.P. He, Y. Kaneko, Y. Okimoto, and Y. Tokura, *Phys. Rev. Lett.* **91**, 056403 (2003).

75. S. Nakatsuji and Y. Maeno, *Phys. Rev. Lett.* **84**, 2666 (2000).
76. L. Balicas, S. Nakatsuji, D. Hall, T. Ohnishi, Z. Fisk, Y. Maeno, and D.J. Singh, *Phys. Rev. Lett.* **95** 196407 (2005).
77. S. Nakatsuji and Y. Maeno, *Phys. Rev. B* **62**, 6458 (2000).
78. M. Udagawa, T. Minami, N. Ogita, Y. Maeno, F. Nakamura, T. Fujita, J.G. Bednorz, and F. Lichtenberg, *Physica B* **219&220**, 222 (1996).
79. S. Sakita, S. Nimori, Z.Q. Mao, Y. Maeno, N. Ogita, and M. Udagawa, *Phys. Rev. B* **63**, 134520 (2001).
80. M. Cuoco, F. Forte, and C. Noce, *Phys. Rev. B* **73**, 094428 (2006).
81. U. Fano, *Phys. Rev.* **124**, 1866 (1961).
82. J.S. Smart, *J. Phys. Chem. Solids* **11**, 97 (1959).
83. Z. Fang and K. Terakura, *Phys. Rev. B* **64**, 020509 (2001).
84. J. C. T. Lee, S. Yuan, S. Lal, Y.- I. Joe, Y. Gan, S. Smadici, K. Finkelstein, Y. Feng, A. Rusydi, P. M. Goldbart, S. L. Cooper, and P. Abbamonte, cond-matxxx (2011).
85. P.B. Allen and V. Perebeinos, *Phys. Rev. Lett.* **83**, 4828 (1999).
86. V. Perebeinos and P.B. Allen, *Phys. Rev. B* **64**, 085118 (2001).
87. R. Kruger, B. Schulz, S. Naler, R. Rauer, D. Budelmann, J. Backstrom, K.H. Kim, S-W. Cheong, V. Perebeinos, and M. Rubhausen, *Phys. Rev. Lett.* **92**, 097203 (2004).
88. L. L. Kruschinskii and P. P. Shorygin, *Opt. Spectrosc. (USSR)* **11**, 12 (1961); *ibid.* **11**, 80 (1961).
89. S. Okamoto and A.J. Millis, *Phys. Rev. B* **70**, 195120 (2004).
90. F. Nakamura, T. Goko, M. Ito, T. Fujita, S. Nakatsuji, H. Fukazawa, Y. Maeno, P. Alireza, D. Forsythe, and S.R. Julian, *Phys. Rev. B* **65**, 220402 (2002).
91. M.C. Aronson, S.B. Dierker, B.S. Dennis, S-W. Cheong, and Z. Fisk, *Phys. Rev. B* **44**, 4657 (1991).
92. M. J. Massey, N.H. Chen, J.W. Allen, and R. Merlin, *Phys. Rev. B* **42**, 8776 (1990).
93. S. McCall, G. Cao, and J. E. Crow, *Phys. Rev. B* **67**, 094427 (2003).
94. Y. Yoshida, S.-I. Ikeda, H. Matsuhata, N. Shirakawa, C.H. Lee, and S. Katano, *Phys. Rev. B* **72**, 054412 (2005).
95. B. Bohnenbuck, I. Zegkinoglou, J. Strempfer, C. Schüßler-Langeheine, C. S. Nelson, Ph. Leininger, H.-H. Wu, E. Schierle, J. C. Lang, G. Srajer, S. I. Ikeda, Y. Yoshida, K. Iwata, S. Katano, N. Kikugawa, and B. Keimer, *Phys. Rev. B* **77**, 224412 (2008).
96. V. I. Anisimov, I. A. Nekrasov, D. E. Kondakov, T. M. Rice, and M. Sigrist, *Eur. Phys. J. B* **25**, 191 (2002).
97. Y. Yoshida, M. Hedo, S.I. Ikeda, N. Shirakawa, and Y. Uwatoko, *Physica B* **403**, 1213 (2008).
98. W.J. Duncan, O.P. Welzel, D. Moroni-Klementowicz, C. Albrecht, P.G. Niklowitz, D. Gruner, M. Brando, A. Neubauer, C. Pfleiderer, N. Kikugawa, A.P. Mackenzie, and F.M. Grosche, *Phy. Status Solidi B* **247**, 544 (2010).
99. F. Baumberger, N. J. C. Ingle, N. Kikugawa, M. A. Hossain, W. Meevasana, R. S. Perry, K. M. Shen, D. H. Lu, A. Damascelli, A. Rost, A. P. Mackenzie, Z. Hussain, and Z.-X. Shen, *Phys. Rev. Lett.* **96**, 107601 (2006).

100. G. Cao, L. Balicas, W.H. Song, Y.P. Sun, Y. Xin, V.A. Bondarenko, J.W. Brill, S. Parkin, X.N. Lin, *Phys. Rev. B* **68**, 174409 (2003).
101. D. J. Lockwood and M. G. Cottam, *J. Appl. Phys.* **64**, 5876 (1988).
102. X. K. Chen, J. C. Irwin, and J. P. Franck, *Phys. Rev. B* **52**, R13130 (1995).
103. A. H. Morrish, *Physical Principles of Magnetism* (Wiley, New York, 1965).
104. A. B. Sushkov, O. Tchernyshyov, W. Ratcliff, S.W. Cheong, and H.D. Drew, *Phys. Rev. Lett.* **94**, 137202 (2005).
105. Z.Q. Mao, M. Zhou, J. Hooper, V. Golub, and C.J. O'Connor, *Phys. Rev. Lett.* **96**, 077205 (2006).
106. D. Fobes, T.J. Liu, Z. Qu, M. Zhou, J. Hooper, M. Salamon, and Z.Q. Mao, *Phys. Rev. B* **81**, 172402 (2010).
107. M.W. Haverkort, I.S. Elfimov, L.H. Tjeng, G.A. Sawatzky, and A. Damascelli, *Phys. Rev. Lett.* **101**, 026406 (2008).
108. K.K Ng and M. Sigrist, *Europhys. Lett.* **49**, 473 (2000).
109. W.-C. Lee and C. Wu, *Phys. Rev. B* **80**, 104438 (2009); W.-C. Lee, D.P. Arovas, and C. Wu, *Phys. Rev. B* **81**, 184403 (2010).
110. V. Oganesyan, S.A. Kivelson, and E. Fradkin, Phys. Rev. B 64, 195109 (2001).

Chapter 5

METAL-INSULATOR TRANSITIONS IN N=1 RUDDLESDEN-POPPER RUTHENATES

Rongying Jin

Department of Physics & Astronomy, Louisiana State University
Baton Rouge, LA 70803, USA
E-mail: rjin@lsu.edu

This review discusses the electronic properties of single-layered (n=1) Ruddlesden-Popper ruthenates with emphasis on the metal-insulator transitions. Such transitions can be induced by either temperature, or pressure, or chemical doping. Through the evaluation of the existing experimental properties and theoretical calculations, we discuss the coupling between several degrees of freedom: charge, spin, lattice and orbital.

Contents

5.1. Introduction

Many chapters in this book deal with Ruddlesden-Popper (RP)[1,2] ruthenates with general formula $A_{n+1}Ru_nO_{3n+1}$ (n = 1, 2,... ∞), where A is an alkali metal. As shown in Fig. 5.1, the structure consists of n-layers of corner-sharing RuO_6 octahedra in an unit cell interleaved with AO layers, i.e., $AO(ARuO_3)_n$. The Ru atoms are located in the center of each octahedron. This is a prototype system, in which physical properties are extremely susceptible to turning parameters such as temperature, magnetic field, pressure, and chemical doping. As the result, the RP ruthenates cover a spectrum of ground-state properties, including an unconventional superconductor, ferromagnetic conductors, metamagnetic conductors, and antiferromagnetic insulators. In this chapter, the aim is to review fascinating physical properties occurring in the single-layered (n=1) ruthenates, that are related to metal-insulator transitions.

Metal-insulator transitions are observed in many materials, accompanied by the resistivity change in both magnitude and temperature dependence. There is a comprehensive review on the possible origins of metal-insulator transitions,[3] what is emphasized in this chapter is the metal-insulator transitions that are driven by electron-electron correlation

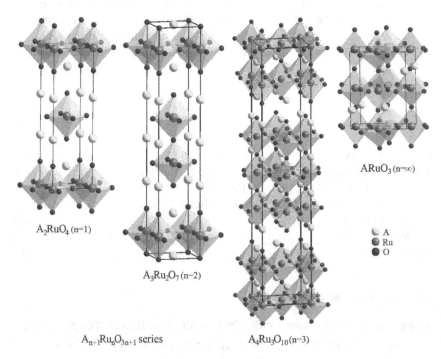

A_2RuO_4 (n=1)

$A_3Ru_2O_7$ (n=2)

$ARuO_3$ (n=∞)

$A_{n+1}Ru_nO_{3n+1}$ series

$A_4Ru_3O_{10}$ (n=3)

○ A
● Ru
● O

Fig. 5.1. A schematic illustration of Ruddleston-Popper series.

effects. The fundamentals of the subject are well described in Ref. [3]. Here, we discuss the possible origins of metal-insulator transitions of single-layered ruthenates through the evaluation of the existing experimental properties and theoretical calculations. X-ray and Roman scattering studies of the ruthenates are presented in Chapters 3 and 4.

5.2. Single-Layered A_2RuO_4 with A = Ca, Sr, Ba

Single-layered ruthenates with general formula A_2RuO_4 form K_2NiF_4-type structure [see Fig. 5.1(left)]. It has been reported that A can be either Ca or Sr or Ba. While Sr_2RuO_4 is the most studied compound because of its unconventional superconductivity[4], little is known about Ba_2RuO_4. This is in part due to the difficulty in synthesizing Ba_2RuO_4 with layered perovskite structure at atmospheric or low pressure.[5] With the help of interfacial strain, Ba_2RuO_4 epitaxial film with desired structure (isostructural with Sr_2RuO_4) has been successfully grown.[6] Figure 5.2 shows the temperature dependence of its electrical resistivity (ρ), which reveals metallic behavior ($d\rho_{ab}/dT > 0$) in a wide temperature range. But an upturn ($d\rho_{ab}/dT < 0$) develops at low temperatures, where the onset temperature varies non-monotonically with the background

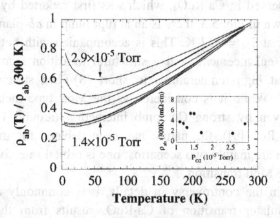

Fig. 5.2. In-plane electrical resistivity normalized by its value at 300 K [$\rho_{ab}(300K)$] for epitaxial Ba_2RuO_4 films grown in various oxygen background pressure (P_{O2}). From top to bottom, the P_{O2} values in units of μTorr are: 29, 17, 15, 12, 25, 11, 13, and 14. The inset shows $\rho_{ab}(300K)$ vs. P_{O2} (from Ref. [6]).

pressure (oxygen).[6] At present, the origin of the metal-insulator transition is unknown. Given the fact that $d\rho_{ab}/dT$ is larger for the film under lower oxygen pressure, the resistivity upturn may be associated with excess oxygen induced by high oxygen pressure. The excess oxygen acts as scattering centers.

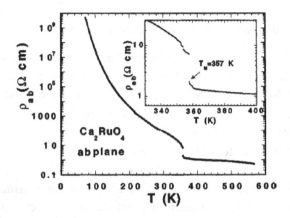

Fig. 5.3. In-plane electrical resistivity as a function of temperature. Inset is the detail of the abrupt jump in ρ_{ab} near the transition T_M (from Ref. [7]).

Nevertheless, a metal-insulator transition with the first-order characteristic is observed in Ca_2RuO_4, which was first reported by Alexander et al.[7] As shown in Fig. 5.3, there is an abrupt jump in *ab*-plane electrical resistivity ρ_{ab} at $T_M = 357$ K. This is accompanied with both magnetic susceptibility enhancement[7] and a structural transition from long (L)-Pbca phase at high temperatures to short (S)-Pbca structure at low temperatures.[8] While it is considered as the Mott type metal-insulator transition driven by strong Coulomb interaction (around $1.1 - 1.2$ eV according to Ref. [9]), the nature has been debated for more than a decade. There are mainly two scenarios: one is orbital selective[10] and the other is single Mott transition.[11-12]

Aside from the controversy in details, it is commonly agreed that the metal-insulator transition of Ca_2RuO_4 results from the structure distortion. According to local-density approximation (LDA) plus dynamic mean-field theory (DMFT), the structure change leads to larger crystal-field splitting between $d_{xz/yz}$ and d_{xy} subbands and narrower bandwidths ($W_{xz/yz}$ and W_{xy}).[10-16] As shown in Fig. 5.4, electrons redistribute when the system undergoes the structure transition.[13] Both x-ray absorption[17] and optical conductivity[18] studies reveal the change of orbital population of the Ru t_{2g} band when Ca_2RuO_4 undergoes metal-insulator transition. As shown in Fig. 5.5, there is sharp decrease of effective number of electrons at T_{MIT} accompanied with the opening of

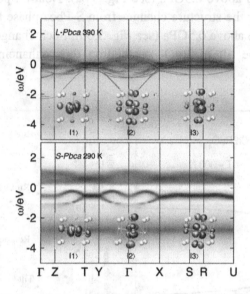

Fig. 5.4. The Mott transition: correlated bands for L-Pbca (top panel) and S-Pbca (bottom panel). Positive (negative) polarization is indicated by red (blue). In the case of S-Pbca, $|1>$ (d_{yz}) is full, $|2>$ (d_{yz}) and $|3>$ (d_{xy}) are half-filled (from Ref. [13]).

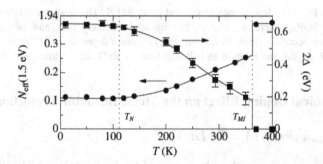

Fig. 5.5. Effective number of electrons N_{eff}(1.5 eV) and optical gap 2Δ of Ca_2RuO_4 (from Ref. [18]).

optical gap.[18] Interestingly, both N_{eff} and Δ saturate below the anti-ferromagnetic transition at T_N.

The scenario that the Mott metal-insulator transition is driven by the structure change is further supported by the results obtained from high-pressure resistivity[19] and structure[20] studies of Ca_2RuO_4. As shown in Fig. 5.6, there is a first-order phase transition from insulator below 0.5GPa to metal above 0.5GPa (see Fig. 5.6a). Neutron powder diffraction reveals that the structure changes from S-Pbca phase below 0.5GPa to L-Pbca phase above 0.5GPa (see Fig. 5.6b). Such change involves the reduction of the tilt of RuO_6 octahedron thus enhancing the orbital overlapping.[20]

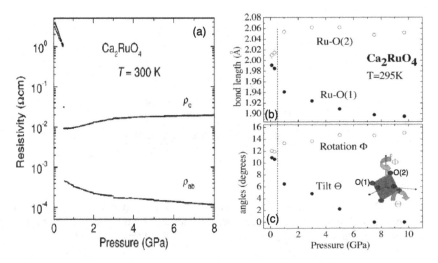

Fig. 5.6. (a) Pressure dependence of ρ_{ab} and ρ_c at 300 K. The insulator-metal transition occurs at 0.5GPa, where ρ_{ab} and ρ_c fall by approximately four and two orders of magnitude, respectively (from Ref. [19]). (b-c) Pressure dependence of the Ru-O(1) and Ru-O2 bond lengths, and rotation and tilt angles of the RuO_6 octahedron (from Ref. [20]).

5.3. Chemical Doping Effect on the Metal-Insulator Transition

5.3.1. $Ca_{2-x}A'_xRuO_4$ (A' = Sr, La, Y)

In many ways, the partial replacement of Ca by Sr or La can be regarded as the application of negative pressure, because either Sr^{2+} (1.32 Å) or

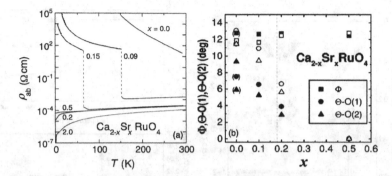

Fig. 5.7. (a) Temperature dependence of the in-plane electrical resistivity ρ_{ab} of $Ca_{2-x}Sr_xRuO_4$ (from Ref. [21]). (b) Sr concentration dependence of the rotation (Φ) and tilt angles (θ-O(1) and (θ-O(2)) of the RuO_6 octahedron (from Ref. [22]).

La^{3+} (1.17 Å) has larger radius than that of Ca^{2+} (1.12 Å). It is thus expected to favor the formation of L-Pbca phase (metallic phase) in either $Ca_{2-x}Sr_xRuO_4$ or $Ca_{2-x}La_xRuO_4$. Indeed, experimental investigation reveals that the substitution of Sr for Ca leads to the drastic suppression of metal-insulation transition temperature and becomes completely metallic with 10% Sr doping (x = 0.2), as demonstrated in Fig. 5.7a.[21] Interestingly, the response of RuO_6 octahedron to Sr doping is very similar to that observed under hydrostatic pressure: the tilt angle decreases with increasing x while the rotation angle remains more or less constant, as shown in Fig. 5.7b.[20] Note that the tilt angle is ~ 6° when $Ca_{2-x}Sr_xRuO_4$ has metallic grand state (x ~ 0.2). At exactly the same tilt angle (see Fig. 5.6b), Ca_2RuO_4 becomes metallic under pressure. This indicates that the structure change is the driving force for the metal-insulator transition in this system.

In principle, an inherent Mott metal-insulator transition should be purely electronic in origin and not assisted by the structure transition.[23,24] This was indeed observed on the surface of $Ca_{1.9}Sr_{0.1}RuO_4$.[25] Combining the several surface-sensitive techniques including scanning tunneling microscopy (STM) and low-energy electron diffraction (LEED), both electronic and structural properties can be measured. As shown in the inset of Fig. 5.8a, the layered nature of $Ca_{1.9}Sr_{0.1}RuO_4$ allows one to create a clean surface by cleaving the single crystal under ultra-high vacuum. For $Ca_{1.9}Sr_{0.1}RuO_4$, the scanning tunneling spectroscopy

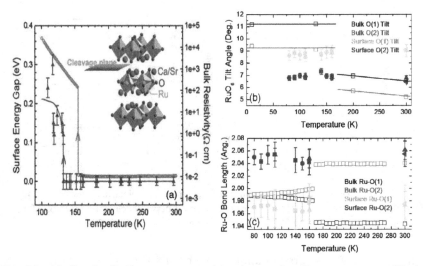

Fig. 5.8. (a) $Ca_{1.9}Sr_{0.1}RuO_4$: temperature dependence of surface energy gap (blue) measured by STM and the in-plane electrical resistivity ρ_{ab} (red). Inset shows the cleavage plane (from Ref. 25). (b) and (c) are temperature dependence of RuO_6 tilt angle and Ru-O bond length of $Ca_{1.9}Sr_{0.1}RuO_4$ on both the surface (from Ref. 26) and bulk (from Ref. 22).

indicates the metal-insulator transition at the surface at T^s_{MIT} ~ 130 K, indicated by the opening of energy gap at the Fermi level (see the blue curve in Fig. 5.8a), which is about 25 K below that of bulk T_{MIT} (see the red curve in Fig. 5.8a).[25] While there is corresponding structure change in bulk as reflected in either RuO_6 tilt angle (see Fig. 5.8b) or Ru-O bond length (see Fig. 5.8c),[22] the surface structure remains intact across the surface MIT.[25-26] It is likely due to the break of the symmetry at the surface. Note that the tilt angles and Ru-O bond lengths are higher at the surface than that in the bulk (Figs. 5.8b and 5.8c). In view of the structural information shown in Figs. 5.6c, 5.7b and 5.8b, we note the common feature is that the Ru-O(2) tilt angle is about or less than 6° in the metallic state.

Regarding the electronic structure of insulating $Ca_{2-x}Sr_xRuO_4$ ($0 \leq x \leq 0.2$), angle-resolved photoemission spectroscopy (ARPES) measurements provide controversial results. Both Figs. 5.9a and 9b are the Fermi surface mapping of $Ca_{1.8}Sr_{0.2}RuO_4$, with one having two bands (α and β only)[27] and one having three (α, β and γ)[28]. One possibility is that the

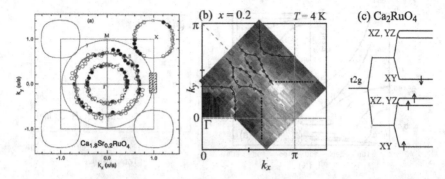

Fig. 5.9. (a) Fermi surface and electron occupancy of different orbitals of $Ca_{1.8}Sr_{0.2}RuO_4$ measured by ARPES at T = 40 K: green contours centered at X are α band, red contours centered at Γ are β band, and blue dashed contours centered at Γ are folded α' band, along with the Fermi crossing points from the fourfold crystal symmetry (red open dots). The black dotted contour centered at Γ is the derived γ Fermi surface according to the Luttinger theorem (from Ref. 27). (b) Mapping of ARPES spectral weight at the Fermi level of $Ca_{1.8}Sr_{0.2}RuO_4$, for which all three bands (α, β, γ) are seen at 4 K (from Ref. 28). (c) Theoretical proposal about the electronic structure of Ca_2RuO_4 and lightly Sr doped compound (from Ref. 10).

sample used in Ref. [27] has actually high Ca concentration than 1.8, thus measuring (T = 40 K) the electronic structure in the insulating region. On the other hand, the work reported in Ref. [28] may be done in the sample with less Ca than 1.8, with the absence of metal-insulator transition as it ends at x = 0.2.[21] Nevertheless, the orbital-selective (γ band) Mott transition is not as predicted by Anisimov et al.[10] As shown in Fig. 5.9c, they proposed Mott transition of $d_{yz/zx}$ ((α and β) bands.[10]

The impact of the substitution of La^{3+} for Ca^{2+} is in two folds: in addition to the structure modification (bandwidth control), it fills t_{2g} band with an electron added from La ion (filling control). As shown in the inset of Fig. 5.10a, the La doping is much more effective than Sr, even though $r_{La3+} < r_{Sr2+}$. With roughly 5% La doping (x = 0.1), $Ca_{2-x}La_xRuO_4$ becomes completely metallic (see Fig. 5.10a), which is also confirmed by Fukazawa and Maeno.[30] Note that, as shown in Fig. 5.10b, Hall coefficient R_H of $Ca_{1.9}La_{0.1}RuO_4$ film changes sign from positive at high temperatures to negative at low temperatures.[31] This reflects increased electron filling by the partial substitution of La for Ca.

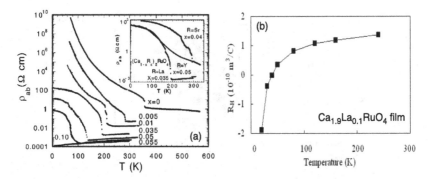

Fig. 5.10. (a) Temperature dependence of the in-plane electrical resistivity ρ_{ab} of $Ca_{2-x}La_xRuO_4$. Inset shows three different dopants (R=La, Y, and Sr) (from Ref. 29). (b) Temperature dependence of Hall coefficient R_H of $Ca_{1.9}La_{0.1}RuO_4$ film (from Ref. 31).

If filling control is the effective way to tune the properties of Ca_2RuO_4, it is expected that Y plays the same role as La. Surprisingly, Y is least effective compared to La and Sr, as demonstrated in the inset of Fig. 5.10a.[29] It is most likely due to the smaller radius of Y^{3+} (1.04 Å) than Ca^{2+}, which is in favor of the insulting phase.

5.3.2. $Sr_2Ru_{1-x}T_xO_4$ (T = Ti, Mn, Fe, Mo, Ir)

The electronic structure of ruthenates can also be altered by partial chemical substitution on Ru site. According to local-density-approximation (LDA) calculations, the band structure of Sr_2RuO_4 is essentially described by three hybridized Ru $4d$ and O $2p$ bands cross the Fermi energy.[32,33] Thus, it is anticipated that any doping to Ru site would change the band structure of Sr_2RuO_4.

Experimentally, it was reported that the partial replacement of Ru by Ti leads to negative resistivity slope ($d\rho_{ab}/dT < 0$),[34] as shown in Fig. 5.11. This is considered as an isovalent doping, i.e., the oxidation state of Ti is 4+. Therefore, there is no change in carrier concentration. On the other hand, the ionic radius of Ti^{4+} (~ 61 Å) is very close to that of Ru^{4+} (~ 62 Å) in the six-fold coordination, expecting little modification of crystal srtucture.[34] This suggests that Ti impurities can be treated as potential scatters without serious effect in inelastic part. Since Ti doping results in small local magnetic moment (~ $0.5\mu_B$/Ti for $0 < x < 0.25$), the

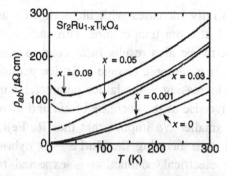

Fig. 5.11. (a) Temperature dependence of the in-plane electrical resistivity ρ_{ab} of $Sr_2Ru_{1-x}Ti_xO_4$ (from Ref. [34]).

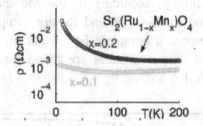

Fig. 5.12. Temperature dependence of the electrical resistivity ρ of $Sr_2Ru_{1-x}Mn_xO_4$ (from Ref. [35]).

metal-nonmetal transition may result from the Kondo effect – itinerant electrons in Ru interact with Ti-induced local moment.

The slope of the electrical resistivity, $d\rho/dT$, also changes sign from positive at high temperatures to negative at low temperatures when introducing Mn into Ru site in $Sr_2Ru_{1-x}Mn_xO_4$.[35] Note that, in Fig. 5.12, such a transition is less dramatic than that occurred in double-layered ruthenate.[36-37] This may be due to the fact that the RuO_6 octahedron in undoped Sr_2RuO_4 bulk is not rotated or tilted. This may not be changed by the partial substitution of Mn. While there is lack of information about Mn dopant in $Sr_2Ru_{1-x}Mn_xO_4$, it is likely the isovalent doping, i.e., Mn^{4+} as that in double-layered ruthenate system.[36,37] With smaller ionic radius (~ 53 Å for Mn^{4+}), the Jahn-Teller distortion of $(Ru/Mn)O_6$ may be reduced, as seen in double-layered ruthenate case.[37]

Back in 1980, Greatrex et al. studied polycrystalline $Sr_2Ru_{1-x}Fe_xO_4$.[38] Although the temperature dependence of the electrical resistivity was not reported, the resistivity increases with increasing Fe and $\rho(x = 0.5)$: $\rho(x = 0) \sim 10^6$: 1 at room temperature. This indicates that the partial substitution of Ru by Fe also results in a metal-to-insulator transition. Structurally, there is monotonic change in lattice parameters. As shown in Fig. 5.13, with increasing x, the lattice parameter a increases while c decrease. These give rise to the decrease of c/a and increase of the unit-cell volume. The smaller c/a implies that the $(Ru/Fe)O_6$ octahedron is more compressed, thus favoring the Ru/Fe – O hybridization. In this circumstance, the electrical conduction is expected to increase with increasing x, in contrast with the experimental observation.[38] One possibility is the Fe doping results in not only the change of bandwidth but also in band filling. According to the Mössbauer measurement results, the oxidation states of Fe and Ru are complicated, both vary with x. In low x region, it is mainly Fe^{3+} while Ru remains in 4+. At x = 0.5, the mixed oxidation states are found for both Fe and Ru: $Sr_2Fe^{3+}_{0.4}Fe^{4+}_{0.1}Ru^{4+}_{0.2}Ru^{5+}_{0.3}O_{3.95}$.[38]

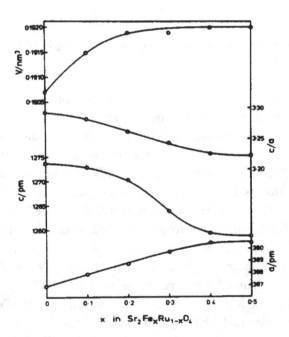

Fig. 5.13. $Sr_2Ru_{1-x}Fe_xO_4$: doping dependence of the lattice parameters a and c, ratio c/a, and the unit-cell volume (from Ref. [38]).

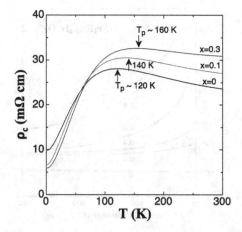

Fig. 5.14. Temperature dependence of the out-of-plane electrical resistivity ρ_c of $Sr_2Ru_{1-x}Mo_xO_4$.

The dopants discussed above are $3d$ metals (Ti, Mn, and Fe). Much less explored is the doping effect on the Ru site by $4d$ metals. We selected Mo to partially replace Ru, as the former is the next nearest neighbor of the latter. It turns out that $Sr_2Ru_{1-x}Mo_xO_4$ exhibits unique features: the in-plane resistivity ρ_{ab} remains metallic for $x \leq 0.4$, even though the magnitude of ρ_{ab} increases with increasing x (not shown). Along the c direction, the coherent $(d\rho_c/dT > 0)$ - incoherent $(d\rho_c/dT < 0)$ crossover temperature T_p increases with increasing x as indicated in Fig. 5.14. This implies that there is enhanced inter-layer coupling upon Mo doping, which leads to better conduction along the c direction. This may be due to larger Mo^{4+} (~ 65 Å) compared to Ru^{4+}. Interestingly, our LEED measurements of $Sr_2Ru_{0.9}Mo_{0.1}O_4$ show that it remains the identical surface structure as that of undoped case,[39] suggesting the bulk has similar structure as well in the doped compounds. However, the STM reveals electronic inhomogeneity in the doped $Sr_2Ru_{0.9}Mo_{0.1}O_4$.[39] Thus, the Mo dopant acts as a scattering center, results in overall increased in-plane resistivity, which increases with increasing x. □

There is also an example of doping in Sr_2RuO_4 using a $5d$ metal. Cava et al. investigated the effect of Ir in Ru site.[40] This results in the itinerant-to-localized electron transition in $Sr_2Ru_{1-x}Ir_xO_4$. Shown in Fig. 5.15 is the electrical resistivity at various doping levels. When $x \geq$

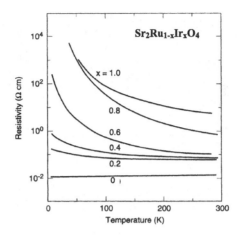

Fig. 5.15. Temperature dependence of the electrical resistivity for polycrystalline $Sr_2Ru_{1-x}Ir_xO_4$ (from Ref. [40]).

0.2, an upturn ($d\rho/dT < 0$) in the resistivity emerges. Since the ionic radius of Ir (~ 77 Å for Ir^{4+} in the six-fold coordination) is considerably larger then that of Ru^{4+}, it is not surprising that Ir doping results in the increase of lattice parameters, as seen experimentally.[40] There is also evidence for the enhanced $(Ru/Ir)O_6$ octahedron distortion such as rotation and/or tilt at high doping levels.[40]

5.4. Summary and Outlook

Ruddlesden-Popper ruthenates series is a unique material system, where charge, spin, lattice and orbital degrees of freedom are strongly coupled. Among single-layered ruthenates, most attention had been paid to Sr_2RuO_4 as there is strong evidence for the p-wave superconductivity.[4] This chapter has focused almost exclusively on the electronic properties of single-layered ruthenates other than Sr_2RuO_4. In particularly, the metal-insulator transitions seen in this class of materials are extremely sensitive to various tuning parameters, such as temperature, pressure, surface symmetry break, and chemical doping. For the first-order metal-insulator transition as seen in $Ca_{2-x}A'_xRuO_4$, it is always accompanied with structural and magnetic property change, reflection of strong coupling between charge, spin, lattice and orbital. Unlike layered cuprate system that has single band, part of complexity of ruthenate physics is due to its multiple band nature. For example, whether or not the metal-insulator transition in $Ca_{2-x}Sr_xRuO_4$ is orbital selective is a remaining question.

Although there are still large discrepancies between existing experimental and theoretical work, it becomes clear that the metal-insulator transitions in single-layered ruthenates are connected to the structure change, such as surface symmetry break, RuO_6 octahedra distortion (rotation and tilt), Ru-O bond length, and O-Ru-O bond angle. In view of existing experimental work, it seems that doping on A site has more dramatic impact than doping on Ru site.

Acknowledgments

This work was partially supported by the U. S. Natural Science fundation with grant No. DMR-1000262. I would like to thank those authors whose figures are used in this review article, and G. Cao and E. W. Plummer for valuable comments.

References

1. S. N. Ruddlesden, P. Popper, *Acta Cryst.* **10**, 538 (1957).
2. S. N. Ruddlesden, P. Popper, *Acta Cryst.* **11**, 54 (1958).
3. M. Imada, A. Fujimori, Y. Tokura, *Rev. Mod. Phys.* **70**, 1039 (1998).
4. For a review, see A. P. Mackenzie, Y. Maeno, *Rev. Mod. Phys.* **75**, 657 (2003).
5. J. A. Kafalas, J. M. Longo, *J. Solid State Chem.* **4**, 55 (1972).
6. Y. Jia, M. A. Zurbuchen, S. Wozniak, A. H. Carim, D. G. Schlom, L.–N. Zou, S. Briczinski, Y. Liu, *Appl. Phys. Lett.* **74**, 3830 (1999).
7. C. S. Alexander, G. Cao, V. Dobrosavljevic, S. McCall, J. E. Crow, E. Lochner, R. P. Guertin, *Phys. Rev.* B 60, R8422 (1999).
8. M. Braden, G. Andre, S. Nakatsuji, Y. Maeno, *Phys. Rev.* B **58**, 847 (1998).
9. A. V. Puchkov, M. C. Schabel, D. N. Basov, T. Startseva, G. Cao, T. Timusk, Z.–X. Shen, *Phys. Rev. Lett.* **81**, 2747 (1998).
10. V. I. Anisimov, I. A. Nekrasov, D. E. Kondakov, T. M. Rice, M. Sigrist, Eur. Phys. J. B **25**, 191 (2002).
11. A. Liebsch, *Phys. Rev. Lett.* **98**, 216403 (2003).
12. A. Liebsch, H. Ishida, *Phys. Rev. Lett.* **91**, 226401 (2007).
13. E. Gorelov, M. Karolak, T. O. Wehling, F. Lechermann, A. I. Lichtenstein, and E. Pavarini, *Phys. Rev. Lett.* **104**, 226401 (2010).
14. Z. Fang, N. Nagaosa, K. Terakura, *Phys. Rev.* B **69**, 045116 (2004).

15. Z. Fang, N. Nagaosa, K. Terakura, *New J. Phys.* **7**, 66 (2005).
16. T. Hotta, E. Dagotto, *Phys. Rev. Lett.* **88**, 017201 (2002).
17. T. Mizokawa, L. H. Tjeng, G. A. Sawatzky, G. Ghiringhelli, O. Tjernberg, N. B. Brooks, H. Fukazawa, S. Nakatsuji and Y. Maeno, *Phys. Rev. Lett.* **87**, 077202 (2001).
18. J. H. Jung, Z. Fang, J. P. He, Y. Kaneko, Y. Okimoto, Y. Tokura, *Phys. Rev. Lett.* **91**, 056403 (2003).
19. F. Nakamura, T. Goko, M. Ito, T. Fujita, S. Nakatsuji, H. Fukazawa, Y. Maeno, *Phys. Rev.* B **65**, 220402R (2002).
20. P. Steffens, O. Friedt, P. Alireza, W. G. Marshall, W. Schmidt, F. Nakamura, S. Nakatsuji, Y. Maeno, *Phys. Rev.* B **72**, 094104 (2005).
21. S. Nakatsuji, Y. Maeno, *Phys. Rev. Lett.* **84**, 2666 (2000).
22. O. Friedt, M. Braden, G. Andre, P. Adelmann, S. Nakatsuji, Y. Maeno, *phys. Rev.* B **63**, 174432 (2001).
23. N. F. Mott, *Proc. Phys. Soc. London Sect.* A **62**, 416 (1949).
24. N. F. Mott, *Metal Insulator Transitions* (Taylor & Francis, London, 1974).
25. R. G. Moore, J. Zhang, V. B. Nascimento, R. Jin, J. Guo, G. T. Wang, Z. Fang, D. Mandrus, E. W. Plummer, *Science* **318**, 615 (2007).
26. R. G. Moore, V. B. Nascimento, J. Zhang, J. Rundgren, R. Jin, D. Mandrus, E. W. Plummer, *Phys. Rev. Lett.* **100**, 066102 (2008).
27. M. Neupane, P. Richard, Z. -H. Pan, Y. -M. Xu, R. Jin, D. Mandrus, X. Dai, Z. Fang, Z. Wang, H. Ding, *Phys. Rev. Lett.* **103**, 097001 (2009).
28. A. Shimoyamada, K. Ishizaka, S. Tsuda, S. Nakatsuji, Y. Maeno, S. Shin, *Phys. Rev. Lett.* **102**, 086401 (2009).
29. G. Cao, S. McCall, V. Dobrosavljevic, C. S. Alexander, J. E. Crow, R. P. Guertin, *Phys. Rev.* B **61**, R5053 (2000).
30. H. Fukazawa, Y. Maeno, *J. Phys. Soc. Jpn.* **70**, 460 (2001).
31. X. Wang, Ph. D. thesis, Pulsed Laser Deposition Growth and Property Studies of $Ca_{2-x}La_xRuO_4$ and RuO_2 Thin Film (Florida State University, 2004).
32. T. Oguchi, *Phys. Rev.* B **51**, 1385 (1995).
33. D. Singh, *Phys. Rev.* B **52**, 1358 (1995).
34. M. Minakata, Y. Maeno, *Phys. Rev.* B **63**, 180504(R) (2001).
35. K. Hatsuda, M. Sc. Thesis, University of Tokyo (2001).
36. R. Mathieu, A. Asamitsu, Y. Kaneko, J. P. He, X. Z. Yu, R. Kumai, Y. Onose, N. Takeshita, T. Arima, H. Takagi, Y. Tokura, *Phys. Rev.* B **72**, 092404 (2005).
37. B. Hu, G. T. McCandless, V. O. Garlea, S. Stadler, Y. Xiong, Julia Y. Chan, E. W. Plummer and R. Jin, *cond-mat/1108.0392* (2011).
38. R. Greatrex, N. N. Greenwood, M. Lal, *Mat. Res. Bull.***15**, 113 (1980).
39. Ismail, L. Petersen, J. Zhang, R. Jin, D. G. Mandrus, E. W. Plummer, *Surface Science* **529**, 151 (2003).
40. R. J. Cava, B. Batlogg, K. Kiyono, H. Takagi, J. J. Krajewski, W. F. Peck, Jr., L. W. Rupp, Jr., C. H. Chen, *Phys. Rev.* B **49**, 11890 (1994).

Chapter 6

THE CONTRADICTORY PHYSICAL PROPERTIES AND EXTREME ANISOTROPY OF $Ca_3Ru_2O_7$

G. Cao and L. E. DeLong

Center for Advanced Materials
Department of Physics and Astronomy,
University of Kentucky,
Lexington, KY 40506, USA
E-mail: cao@uky.edu

P. Schlottmann

Department of Physics, Florida State University,
Tallahassee, FL 32306, USA

The bilayered Ruddlesden-Popper (RP) compound $Ca_3Ru_2O_7$ displays a wide variety of physical properties and exhibits the signatures of almost every ordered state (except for superconductivity) known in condensed matter physics, including exotic phenomena not found in other materials. For example, $Ca_3Ru_2O_7$ exhibits conflicting hallmarks of both insulating and metallic states. We present a brief introduction to the layered ruthenates, and then discuss transport and thermodynamic properties of $Ca_3Ru_2O_7$ with a focus on the following intriguing phenomena: *(1) colossal magnetoresistance (CMR) via suppression of a ferromagnetic (FM) state, (2) strong spin valve effect in bulk single crystals, (3) quantum oscillations in a nonmetallic state, and (4) oscillatory magnetoresistance periodic in B (rather than 1/B).* We argue that these novel properties derive from the orbital degrees of freedom of the Ru-ions, which drive a complex phase diagram via their couplings to spin (spin-orbit interaction) and the lattice (Jahn-Teller effect).

Contents

6.1. Introduction

A primary characteristic of the 4d- and 5d-electron transition elements is that their d-orbitals are more extended compared to those of their 3d-electron counterparts. Strong p-d hybridization and electron-lattice coupling, along with a reduced (with respect to 3d-transition metals) intra-atomic Coulomb interaction U, are thus anticipated in these systems. Consequently, the magnitudes of U and the 4d- and 5d-electron bandwidth, W, are comparable in the layered 4d- and 5d-transition metal oxides, leaving them precariously balanced on the border between metallic and insulating behavior, and/or on the verge of long-range magnetic order. Small perturbations, such as slight changes in lattice parameters, application of a magnetic field, etc., can readily tip the balance, inducing drastic changes in the ground state. It is therefore not surprising that these materials exhibit nearly every cooperative phase known in solids.

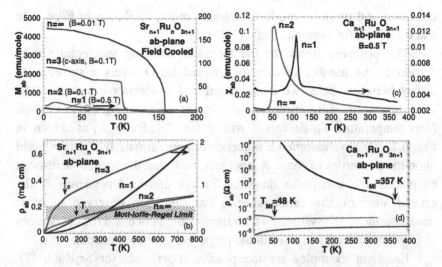

Fig. 6.1. Magnetic susceptibility χ, magnetization M (upper panel) and resistivity ρ (lower panel) as a function of temperature for R-P series $Sr_{n+1}Ru_nO_{3n+1}$ (left column) and $Ca_{n+1}Ru_nO_{3n+1}$ (right column) with n = 1,2,3 and ∞. Note the sharp differences in the ground state between the Ca- and Sr-compounds and n-dependence of M, χ and ρ, and (2) the temperature range for ρ for $Sr_{n+1}Ru_nO_{3n+1}$ is 1.7 K < T < 800 K.

Our early research on ruthenates, iridates and related systems, along with studies by many other authors, has revealed a wide array of intriguing phenomena. One of the most interesting classes of transition metal oxides is the Ruddlesden-Popper (RP) series, $(Sr,Ca)_{n+1}T_nO_{3n+1}$, where T is a 3d, 4d or 5d transition metal element and n is the number of T-O layers per unit cell. We note that the physical properties of the entire RP series, $Ca_{n+1}Ru_nO_{3n+1}$ and $Sr_{n+1}Ru_nO_{3n+1}$, critically depend on the distortions and relative orientations of the corner-shared RuO octahedra, as well as the radius of the alkaline earth cation. This local environment of the Ru-ions determines the crystalline electric field splitting of the 4d-levels and, hence, the band structure of a given compound. Consequently, inter- and intra-layer magnetic couplings and the ground state are critically linked to n and the cation type (Ca or Sr). Specifically, the $Sr_{n+1}Ru_nO_{3n+1}$ compounds are metallic and tend to be ferromagnetic (with Sr_2RuO_4 being an exception) (see **Figs. 6.1a** and **6.1b**), whereas the $Ca_{n+1}Ru_nO_{3n+1}$ compounds are all proximate to a metal-nonmetal transition and prone to antiferromagnetic order, as shown in **Figs. 6.1c** and **6.1d** (see, for example, Refs. [1-20]).

The observed trend for the magnetic ordering temperature with respect to the number of directly coupled Ru-O layers n surprisingly differs between these two isostructural and isoelectronic systems. The Curie temperature T_C increases with n for $Sr_{n+1}Ru_nO_{3n+1}$, whereas the Néel temperature T_N decreases with n for $Ca_{n+1}Ru_nO_{3n+1}$, as shown in **Fig. 6.2**. We have assigned a sequence of approximate *W/U* ratios based upon the properties of these compounds, such that the two RP series can be placed into one phase diagram. Such a drastic dependence of the ground state on the cation species has not been observed in other transition metal RP systems, which implies the lattice and orbital degrees of freedom play critical roles in the properties of these materials.

Excellent examples are the p-wave superconductor Sr_2RuO_4 [7], $Sr_3Ru_2O_7$ [15] and $Ca_3Ru_2O_7$ [1]. The latter compound, the focus of our review, has a bilayer structure (**Fig. 6.3**) and exhibits a large number of intriguing phenomena:

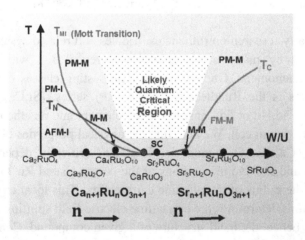

Fig. 6.2. Phase diagram (T vs W/U) qualitatively describing $Ca_{n+1}Ru_nO_{3n+1}$ and $Sr_{n+1}Ru_nO_{3n+1}$. It is generated based on our own studies on single crystals of $(Ca,Sr)_{n+1}Ru_nO_{3n+1}$ except for Sr_2RuO_4 [7] and $Sr_3Ru_2O_7$ [14, 15]. As illustrated, the ground state can be readily changed by changing the cation, and physical properties can be systematically tuned by altering the number of Ru-O layers, n. SC stands for superconductor, FM-M ferromagnetic metal, AFM-I antiferromagnetic insulator, PM-M paramagnetic metal, M-M magnetic metal.

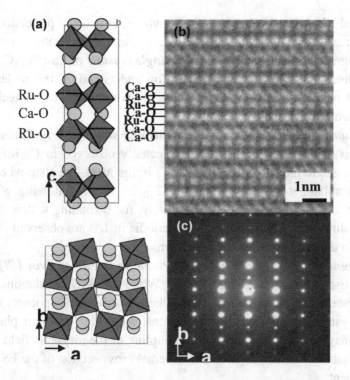

Fig. 6.3. (a) Crystal structure of $Ca_3Ru_2O_7$; (b) TEM image along [010] zone axis; (c) Diffraction pattern of the basal plane. Note that the dark stripes in the TEM image are magnetic Ru-O layers and light gray stripes are insulating Ca-O layers.

(1) Colossal magnetoresistance (CMR) via suppression of a ferromagnetic (FM) state [18-27]: Conventional CMR occurs only within a spin-polarized state with magnetic field applied along the magnetic easy axis. We have discovered that bilayered $Ca_3Ru_2O_7$ (n = 2) exhibits CMR only when the magnetic field is applied *perpendicular* to the easy axis of magnetization (i.e., when the FM state *is suppressed*). This new phenomenon is fundamentally different from those of all other CMR systems that are driven by magnetization saturation.

(2) Bulk spin valve (SV) effect [32-34]: A SV is a device structure consisting of metallic FM layers separated by nonmagnetic insulating layers. The electrical resistance of this device can be manipulated via control of the relative polarization of the FM layers. The SV effect therefore depends upon precision deposition and nanoscale patterning of

artificial **thin-film heterostructures** whose structural perfection and performance characteristics are difficult to control. In contrast, we have discovered a strong SV effect in **bulk single crystals** of $Ca_3(Ru_{1-x}Cr_x)_2O_7$. This discovery opens new avenues for understanding the underlying physics of the SV effect, and enhances the possibilities for its technical exploitation.

(3) Quantum oscillations in the nonmetallic state [22-26]: Shubnikov-de Haas (SdH) oscillations are unexpectedly observed in $Ca_3Ru_2O_7$ at high magnetic induction B (up to 45 T) in the Mott-like state where the metal-insulator transition occurs at T_{MI} = 48 K following a Neel temperature at T_N = 56 K [1]. Particularly, for B rotating within the **bc-**plane, slow, strong SdH oscillations periodic in 1/B are observed for T ≤ 1.5 K in the presence of metamagnetism.

(4) Magnetoresistance oscillations periodic in B (rather than 1/B) [26]: These oscillations persist up to 15 K. While the SdH oscillations are a manifestation of the presence of small Fermi surface (FS) pockets in this Mott system, the B-periodic oscillations, an exotic quantum phenomenon, may be a result of anomalous coupling of the magnetic field to the t_{2g} orbitals that makes the relevant extremal cross-section of the FS field-dependent.

We next review the transport and thermodynamic properties of $Ca_3Ru_2O_7$ with emphasis on the intriguing issues cited above.

6.2. Some Underlying Properties of $Ca_3Ru_2O_7$

In the case of Ru^{4+} ($4d^4$) ions in the layered ruthenates, the Hund's rule energy at each Ru site is not large enough to overcome the e_g-t_{2g} crystalline electric field splitting, so that the e_g levels are not populated. Hence, one t_{2g} orbital is doubly occupied, while the other two host a single electron each. The three t_{2g} levels (d_{xy}, d_{zx} and d_{yz}) are expected to have different energies because the RuO_6 octahedra are deformed. These splittings are generally larger than k_BT and the Zeeman splitting. The RuO_6 octahedra are corner-shared and often tilted. The tilting plays an important role in determining the overlap between the orbitals of neighboring octahedra; therefore, small changes in tilt can result in qualitative changes in physical properties. Band structure calculations for

$Ca_3Ru_2O_7$ demonstrate that the Fermi surface is very sensitive to small structural changes that readily shift the Fermi energy [34, 35], and may even change the topology of the Fermi surface. In view of the strong sensitivity of the electronic structure to crystal field and tilting angle asymmetries, the magnetic properties of $Ca_3Ru_2O_7$ are expected to strongly depend on the orientation of an applied magnetic field.

The bilayered $Ca_3Ru_2O_7$ has room temperature lattice parameters $a = 5.3720(6)$ Å, $b = 5.5305(6)$ Å, and $c = 19.572(2)$ Å [21]. The crystal structure is severely distorted by the tilting of the RuO_6 octahedra (see **Fig. 6.3**). The tilt projects primarily onto the **ac**-plane ($153.22°$), while it only slightly affects the **bc**-plane ($172.0°$) [21]. These bond angles are crucial in determining the crystal field splitting of the 4d-orbitals and the overlap matrix elements between orbitals within the basal **ab**-plane. They directly impact the band structure and are the origin of the anisotropic properties of the compound. In zero magnetic field, $Ca_3Ru_2O_7$ undergoes an antiferromagnetic (AFM) transition at $T_N = 56$ K while remaining metallic, and then a Mott-like transition at $T_{MI} = 48$ K with a dramatic reduction (up to a factor of 20 along the **c**-axis) in the conductivity as temperature decreases below T_{MI} [1, 21-26]. This transition is accompanied by an abrupt shortening of the **c**-axis lattice parameter below T_{MI} [22, 23] (right scale in **Fig. 6.4b**). The resultant, strong magnetoelastic coupling induces Jahn-Teller distortions of the RuO_6 octahedra [9, 10], thus lowering the energy of the d_{xy} orbitals relative to d_{zx} and d_{yx} orbitals, which is a state of orbital order (OO). This yields an orbital distribution with two electrons occupying the lower d_{xy} level, and one electron occupying each of the nearly degenerate, higher-energy d_{zx} and d_{yz} levels [29]. Consequently, AFM and OO coexist below $T_{MI} = 48$ K, which explains the poor metallic behavior for $T < T_{MI}$ and $B < B_c$ (critical field for the metamagnetic transition). The above picture is consistent with Raman-scattering studies of $Ca_3Ru_2O_7$, which reveal that the transition at T_{MI} is associated with the opening of a charge gap, $\Delta_c \sim$ 0.1 eV, and the concomitant softening and broadening of an out-of-phase O phonon mode [27-29].

Intermediate between T_{MI} and T_N there exists an ***antiferromagnetic metallic (AFM-M) state*** (see **Fig. 6.4c**) that is rarely seen in oxides [1]. Secondly, the system is highly anisotropic, as shown in **Fig. 6.4a**. The low field M(T) for the **b**-axis (the magnetic easy axis) exhibits two phase transitions at $T_N = 56$ K and $T_{MI} = 48$ K. In contrast, M(T) for the **a**-axis exhibits no discernable anomaly at T_{MI}, but a sharp peak at T_N, as seen in **Fig. 6.4a** [33, 36]. For the field parallel to the **c**-axis, the two transitions are observed but they are considerably weakened. As **B** is increased parallel to the **b**-axis, T_{MI} shifts slightly downward, whereas T_N

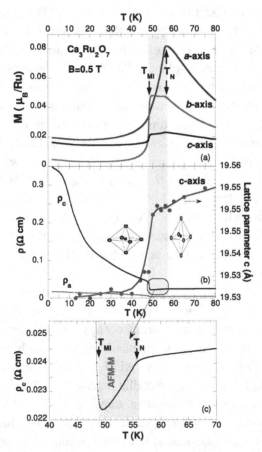

Fig. 6.4. (a) Temperature dependence of magnetization M at B = 0.5 T for B applied along the a-, b- and c-axis, respectively; (b) Temperature dependence of the c-axis resistivity ρ_c (left-hand scale) and the lattice parameter of the c-axis (right-hand scale) and (c) enlarged ρ_c in the shaded area which defines the antiferromagnetic metallic (AFM-M) region between T_{MI} and T_N. Note that the abrupt shorting in the c-axis near T_{MI}, which flattens RuO octahedra.

remains essentially unchanged, but eventually becomes rounded off at higher fields. For B ≥ 6 T, the magnetic state is driven into a ***spin-polarized or field-induced ferromagnetic (FM) state***. In contrast, when magnetic field is applied along the **a**-axis, T_N decreases with increasing B at an approximate rate of 2 K/T. Remarkably, the ground state for **B ∥ a** remains AFM, entirely different from that for **B ∥ b** [21-25]. Unlike the **a**- and **b**-axis magnetizations, the **c**-axis magnetization remains essentially unchanged with applied field, which suggests that intra-**ab**-plane spin couplings are particularly strong.

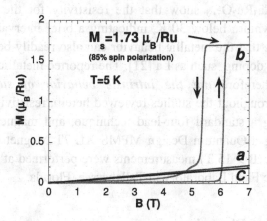

Fig. 6.5. Isothermal magnetization M for B ∥ a-,b-,c-axis at T = 5 K. Note that the first-order metamagnetic transition occurs along the magnetic easy-axis, the b-axis, polarizing the spins by more than 85%.

The anisotropy of the magnetic state is further illustrated by the isothermal magnetization at T = 5 K (**Fig. 6.5**), which displays a first-order metamagnetic transition at B = 6 T applied along the easy **b**-axis, leading to a spin-polarized state with a saturation moment M_s = 1.73 μ_B/Ru. The latter value corresponds to more than 85% of the hypothetical saturation magnetization (2 μ_B/Ru) expected for an S = 1 system. The behavior is completely different if the field is applied along the **a**- or **c**-axis, in part due to a strong anisotropy field of 22.4 T [30].

We next describe the growth and structural characterization of single crystals studied, before we turn to a discussion of the experimental results for their electric, magnetic and thermal properties.

6.3. Single Crystal Growth and Characterization

Single crystals were grown using both flux and floating zone techniques [25]. The high quality of these crystals was confirmed by single-crystal x-ray diffraction, SEM and TEM (see **Figs. 6.3b** and **6.3c**). The highly anisotropic magnetic properties of $Ca_3Ru_2O_7$ were used to orient the magnetic easy **b**-axis and to identify twinned crystals that often show a small kink at 48 K in the **a**-axis susceptibility. There are *no differences* in the magnetic and transport properties and Raman spectra of crystals grown using either the flux or floating zone methods. Our studies on oxygen-rich $Ca_3Ru_2O_{7+\delta}$ show that the resistivity for the basal plane displays a downturn below 30 K, indicating a brief interval of metallic behavior. Note that the metallic behavior can also readily be induced by other impurity doping, such as La [21]. The reported metallic behavior in Ref. 33 is therefore *not the intrinsic behavior of stoichiometric* $Ca_3Ru_2O_7$. Throughout the studies reviewed herein, resistivity data were obtained using a standard four-lead technique, and magnetization was obtained using a Quantum Design MPMS XL 7T Magnetometer. High magnetic field (B > 15 T) measurements were performed at the National High Magnetic Field Laboratory in Tallahassee, Florida.

6.4. Results and Discussion

6.4.1. *Colossal Magnetoresistance via Suppression of a Ferromagnetic State*

The unusual CMR of $Ca_3Ru_2O_7$ is clearly illustrated in **Fig. 6.6a**, which shows the field dependence of the interplane resistivity ρ_c (measuring electric field along **c**-axis) with applied magnetic field along the **c**-axis at T = 0.4 K and $0 \leq B \leq 45$ T with **B** ǁ **a**-, **b**- and **c**-axes. ρ_c is extraordinarily sensitive to the orientation of **B**. For **B** ǁ **b** (magnetic easy-axis) ρ_c shows an abrupt drop by an order of magnitude at 6 T; this is identified with a first-order metamagnetic transition leading to a spin-polarized state with a saturated moment, M_s, of 1.73 μ_B/Ru (more than 85% saturation; see **Fig. 6.6a**) [1, 21-31]. The reduction of ρ_c is attributed to an increase of coherent motion of the electrons between Ru-O

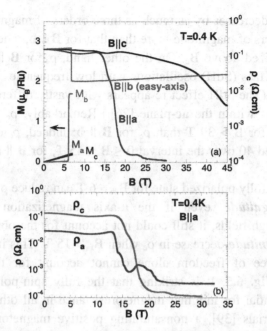

Fig. 6.6. (a) Magnetic field dependence of the c-axis resistivity ρ_c for B ‖ a-, b- and c-axis at T = 0.4 K (right scale) up to 45 T. Isothermal magnetization M for B ‖ a-, b- and c-axis at T = 2 K (left scale) is plotted for comparison. Note that ρ_c for B ‖ b is greatest when B > 39 T. (b) Field dependence of both the a-axis resistivity ρ_a and c-axis resistivity ρ_c for B ‖ a-axis at T = 0.4 K. Note that the inter-plane ρ_c is conspicuously smaller than intra-plane ρ_a when B > B_c [Ref. 31].

planes separated by insulating Ca-O planes; this could reflect a situation similar to "spin-filters", where the probability of tunneling depends on the angle between the spin magnetization of adjacent FM layers. Existing data (**Fig. 6.6a**) indicate that the fully spin polarized state can lower the resistivity by at most a factor of 10. As B is increased further from 6 to 45 T, ρ_c increases linearly with B by more than 30%, which is interesting in its own right, since a quadratic dependence is expected for simple metals [38]. Because spin scattering is apparently reduced to its minimum at B = 6 T (**Fig. 6.6a**), the linear increase of resistivity above 6 T can only arise from an interaction between OO and applied magnetic field.

For **B ‖ a** (magnetic hard-axis), there is no metamagnetic transition and the system remains AFM. In sharp contrast to ρ_c for **B ‖ b**, ρ_c for

B ∥ **a** rapidly decreases by as much as three orders of magnitude at $B_c =$ 15 T, two orders of magnitude more than that for **B** ∥ **b**, where spins can be fully polarized above B_c. On the other hand, ρ_c for **B** ∥ **c**, displays Shubnikov-deHaas (SdH) oscillations with low frequencies of 28 T and 10 T [22-26]. (The SdH effect re-appears with vastly different behavior when **B** rotates within the **ac**-plane [26].) Remarkably, ρ_c for **B** ∥ **c** is much smaller for B > 39 T than ρ_c for **B** ∥ **b**; indeed, ρ_c decreases by factors of 7 and 40 over the interval $0 \leq B \leq 45$ T, for **B** ∥ **a** and **B** ∥ **c**, respectively.

Since the fully polarized state for $B_{\|b} > 6$ T can reduce ρ by *only one order of magnitude*, even if the **a**-axis magnetization were fully polarized at high fields, it still could not account for the observed *three orders of magnitude* decrease in ρ_c when $B_{\|a} > 15$ T. This indicates that the spin degree of freedom alone cannot account for the behavior observed in **Fig. 6.5**. It is striking that the fully spin-polarized state, which is essential for minimal magnetoresistance in all other magneto-resistive materials [39], a nonsaturating positive magnetoresistance is observed above B_c in $Ca_3Ru_2O_7$ (**Fig. 6.6a**)! The **a**-axis resistivity ρ_a behaves very similarly to ρ_c (**Fig. 6.6b**); in particular, for **B** ∥ **b**, the decrease in ρ_a is also one order of magnitude, the same as that of ρ_c, suggesting that the reduction in both ρ_a and ρ_c is driven by the same in-plane spin polarization. For **B** ∥ **a**, ρ_a decreases by two orders of magnitude when B > B_c. This comparison confirms that the spin-polarized state along the **b**-axis is indeed not favorable for electron hopping.

It is striking that ρ_a (~10^{-3} Ω cm) is larger than ρ_c (~10^{-4} Ω cm) when B > B_c. There may be a change in effective dimensionality driven by B that results in a smaller inter-plane ρ_c when B > B_c. This "crossover" presumably has a lesser impact on the intra-plane ρ_a, and could be analogous to that driven by temperature, as discussed in ref. 36.

Shown in **Fig. 6.7** are the magnetization M and ρ_c as a function of B for **B** ∥ **a** and **B** ∥ **b** over the temperature range, $40 \leq T < 56$ K, where B_c along both the **a**- and **b**-axes, falls into the range below 7 T, so that M can be fully characterized using a SQUID magnetometer. Moreover, direct comparisons of ρ, B and M(T,B) permits wider investigations of their correlations and the role of OO in the anomalous physical properties of $Ca_3Ru_2O_7$.

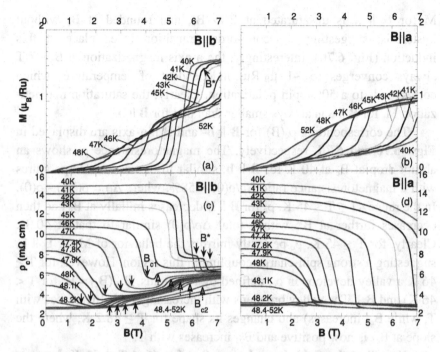

Fig. 6.7. M and ρ_c as a function of B for B ∥ b (panels (a) and (c)) and for B ∥ a (panels (b) and (d)) for $40 \leq T < 56$ K. The solid arrows (empty arrows) indicate B* and B_{c1} (B_{c2}). Note that M_a tends to converge to 1 μ_B/Ru and is always smaller than M_b, but the reduction of ρ_c is always greater for B ∥ a than for B ∥ b [Ref. 31].

Figure 6.7a displays M as a function of B for **B ∥ b**, the magnetic easy axis. At 40 K, M(B) resembles its low-temperature behavior (see **Fig. 6.6**) but with slightly lower $M_s = 1.6$ μ_B/Ru) and $B_c = 5.8$ T. For $41 \leq T \leq 45$ K, a second transition develops at $B^* > B_c$, suggesting an intermediate FM state that is not fully polarized along the **b**-axis for $B_c < B < B^*$. A possible interpretation of this effect is that the spins are rotating away from the **b**-axis due to a shortening of the **c**-axis near T_{MI}, and hence a stronger field (B*) is required to re-align these spins along the **b**-axis. Since the spin rotation tends to become stronger as T approaches T_{MI}, B* increases with T. The easy-axis M ≈ 1 μ_B/Ru at B_c and increases by 0.6 μ_B/Ru at B*. Only half of the ordered spins are thus aligned with the **b**-axis in the spin reorientation (SR) region for $B_c < B < B^*$. B_c decreases with T and vanishes near $T_N = 56$ K. Unlike M for **B ∥ b**,

M for **B ∥ a** is unsaturated at B > B_c, and rounded at B_c without hysteresis, suggesting a second-order transition takes place at this induction (**Fig. 6.7b**). Interestingly, the a-axis magnetization at B = 7 T always converges to ~1 μ_B/Ru, independent of temperature, which corresponds to a 50% spin polarization. Clearly, the saturation magnetization M_s for **B ∥ a** is always smaller than that for **B ∥ b.**

The corresponding $\rho_c(B)$ for **B ∥ b**- and **B ∥ a**-axis are displayed in **Figs. 6.7c** and **6.7d**, respectively. The magnetoresistance ρ_c shows an abrupt drop at B_c at 40 K for **B ∥ b**, similar to that at low temperatures with a magnetoresistance ratio $\Delta\rho/\rho(0)$ = 58%, where $\Delta\rho = \rho(7T) - \rho(0)$. In the range $41 \leq T \leq 45$ K, ρ_c for **B ∥ b** decreases initially at B_c and then decreases further at B* with a total $\Delta\rho/\rho(0)$ similar to that at 40 K. Clearly, for $T \leq 45$ K, ρ_c perfectly mirrors the behavior of M for **B ∥ b**, suggesting a strong spin-charge coupling in this region. However, for $T \geq$ 46 K a valley develops in ρ_c, defined by two fields, B_{c1} ($B_{c1} = B_c$ for T < 46 K) and B_{c2}. The valley broadens with increasing T (B_{c1} decreases with T, while B_{c2} increases) and changes its shape for $T \geq 48.2$ K, where the slope at B_{c1} is now positive and B_{c1} increases with T.

Note that the field dependence of ρ_c for $46 \leq T \leq 52$ K does not track the field dependence of M (compare **Figs. 6.7a** and **6.7c**). We interpret the lack of parallel behavior between M and ρ_c as exemplary of the crucial role that orbital degrees of freedom play in electron hopping for **B ∥ b**. Moreover, *the reduction in ρ_c for B ∥ a (Fig. 6.7d) is always much larger than that for B ∥ b (Fig. 6.7c); and yet M_s for B ∥ a is always smaller than M_s for B ∥ b.* The temperature dependence of M(B = 7T) (left scale) and $\Delta\rho/\rho(0)$ at 7 T (right scale) for **B ∥ a**- and **B ∥ b**-axis is summarized in **Fig. 6.8a**. *Such an inverse correlation between M and $\Delta\rho/\rho(0)$ suggests that the spin-polarized state is indeed detrimental to the CMR.* For T > T_{MI}, the metallic state is recovered for B < B_{c1}. But applying **B ∥ b**-axis leads to a rapid increase in ρ_c (i.e., a positive $\Delta\rho/\rho(0)$ reaching as high as 112% for B > B_{c2}), whereas applying **B ∥ a**-axis results in essentially no changes in ρ_c.

As discussed above, for B < 6 T, a Mott-like state is induced by the onset of OO facilitated by the c-axis shortening at T_{MI} [22, 23]. When $B_{∥b}$ > 6 T the onset of FM order stabilizes the OO. The OO could be either a ferro-orbital (FO) or an antiferro-orbital (AFO) configuration; hence the

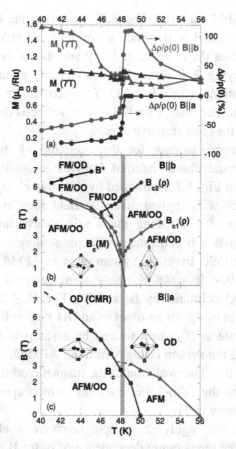

Fig. 6.8. (a) Temperature dependence of M (triangles) and $\Delta\rho/\rho(0)$ (solid circles, right scale) at 7 T for B ∥ a- and b-axis. (b) and (c) phase diagrams plotted as B vs T summarizing various phases for B ∥ a- and b-axis, respectively. Note that in (b) $B_{c1}(\rho)$ and $B_{c2}(\rho)$ indicate the curves are generated based on ρ, and $B_c(M)$ on M [Ref. 31].

system is in either a FM/FO or a FM/AFO state. The former inhibits the hopping of the 4d electrons because of the Pauli exclusion principle, while the latter permits intersite transitions, but at the expense of a Coulomb energy. Therefore, despite of an order-of-magnitude drop in ρ_c due to increased spin-polarization when B > 6 T, a fully metallic state can never be reached for **B ∥ b**. In fact, the linear increase in ρ_c with increasing B for $B_{\parallel b}$ > 6 T shown in **Fig. 6.6** may signal strengthened OO via an enhanced FM state. Conversely, applying **B ∥ a**-axis steadily suppresses the AFM state (see **Fig. 6.9**), removing the OO through the SO interaction when $B > B_c$. Such an orbitally disordered (OD) state drastically increases the electron mobility and drives the CMR. On the other hand, applying **B** ∥ **c**-axis has a noticeable impact on spin and orbital configurations when $B > 35$ T, where ρ_c drops rapidly and becomes much smaller than ρ_c for **B** ∥ **b**. This suggests that the electronic state for **B** ∥ **b** is the most resistive one.

Complementary evidence for the evolution of the field-induced magnetic and orbital phases inferred from the magnetic and transport behavior shown in **Fig. 6.7** is provided by the rapid changes of the Ru-O phonon frequency with applied magnetic field seen in Raman studies [29]. Application of field along the **a**-axis clearly favors the CMR, whereas applying **B** ∥ **b** generates a rich phase diagram. Below 40 K, applied magnetic field drives the system from an AFM/OO to a FM/OO state, and for $40 < T < 48$ K, the system enters a region of spin reorientation (SR) delimited by B_c and B^*. For $46 \leq T < 48.2$ K, the valley seen only in ρ_c signals an onset of an OD state at B_{c1} and then a re-occurring OO state at B_{c2} characterized by a sharp increase in ρ_c. For $48.2 < T < 56$ K, the system changes from an AFM/OD to an AFM/OO phase when $B > B_{c1}$. The evolution of the magnetic/orbital configuration is associated with the Jahn-Teller coupling, which appears to be most drastic in the vicinity of T_{MI}.

In addition, the highly anisotropic behavior is also reflected in **Fig. 6.9**, where the temperature dependence of ρ_c for **B** ∥ **a** and **B** ∥ **b** is displayed. For **B** ∥ **b**, at low temperatures ρ_c increases slightly with increasing B when $B < 6$ T, and decreases abruptly by about an order of magnitude when $B \geq 6$ T, at which the first-order metamagnetic transition leads to the spin-polarized state, as seen in **Figs. 6.5** and **6.6**. A further increase in B only results in slightly higher resistivity at low temperatures. In fact, ρ_c at $B = 28$ T still shows nonmetallic behavior. In sharp contrast, when **B** ∥ **a**, T_{MI} decreases systematically at a rate of 2K/T, and disappears at $B > 24$ T, as seen in **Fig. 6.9b**. These results once again suggest a strong magneto-orbital coupling. It is also striking that at $B = 30$ T, where the metallic state is recovered and the AFM is completely suppressed; as shown in **Fig. 6.9c**, $\rho_c \sim T^{1.2}$ for $T < 56$ K. A near-linear power law possibly signals the proximity to a quantum critical point.

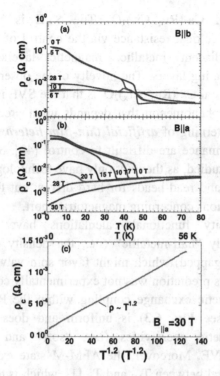

Fig. 6.9. Temperature dependence of ρ_c at a few representative B up to 30 T applied along (a) the b-axis and (b) a-axis for 1.2 < T < 80 K. Note that T_{MI} is effectively suppressed by B and eventually vanishes for B > 24 T when B ∥ a. (c) ρ_c at $B_{\parallel a} = 30$ T. Note that $T^{1.2}$-dependence implies the proximity to a quantum critical point.

It is clear that the orbital degrees of freedom, and their coupling to spin and lattice degrees of freedom, play a critical role in the remarkable physical properties of Ca₃Ru₂O₇. As a consequence, the anisotropies observed by applying **B** along the **a**-, **b**-, or **c**-axis define novel and contrasting properties. Most notably, the CMR achieved by avoiding the spin-polarized state is fundamentally different from that in all other magnetoresistive materials.

6.4.2. *Bulk Spin Valve Effect*

The exotic CMR discussed above is not the only peculiar property of Ca₃Ru₂O₇. A strong *spin valve effect* (SVE) has been identified in *bulk* single crystals of $Ca_3(Ru_{1-x}Cr_x)_2O_7$. The SVE is based upon the manipulation of electrical resistance via the control of the relative spin alignment of adjacent metallic, magnetic layers separated by nonmagnetic insulating layers. The novelty of the observation of a SVE in bulk, single-crystal $Ca_3(Ru_{1-x}Cr_x)_2O_7$ is that the SVE is thought to be a delicate quantum phenomenon that depends upon precision deposition and nanoscale patterning of *artificial thin-film heterostructures* whose quality and performance are difficult to control [40]. Spin valves have been intensively studied, as they not only have technological potential as magnetic sensors and read-heads for hard drives, but they also present fundamental questions concerning magnetotransport.

Recent density functional calculations have suggested that $Ca_3Ru_2O_7$ is nearly "half-metallic" (i.e., the density of states of the minority spins is gapped) which might favor spin valve (SV) behavior [34]. However, this prediction was not experimentally confirmed, simply because the magnetic exchange coupling within the Ru-O bilayers of pure $Ca_3Ru_2O_7$ (see **Fig. 6.3**) is uniform and does not render the contrasting magnetic coercivities (i.e., "soft" and "hard" layers) necessary for a SVE. Moreover, the AFM-M state exists only over a narrow, 8 K-interval between T_{MI} and T_N [1], which is too restrictive for the underlying physics of a SV phenomenon to be fully investigated.

We have pursued transport and thermodynamic studies of single crystals of $Ca_3(Ru_{1-x}Cr_x)_2O_7$ with $0 \leq x \leq 0.20$, since Cr substitution is uniquely effective for widening the stability range of the AFM-M state while preserving the most intriguing properties of pure $Ca_3Ru_2O_7$. This is particularly evident in the magnetization data for $x = 0.20$ and **B ∥ b** (M_b), which clearly show that Cr substitution lowers T_{MI} and rapidly raises T_N, greatly extending the AFM-M state from an 8 K interval to one over 70 K (**Fig. 6.10a**). The magnetization for **B ∥ a** (M_a) reflects the same increase in T_N as M_b, but with an upturn below T_{MI} for $x > 0$, suggesting that spin canting is present in the AFM-M state for **B ∥ a** (**Fig. 6.10b**). The decrease of T_{MI} implies delocalization of d-electrons and the increase in T_N signals an enhanced exchange coupling between nearest-neighbor spins. This is similar to $SrRu_{1-x}Cr_xO_3$ and $CaRu_{1-x}Cr_xO_3$, where the FM state is either strongly enhanced [42-45] or induced [46] by Cr doping, respectively.

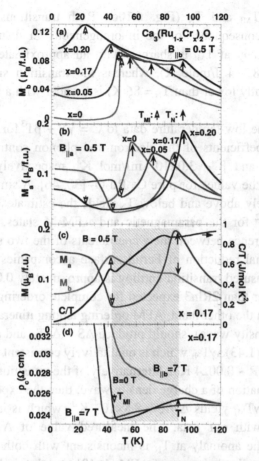

Fig. 6.10. The magnetization M as a function of temperature T for (a) the b-axis M_b, (b) a-axis M_a at B = 0.5 T for various Cr content x, (c) M_a and M_b at B = 0.5 T for x = 0.17; right scale: C/T vs. T for x = 0.17. (d) Temperature dependence of the c-axis resistivity ρ_c for x = 0.17 for B ‖ a and B ‖ b at B = 0 and 7 T. Note that the shaded area is the range of the AFM-M state [Ref. 32].

6.4.2.1. *High-Field Stability of the Antiferromagnetic Metallic State*

We now focus on probing the AFM-M state of a representative composition, x = 0.17, as a function of temperature, magnetic field strength and orientation. The magnetizations M_a and M_b of $Ca_3(Ru_{0.83}Cr_{0.17})_2O_7$ indicate T_{MI} = 42 K and T_N = 86 K at B = 0.5 T, as shown in **Fig. 6.10c**. The specific heat C(T) exhibits two mean-field-like, 2nd-order phase transitions at T_{MI} and T_N (**Fig. 6.10c**). Both transitions are slightly smeared as a consequence of the inhomogeneous Cr distribution. The transition anomaly at T_{MI} is sharper, with an approximate jump $\Delta C \sim$ 0.31 R (R = 8.314 J/mole K), whereas the transition signature near T = 83 K (slightly lower than T_N = 86 K) is broader with a smaller $\Delta C \sim$ 0.28 R.

A fit of the low-temperature data to $C = \gamma T + \beta T^3$ for 1.7 K < T < 30 K yields coefficients of the electron and phonon contributions, $\gamma \sim$ 31 mJ/mol K and $\beta \sim 3.0 \times 10^{-4}$ mJ/mol K^3, respectively, which are comparable to the values for pure $Ca_3Ru_2O_7$ [30, 46]. A similar fit to the data immediately above and below T_N yields the estimates γ = 680 and 79 mJ/mole K^2 for the paramagnetic and AFM-M states, respectively. The large difference between the γ coefficients of the two states implies that a substantial reduction of Fermi surface accompanies the onset of AFM. The measured transition entropy is approximately 0.037 R, which is much smaller than 2Rln3 expected for complete ordering of localized S = 1 spins. On the other hand, AFM ordering among itinerant spins (e.g., as in a spin-density wave), should produce $\Delta S \sim \Delta \gamma T_N$ and a mean-field-like step $\Delta C \sim (1.43) \Delta \gamma T_N$, which is qualitatively consistent with our data, but implies $\Delta \gamma / R \sim 0.0024$ K^{-1}. Alternatively, if the transition at T_{MI} were due to the formation of a charge density wave, then the expected entropy change $\Delta S \sim \Delta \gamma T_{MI}$ yields $\Delta \gamma / R \sim 8.8 \times 10^{-4}$ K^{-1}, which is in still greater disagreement with our data. The measured value of $\Delta \gamma / R$ therefore suggests that the anomaly at T_N is inconsistent with both conventional itinerant and localized pictures of AFM [1, 48], and thus indicates a more complex spin ordering within the AFM-M state, which should be reflected in anomalous high-field behavior.

6.4.2.2. *Anomalous Anisotropy of Magnetic and Magnetotransport Properties*

Indeed, $Ca_3(Ru_{0.83}Cr_{0.17})_2O_7$ is driven toward a spin-polarized state at B > 5 T for **B** ‖ **b**, as T_N decreases at a rate of $\Delta T_N / \Delta B$ = -1.4 K/T. However, T_N is readily suppressed at an astonishing rate $\Delta T_N / \Delta B$ = -7.5 K/T for **B** ‖ **a**-axis, suggesting that the magnetic lattice softens. The conductivity in the AFM-M state is strongly spin-dependent, as reflected in the **c**-axis resistivity ρ_c at B = 0, which sharply drops for T < T_N (**Fig. 6.10d**), despite the reduced Fermi surface that must accompany the reduction of γ at T_N. These considerations point to a reduction of spin-scattering with increasing magnetic order. This behavior differs from that of other AFM-M systems such as Cr and Mn [44] where the initial onset of AFM order at T_N is accompanied by a rise in resistivity. Moreover, the conductivity anisotropy of the AFM-M state is strongly anisotropic, as reflected in ρ_c for **B ∥ b**, where the AFM state becomes semiconducting, despite the emerging field-induced FM state for B > 5 T. Conversely, for **B ∥ a**, ρ_c *decreases* with applied field, as shown in **Fig. 6.2d**. All in all, in the range T_{MI} < T < T_N the system remains an AFM-M for **B ∥ a**, but becomes FM and semiconducting for **B ∥ b**. This behavior indeed bears a resemblance to that predicated for half-metallic systems [49].

The field dependences of the magnetoresistivity ratio $[\rho_c(B)-\rho_c(0)]/\rho_c(0)$ and magnetization M(B,T) exhibit two important trends:

(1) **Figs. 6.11a** and **6.11b** reveal a metamagnetic transition that occurs at T < 30 K for M_b but at T > 30 K for M_a, respectively; the reversed anisotropy implies that the magnetic easy axis rotates from the **b**-axis to the **a**-axis for T > 30 K; (2) Although M_a is notably *smaller* than M_b at T ≤ 30 K, $[\rho_c(7T)-\rho_c(0)]/\rho_c(0)$ for **B ∥ a** is greater in magnitude than that for **B ∥ b** (**Figs. 6.11c** and **6.11d**).

As we observed for pure Ca₃Ru₂O₇, an *inverse relation* exists between M(B) and $[\rho_c(B)-\rho_c(0)]/\rho_c(0)$ in Cr-doped Ca₃Ru₂O₇, and supports the existence of a novel magnetotransport mechanism based on orbital order rather than a spin-polarized state, as discussed in Section A (also see **Fig. 6.7**).

The central finding in Cr doped Ca₃Ru₂O₇ is that $\rho_c(B)$ for **B ∥ a** *strongly peaks* at a critical field B_{C2} that marks a sharp change in slope of $M_a(B)$. M_a increases in the range 35 K < T < 45 K via two distinct transitions at B_{C1} and B_{C2} < B_{C1}, culminating at a saturation magnetization M_s ~3 μ_B/f.u. at 7 T, as shown in **Figs. 6.11b** and **6.12b**. A rapid increase in magnetization normally implies a monotonic reduction of spin scattering with increasing field, as observed for x = 0 for **B ∥ a** (see **Fig. 6.12a**), x = 0.17 for **B ∥ b** (**Figs. 6.11a** and **6.11c**), and most other

materials. This expectation clearly disagrees with the peak in $\rho_c(B)$
(**Figs. 6.11d, 6.12b** and **6.12c**) observed over an extended range 35 K <
T < 65 K for **B ∥ a**.

6.4.2.3. *The Bulk Spin Valve Scenario*

These singular physical properties of $CaRu_{1-x}Cr_xO_3$ can be explained
within a SV scenario, as sketched in **Fig. 6.12b**. The AFM coupling
between bilayers is far weaker than the FM interaction within a bilayer
for x = 0. The existence of a robust SVE demands that Cr substitution is
not uniform, but results in *some* Ru-O layers being replaced by a Cr-rich
layer; for x = 0.17, the replacement of a Ru-O layer by a Cr-O-rich layer
is likely to occur every 2 or 3 bilayers. Extremely weak superlattice
peaks expected in single-crystal x-ray diffraction data for a highly
ordered stacking of Cr-O layers have not been observed, but an ordered
stacking of the Cr-O layers is not necessary for our discussion. The
strong anomalies seen in C(T), along with the drastically enhanced T_N,
are consistent with an ***inhomogeneous, but non-random, distribution*** of
Cr ions.

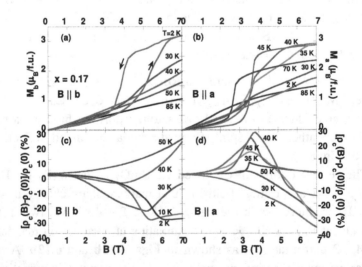

Fig. 6.11. Field dependence of M for (a) M_b and (b) M_a, and field dependence of
(c) $[\rho_c(B)-\rho_c(0)]/\rho_c(0)$ for B ∥ b and (d) B ∥ a for $2 \leq T \leq 85$ K for the compound
$Ca_3(Ru_{0.83}Cr_{0.17})_2O_7$. Note the peaks in (d) [Ref. 32].

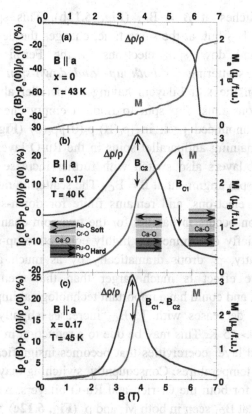

Fig. 6.12. Field dependence of $[\rho_c(B)-\rho_c(0)]/\rho_c(0)$ and M (right scale) for B ∥ a at (a) T = 43 K for x = 0, (b) 40 K for x = 0.17, and (c) 45 K for x = 0.17. The thin (thick)-line arrow indicates the spin direction in the Cr-O rich (RuO) layer. Note that the antiparallel alignment in layers having both Cr and Ru spins corresponds to the peak in ρ_c; this peak is absent in x = 0 [Ref. 32].

The presence of the Cr-O layer causes a spin canting at low fields and temperatures, which gives rise to the upturn in M_a at T_{MI} (**Figs. 6.10b** and **6.10c**). The magnetization of each un-substituted Ru-O bilayer, or **hard magnetic bilayer**, is pinned due to the strong exchange coupling within the bilayer. In contrast, the magnetization of a Cr-rich layer, which constitutes a **soft magnetic bilayer**, is relatively free to rotate toward the applied field because of its interrupted or weakened exchange coupling, as compared to the undoped Ru-O bilayers. **Antiparallel alignment** in the soft bilayer is achieved when the spin in the Cr-rich layer fully switches at $B = B_{C2}$ (**Fig. 6.12b**). This spin switching enhances M_a at B_{C2} but, at the same time, changes the density of states for the up- and down-spin electrons at the Fermi surface. The probabilities of scattering for *both up- and down-spin electrons* are enhanced within the soft bilayers having antiparallel alignment since transport electrons must flip spin to find an empty energy state; this explains the pronounced peak in $[\rho_c(B)-\rho_c(0)]/\rho_c(0)$ (**Figs. 6.12b** and **6.12c**). The remaining antiparallel spins in the Ru-O layers in both the soft and hard bilayers also switch with further increase of B, finally completing the spin alignment at $B = B_{C1}$. The scattering rate is then zero for the up-spin electrons, and remains finite for down-spin electrons. Since conduction occurs in parallel for the two spin channels, the total resistivity is chiefly determined by highly conductive up-spin electrons and, consequently, ρ_c drops dramatically by as much as 40%. This magnetoresistive effect is much larger than that seen in thin-film mulilayers [41], and could have important technological implications.

While B_{C1} decreases with T, B_{C2} increases slightly with T and disappears at T > 45 K. This may be due to a reduction in the difference in soft and hard layer coercivities that becomes insignificant in applied fields at higher temperatures. Consequently, switching may occur almost simultaneously for both the Cr-rich and Ru-O bilayers, resulting in one sharp transition at B_{C1} seen in both M_a and ρ_c (**Fig. 6.12c**), which persists up to 70 K. The B-T phase diagram given in **Fig. 6.13** shows that all transitions (T_N, T_{MI}, B_{C1} and B_{C2}) meet at a tetracritical point at B = 3.8 T and T = 45 K when **B** is applied along the **a**-axis. It is in the vicinity of this point that prominent SV behavior exists.

Previous to this work, the SV phenomenon was realized solely in heterostuctures and multilayer films [41]. Cr substitution apparently alters the density of states in the AFM-M state and creates soft and hard bilayers having antiparallel spin alignments that induce SV behavior for **B ∥ a**. Additional studies will be required to identify the exact nature of the magnetic microstructure that governs the striking magnetoresistive behavior of this system.

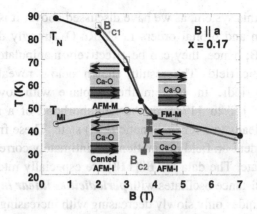

Fig. 6.13. T-B diagram for x = 0.17 based on the data or B ∥ a. The thin (thick)-line arrow indicates the spin direction in the Cr-O rich (RuO) layer. The shaded area is the temperature range of interest for the spin-valve effect [Ref. 32].

6.4.3. *Oscillatory Magnetoresistance Periodic in 1/B and B*

Quantum oscillations, such as the Shubnikov-deHaas (SdH) effect, are a manifestation of oscillations of the density of states due to the magnetic field evolution of Landau levels, and therefore provide a direct probe of the Fermi surface. This probe is particularly important for investigations of correlated electron systems whose quantum oscillations are experimentally observable. Observations of the SdH effect are common in metals where the carrier mean-free path is sufficiently long and resistivity is sufficiently low. For most transition element oxides, the mean-free paths may be comparable with the lattice spacing and the resistivity is much higher. Therefore, observations of the SdH effect in oxides are not as common as in metals, and certainly unexpected in Mott systems such as $Ca_3Ru_2O_7$. Besides the SdH effect, which is defined by resistivity oscillations periodic in 1/B, a striking and intriguing oscillatory resistance that is *periodic in B*, are both observed in $Ca_3Ru_2O_7$. The latter phenomenon, never seen in bulk materials before, reflects unique new coupling of the Fermi surface to the magnetic field.

Since the tilting of the RuO_6 octahedra determines the overlap between the three t_{2g} orbitals (d_{xy}, d_{zx} and d_{yz}) of neighboring corner-shared octahedra, small changes in this tilt can induce complex phenomena in this system, as we have discussed above. It is clear that the long-range spin and orbital orders in $Ca_3Ru_2O_7$ strongly depend on the orientation of B; hence, they can be effectively manipulated and probed with a magnetic field. The results shown below reveal strong SdH oscillations (periodic in 1/B) in the **bc**-plane with low frequencies ranging from 30 T to 117 T in the neighborhood of a metamagnetic transition that leads to a ferromagnetic (FM) state. These frequencies are strongly dependent on field orientation and intimately correlated with the induced FM state. The data for **B∥[110]** are especially interesting, since the magnetoresistance oscillates with *periodicities linear in B up to 15 K*, and with amplitudes only slowly decreasing with increasing temperature.

6.4.3.1. *Shubnikov-deHaas Effect for B ∥ c*

As seen above, for **B ∥ c**, ρ_c displays slow SdH oscillations in the absence of the metamagnetism, signaling the existence of very small Fermi surface cross-sectional areas. It is plausible that the d_{xy} orbitals give rise to small lens-shaped Fermi surface pockets that are very sensitive to slight structural changes. The observed oscillations must then be associated with the motion of the electrons in the **ab**-plane---i.e., with the d_{xy} orbitals. The oscillations in ρ_c correspond to two extremely low frequencies, $f_1 = 28$ T and $f_2 = 10$ T, that, based on crystallographic data [21] and the Onsager relation $F_0 = A(h/4\pi^2 e)$ (e is the electron charge, h Planck's constant), correspond to a cross-sectional area A of only 0.2% of the first Brillouin zone. From the temperature dependence of the SdH amplitude, the cyclotron effective mass is estimated to be $\mu_c = (0.85 \pm 0.05)m_e$, where m_e is the free-electron mass. This is markedly smaller than the thermodynamic effective mass ($m^* \sim 3m_e$) estimated from the electronic contribution γ to the specific heat [30, 47]. There are three possible sources for this discrepancy: (1) The cyclotron effective mass is measured in a large magnetic field that quenches correlations, while the specific heat is a zero-field measurement. (2) μ_c only refers to one closed orbit, while the thermodynamic effective mass reflects an average over the entire Fermi surface. (3) The Dingle temperature, $T_D = h/4\pi^2 k_B \tau$ is a measure of scattering at the Fermi surface (estimated to be 3 K for the observed oscillations, comparable to those of good organic metals); and the Dingle temperatures of the highest mass electron orbits may be too large to permit resolved oscillations.

6.4.3.2. *Shubnikov-deHaas Effect in the bc-Plane*

The field dependence of ρ_c (on a logarithmic scale) for **B** rotating in the **bc**-plane (**B** ∥ **b**, $\theta = 0°$; and **B** ∥ **c**, $\theta = 90°$) at T = 0.4 K is shown in **Fig. 6.14a** for B ranging from 11 T to 45 T. B_c occurs at 6 T for **B** ∥ **b**, and increases with increasing θ (i.e., as **B** rotates towards to the **c**-axis). The striking finding is that strong SdH oscillations are observed to be qualitatively different for $11° \leq \theta < 56°$ and $56° < \theta \leq 90°$. It is then likely that the value $\theta = 56°$ marks the suppression of the OO state as **B** rotates further away from the easy-axis of magnetization. This also destabilizes the spin-polarized state, and thus the OO state via direct coupling to the field or the spin-orbit interaction. (This is only possible perturbatively, because the spin-orbit interaction is quenched by crystalline fields). Consequently, the electron mobility increases drastically, which explains the largely enhanced conductivity for $56° < \theta \leq 90°$.

Strong oscillations occur at $\theta < 56°$ only for B > B_c, and with frequencies significantly larger than the ones observed for **B** ∥ **c**. For $0° < \theta < 56°$ and B > B_c, ρ_c increases with both B and θ and displays oscillatory behavior only for $11° \leq \theta < 56°$. While the extremal orbits responsible for the oscillations are facilitated by the FM state, it is remarkable that no oscillations are seen when $\theta = 0$ (**B** ∥ **b**) where the field-induced FM state is fully established at $B_c = 6$ T.

Remarkably, no oscillations were discerned for **B** rotating within the **ac**-plane for B < 45 T. The bumps seen in ρ for **B** ∥ **a** (**Fig. 6.13a**) are not oscillatory at higher B. The **ac**-plane is perpendicular to the easy **b**-axis of magnetization and has no FM component, suggesting a critical link of the SdH oscillations to the fully spin-polarized state. The spin-polarized state and the different projections of the tilt angles of the RuO_6 octahedra onto the **ac**- and **bc**-planes are expected to affect the Fermi surface.

On the other hand, for $56° < \theta \leq 90°$, the oscillations disappear for B > B_c but are present for B < B_c, accompanying the much more conducting phase at high fields, as shown in **Figs. 6.14a** and **6.14b**. The frequency of the oscillations seen for B < B_c remains essentially

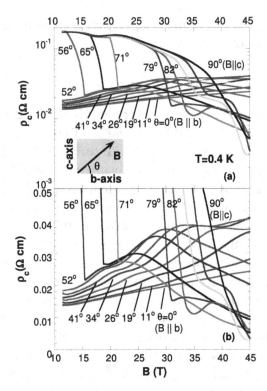

Fig. 6.14. (a) ρ_c for B rotating in the bc-plane with $\theta = 0$ and $90°$ corresponding to B || b and B || c, respectively, and (b) enlarged ρ_c on a linear scale for clarity. Note that the range of B is from 11 to 45 T [Ref. 26].

independent of θ for $65° \le \theta \le 90°$. Since the d_{xy} orbitals are believed to be responsible for the oscillations **B || c**, the insensitivity of the frequency with tilting of **B** suggests that the oscillations in the absence of the metamagnetism originate from a nearly spherical pocket of the same d_{xy} orbitals. Conversely, the oscillations for $11° \le \theta < 56°$ and B > B_c could be associated with a spin-polarized state and ordered d_{zx} and/or d_{yz} orbitals. These orbitals permit only limited electron hopping (as confirmed by a larger ρ_c), and thus lower density of charge carriers and longer mean free path which in turn facilitates electrons to execute their circular orbits.

Figure 6.15 shows the amplitude of the SdH oscillations as a function of 1/B for several representative angles θ at (a) T = 0.4 K

(Fig. 6.15a) and (b) 1.5 K (Fig. 6.15b). The SdH signal is defined as $\Delta\rho/\rho_{bg}$, where $\Delta\rho = (\rho_c - \rho_{bg})$ and ρ_{bg} is the background resistivity that is obtained by fitting the experimental ρ_c to a polynomial. The Fast Fourier Transformation (FFT) analysis yields the same frequencies as those determined from **Figs. 6.15a** and **6.15b**. The result from the FFT is shown as **Fig. 6.15c** for a representative ρ_c at T = 1.5 K and $\theta = 26°$. Clearly, the oscillations are strong and slow, and their phase and frequency shift systematically with changing θ. The oscillations vanish for $\theta > 56°$, suggesting that the extremal cross-section responsible is highly susceptible to the orientation of **B**. SdH oscillations are usually

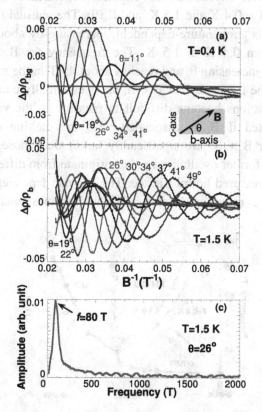

Fig. 6.15. The amplitude of the SdH oscillations defined as $\Delta\rho/\rho_{bg}$ as a function of inverse field B^{-1} for various θ and for (a) T = 0.4 K and (b) 1.5 K; (c) the SdH amplitude as a function of frequency obtained from the Fast Fourier Transformation for a representative ρ_c at T = 1.5 K and $\theta = 26°$ [Ref. 26].

rather weak in metals [38]; the remarkably strong oscillatory behavior for $11° \le \theta < 56°$ may arise from an extremal orbit with a flat dispersion perpendicular to the cross-section, so that a large constructive interference can occur. It is also noted that the $1/\cos\theta$ dependence seen in **Figs. 6.15a** and **6.15b** may derive from a cylindrical Fermi surface elongated along the **c**-axis, which approximates two-dimensional conductivity. With further increase of $\theta \ge 56°$ the impact of **B** on the Fermi surface becomes even more dramatic and the closed orbit is no longer observed, and may be replaced by open orbits that do not support SdH oscillations.

Figure 6.16 illustrates the angular dependences of the SdH frequency for T = 0.4 K and 1.5 K, and $B_c(\theta)$. The unusual feature is that the *frequency* is temperature-dependent, increasing by about 15% when T is raised from 0.4 K to 1.5 K. The frequency for $B > B_c$ rapidly decreases with increasing θ, and reaches about 45 T in the vicinity of $\theta = 56°$; whereas the frequency for $B < B_c$ stays essentially constant for $\theta > 56°$. The oscillations become difficult to measure in the vicinity of B_c. This is expected if B_c is associated with the melting of OO. The frequencies for $B > B_c$ are significantly larger than those for $B < B_c$, suggesting the former oscillations either originate from different electron orbits or a restructured Fermi surface. The angular dependence of B_c, on the other hand, is rather weak for $\theta < 56°$, but becomes much stronger for

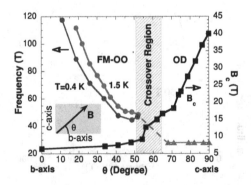

Fig. 6.16. The angular dependence of the frequency (solid circles for $B > B_c$, and triangles for $B < B_c$) for T = 0.4 K and 1.5 K (left scale) and B_c (solid squares) (right scale). Note that the shaded area in the vicinity of 56° marks the melting of OO as θ increases [Ref. 26].

$\theta > 56°$. Note that ρ_c displays a weak plateau at high magnetic fields and approaching $90°$, which disappears for **B** ‖ **c** because the FM state is no longer energetically favored. Such an inverse correlation between the frequency and B_c reinforces the point that the field-induced FM state reconstructs the Fermi surface and facilitates the oscillatory effect. It is evident that the angular interval $52° < \theta < 65°$ about $\theta = 56°$ marks a crossover region between the FM-OO state and the orbitally degenerate state, as shown in **Fig. 6.16**.

6.4.3.3. *Magnetoresistance Oscillations Periodic in B*

For **B**‖[110], ρ_c also exhibits oscillations in the magnetoresistance as displayed in **Fig. 6.17**. *However, these oscillations are periodic in B* (rather than 1/B) with a period $\Delta B = 11$ T, and are persistent up to an *unusually high temperature T = 15 K*.

This highly unusual observation is corroborated by plotting the data, both as a function of 1/B (**Fig. 6.17a** and **6.17b**) and B (**Fig. 6.17c** and **6.17d**). The oscillations die off rapidly if B slightly departs (e.g., by $\pm 5°$) from the [110] direction.

Magnetoresistance oscillations periodic in 1/B (SdH effect) are a manifestation of the constructive interference of quantized extremal orbits of Fermi surface cross-sections perpendicular to the field. Due to the Pauli principle the electrons are constrained to move about the Fermi surface. The projection of the real-space trajectory of a nearly-free electron onto a plane perpendicular to **B** reproduces the **k**-space trajectory, but rotated by $\pi/2$ and scaled by a factor $c\hbar/|e|B$. Hence, trajectories with constructive interference in real space are expected to be periodic in B rather than 1/B (the frequency is proportional to the cross-sectional area in reciprocal space, so that the relation to the real space is B^2). Oscillations in the magnetoresistivity periodic in B are realized in some mesoscopic systems and are always related to finite size effects. Examples are (i) the Aharanov-Bohm (AB) effect [38, 50, 51], (ii) the Sondheimer effect [38, 52], and (iii) the edge states in quantum dots [53]. Note that each of the cases involves a geometrical confinement.

The AB interference occurs when a magnetic flux threading a metallic loop changes the phase of the electrons, generating oscillations in the magnetoresistance and is observed only in mesoscopic conductors, but not in bulk materials. The Sondheimer effect requires a thin metallic film with the wavefunction vanishing at the two surfaces. The thickness of the film has to be comparable with the mean free path. This gives rise to boundary scattering of the carriers that alters the free electron trajectories and the possibility of interference. Finally, the quantum Hall edge states require real-space confinement [52].

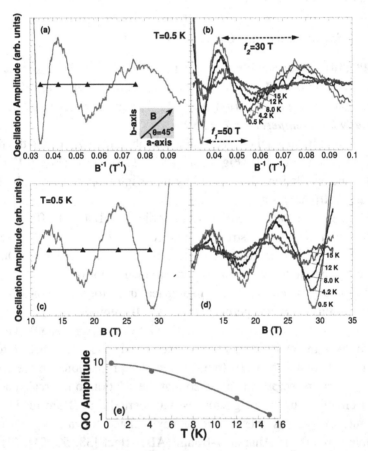

Fig. 6.17. (a) The amplitude of the oscillations as a function of B^{-1} for B‖[110] and T = 0.5 K and (b) for various temperatures up to 15 K (Note oscillations are not periodic in 1/B). (c) The amplitude of the oscillations as a function of B for B‖[110] and T = 0.5 K and (d) for various temperatures up to 15 K (Note oscillations are periodic in B instead). (e) The amplitude of the oscillations as a function of temperature [Ref. 26].

Since a bulk, three-dimensional material has no real space confinement for the orbits of the carriers, the most likely explanation for the periodicity as a function of B is a Fermi surface cross-section that changes with field. The t_{2g}-orbitals have off-diagonal matrix elements arising from the orbital Zeeman effect, and hence couple directly to the magnetic field. It is therefore conceivable that the magnetic field could lead to a dramatic change of the Fermi surface if it points along a special direction. Note that the Fermi surface pockets involved are very small (low frequencies as a function of 1/B) and susceptible to external influences. If there is more than one conducting portion of the Fermi surface, occupied states can be transferred from one pocket to another with relatively small changes in the external parameters. This is also consistent with the 15% of change in the frequency when T is raised from 0.4 K to 1.5 K, as shown in **Fig. 6.3**. Indeed, the amplitude of the oscillations follows the Lifshitz-Kosevich behavior expected for SdH oscillations (**Fig. 6.17e**). We also must note that the AB effect at finite T would show the same amplitude dependence [54]. It is still perplexing that the cross-section of the observed pocket is only 0.2% of the Brillouin zone, so the position of the Fermi energy is fixed at the non-quantized level of other Fermi surface branches. In such a situation, the density of states oscillates only against 1/B. In addition, if the origin of the oscillations periodic in B is ascribed to the Landau quantization, it is then perplexing as to why there are no SdH oscillations for **B∥[110]**, simultaneous with the oscillations periodic in B.

The observations of the oscillations in the magnetoresistance in Ca$_3$Ru$_2$O$_7$ periodic both in B and 1/B certainly reflect a crucial dependence of the quantized orbits on the orientation of **B**. These novel phenomena highlight the critical role of the orbital degrees of freedom embodied via the coupling of the t_{2g}-orbitals to the magnetic field and certainly merit more experimental and theoretical efforts.

6.5. Conclusions

In summary, the unusual and sometimes spectacular physical properties of Ca$_3$Ru$_2$O$_7$ discussed above not only span almost every ordered state (except for superconductivity) known in condensed matter physics, but include unusual phenomena not found in other materials. It has become increasingly clear that the orbital degrees of freedom drive the complex phase diagram of $Ca_3Ru_2O_7$ via (spin-orbit) coupling of the orbital degrees of freedom to the spin, and also (via a Jahn-Teller effect) to lattice degrees of freedom. These phenomena present profound intellectual challenges, and pose a set of intriguing questions whose answers will eventually establish a deeper understanding of highly correlated transition element oxides.

Acknowledgements

This work was supported by NSF through grants DMR-0856234 and EPS-0814194. P.S. is supported by DoE Grant #DE-FG02-98ER45707. L.D. is supported by DoE Grant #DE-FG02-97ER45653.

References

1. G. Cao, S. McCall, J. E. Crow and R. P. Guertin, Phys. Rev. Lett. **78** (1997) 1751
2. P. Allen, H. Berger, O. Chauvet, L. Forro, T. Jarlborg, A. Junod, B. Revaz, and G. Santi, Phys. Rev. B **53** (1996) 4393
3. G. Cao, S. McCall, J. Bolivar, M. Shepard, F. Freibert, P. Henning, J.E. Crow and T. Yuen, Phys. Rev. B **54** (1996) 15144
4. G. Cao, S. McCall, M. Shepard, J.E. Crow and R.P. Guertin, Phys. Rev. B **56** (1997) 321
5. I.I. Mazin, D.J. Singh, Phys. Rev. B **56** (1997) 2556
6. L. Klein, J.S. Dodge, C.H. Ahn, G.J. Snyder, T.H. Geballe, M.R. Beasley and A. Kapitulnik, Phys. Rev Lett. **77** (1996) 2774
7. Y. Maeno, H. Hashingmoto, K. Yoshida, S. Ishizaki, T. Fujita, J.G. Bednorz, and F. Lichtenberg, Nature (London) 372 (1994) 532
8. Y. Maeno, T.M. Rice, and M. Sigrist, Physics Today 54 (2001) 42
9. G. Cao, S. McCall, M. Shepard, J.E. Crow and R.P. Guertin, Phys. Rev. B **56** (1997) **R2916**
10. C.S. Alexander, G. Cao, V. Dobrosavljevic, E. Lochner, S. McCall, J.E. Crow and P.R. Guertin, Phys. Rev. B **60** (1999) **R8422**
11. G. Cao, S. McCall, V. Dobrosavljevic, S.C. Alexander, and J.E. Crow, and R.P. Guertin, Phys. Rev. B **61** (2000) **R5053**
12. G. Cao, S. McCall, J. E. Crow, and R. P. Guertin, Phys. Rev. B 56 (1997) 5387
13. A.V. Puchkov, M.C. Schabel, D.N Basov, T. Startseva, G. Cao, T. Timusk, and Z.-X. Shen, Phys. Rev. Lett. **81** (1998) 2747.
14. Shin-Ichi Ikeda, Yoshiteru Maeno, Satoru Nakatsuji, Masashi Kosaka, and Yoshiya Uwatoko, *Phys. Rev. B* **62** (2000) **R6089**

15. S.A. Grigera, R.P. Perry, A.J. Schofield, M. Chiao, S.R. Julian, G.G. Lonzarich, S.I. Ikeda, Y. Maeno, A.J. Millis and A.P. Mackenzie, Science **294** (2001) 329
16. M.K. Crawford, R.L. Harlow, W. Marshall, Z. Li, G. Cao, R.L. Lindstrom, Q. Huang, and J.W. Lynn. Phys. Rev. B **65** (2002) 214412
17. G. Cao, L. Balicas, W. H. Song, Y.P. Sun, Y. Xin, V.A. Bondarenko, J.W. Brill, S. Parkin, and X.N. Lin, Phys. Rev. B **68** (2003) 174409
18. Rajeev Gupta, M. Kim, H. Barath, S.L. Cooper and G. Cao, Phys. Rev Lett. **96** (2006) 067004
19. S. Chikara , V. Durairaj, W.H. Song, Y.P. Sun, X.N. Lin, A. Douglass, G. Cao, P. Schlottmann, and S. Parkin, Phys. Rev. B **73** (2006) 224420
20. G. Cao, S. Chikara, J. W. Brill, and P. Schlottmann, Phys. Rev. B **75** (2007) 024429
21. G. Cao, K. Abbound, S. McCall, J.E. Crow and R.P.Guertin, Phys. Rev. B **62** (2000) 998
22. G. Cao, L. Balicas, Y. Xin, E. Dagotto, J.E. Crow, C.S. Nelson, and D.F. Agterberg, Phys. Rev. B **67** 060406(R) (2003)
23. G. Cao, L. Balicas, Y. Xin, J.E. Crow, and C.S. Nelson, Phys. Rev. B **67**, 184405 (2003)
24. G. Cao, L. Balicas, X.N. Lin, S. Chikara, V. Durairaj, E. Elhami, J.W. Brill, R.C. Rai and J. E. Crow, Phys. Rev. B **69** (2004) 014404
25. G. Cao, X.N. Lin, L. Balicas, S. Chikara, J.E. Crow and P. Schlottmann, New Journal of Physics **6** (2004) 159
26. V. Durairaj, X. N. Lin, Z.X. Zhou, S Chikara, E. Elhami, P. Schlottmenn and G. Cao, Phys. Rev. B **73** (2006) 054434
27. H.L. Liu, S. Yoon, S.L. Cooper, G. Cao and J.E. Crow, Phys. Rev. B **60** (1999) R6980,
28. C.S. Snow, S.L. Cooper, G. Cao, J.E. Crow, S. Nakatsuji, Y. Maeno, Phys. Rev. Lett. **89** (2002) 226401
29. J.F. Karpus, R. Gupta, Barath, S.L. Cooper and G. Cao, Phys. Rev. Lett. **93** (2004) 167205
30. S. McCall, G. Cao, and J.E. Crow, Phys. Rev. B **67,** (2003) 094427
31. X. N. Lin, Z.X. Zhou, V. Durairaj, P. Schlottmenn and G. Cao, Phys. Rev. Lett. **95** (2005) 017203
32. G. Cao, V. Durairaj, S. Chikara, and L.E. DeLong and P. Schlottmenn, Phys. Rev. Lett. **100** (2008) 016604
33. G. Cao, V. Durairaj, S. Chikara, and L.E. DeLong and P. Schlottmenn, Phys. Rev. Lett. **100** (2008) 0159902
34. D.J. Singh and S. Auluck, Phys. Rev. Lett. **96** (2006) 097203
35. Guo-Qiang Liu, Phys. Rev. B **84**, 235137 (2011)
36. In text and figures in our earlier papers, the terms, "**a**-axis" and **b**-axis" should be reversed.
37. E. Ohmichi, Y. Yoshida, S.I. Ikeda, N Shirakawa and T. Osada, Phys. Rev. B. **70** (2004) 104414.
38. A.B. Pippard, *Magnetoresistance in Metals* (Cambridge University Press, Cambridge, 1989)

39. For example, Yoshinori Tokura, *Colossal Magnetoresistive Oxides* (Gordon and Beach Science Publishers, Australia, 2000)
40. T. Valla, P. D. Johnson, Z. Yusof, B. Wells, Q. Li, S. M. Loureiro, R. J. Cava, M. Mikami, Y. Mori, M. Yoshimura and T. Sasaki, Nature **417** (2002) 627
41. For example, E.Y. Tsymbal and D.G. Pettifor, p. 113, *Solid State Physics* v. 56, ed. Henry Ehrenreich and Frans Spaepen (Academic Press, New York, 2001)
42. L.Pi, A Maignan, R. Retoux and B. Raveau, J. Phys.: Condensed Matter **14** (2002) 7391
43. Z.H. Han, J.I. Budnick, W.A. Hines, B. Dabrowski, S. Kolesnik, and T. Maxwell, J. Phys.: Condensed Matter **17** (2005) 1193
44. B. Dabrowski, O. Chmaissem, P.W. Klamut, S. Kolesnik, M. Maxwell, J. Mais, Y. Ito, B. D. Armstrong, J.D. Jorgensen, and S. Short, Phys. Rev. B **70** (2004) 014423
45. B. Dabrowski, O. Chmaissem, S. Kolesnik, P.W. Klamut, M. Maxwell, M. Avdeev, and J.D. Jorgensen, Phys. Rev. B **71** (2005) 104411
46. V. Durairaj, S. Chikara, X.N. Lin, A. Douglass, G. Cao, P. Schlottmann, E.S.Choi and R.P. Guertin, Phys. Rev B **73** (2006) 214414
47. V. Varadarajan, S, Chikara, V. Durairaj, X.N. Lin, G. Cao, and J.W. Brill, Solid State Comm. **141** (2007) 402
48. Eric Fawcett, Rev. Mod. Phys. **60** (1988) 209
49. R.A. de Groot, F. M. Mueller, P. G. van Engen and K. H. J. Buschow Phys. Rev. Lett. **50** (1983) 2024
50. C.W. J. Beenakker and H. van Houten, *Solid State Physics* (Academic Press, New York, 1991), edited by H. Ehrenreich and D. Turnbull, vol. 44, p. 65.
51. R.A. Webb, S Washburn, C.P. Umbach, R.B. Laibowitz, Phys. Rev. Lett. **54**, (1985) 2696
52. E.H. Sondheimer, Phys. Rev. **80** (1950) 401
53. A. Yacobi, R. Schuster, and M. Heilblum, Phys. Rev. B **53** (1996) 9583
54. P. Schlottmann and A.A. Zvyagin, J. Appl. Phys. **79**, 5419 (1996); J. Phys. Cond. Matter 9 (1997) 7369

Chapter 7

SURFACES OF TRANSITION-METAL COMPOUNDS: THE INTERPLAY BETWEEN STRUCTURE AND FUNCTIONALITY

Xiaobo He[1], Jing Teng[1], Von Braun Nascimento[1], R. G. Moore[2,3], Guorong Li[1], Chen Chen[1], Jiandi Zhang[1], E. W. Plummer[1]*

[1]*Department of Physics and Astronomy, Louisiana State University, Baton Rouge LA 70803, USA.*
[2]*Geballe Laboratory for Advanced Materials, Departments of Physics and Applied Physics, Stanford University, Stanford, CA 94305, USA.*
[3]*Stanford Institute for Materials and Energy Sciences, SLAC National Accelerator Laboratory, Menlo Park, CA 94025, USA.*
E-mail: wplummer@phys.lsu.edu

The exotic behavior of correlated electron systems has both excited and challenged material scientists for decades. Better experimental techniques, higher quality samples, the discovery of new functional materials, and a focus on complexity have all driven the current flurry of activity. This volume features a series of articles describing the *Frontiers of 4d- and 5d- Transtion Metal Oxides.* In this article we describe *Frontiers of 4d- and 5d- Transition Metal Compounds (TMCs) in an Environment of Broken Symmetry, i.e. a Surface.* The unique environment of a surface, with its inherently broken translational symmetry, provides an exquisite playground for probing the physics of TMCs. The surface structure, surface property relationship will be demonstrated, where small structural changes effect physical properties such as the metal-to-insulator transition temperature, the presence of superconductivity, magnetic ordering, electron-phonon coupling, spin-phonon coupling, and even quantum critical behavior.

Contents

Acronyms

AFM	antiferromagnetic
ARPES	angle-resolved photoemission spectroscopy
CAF	canted antiferromagnet
CEM	correlated electron material
CMR	colossal magnetoresistance
CSRO	$Ca_{2-x}Sr_xRuO_4$
CTR	crystal truncation rod
DAS	Dimer-Adatom-Stacking
FM	ferromagnetic
GIXS	grazing incidence X-ray scattering
HREELS	high resolution electron energy loss spectroscopy
HREM	high resolution electron microscopy
HTT	high-temperature tetragonal
LDOS	local density of states
LEED	low energy electron diffraction
LTO	low-temperature orthorhombic
MIT	metal-to-insulator transition
MT	muffin-tin
QCP	quantum critical point
RFA	retarding field analyzer
R-factor	reliability factor
RHEED	reflection high energy electron diffraction
RP	Ruddlesden-Popper
SSR	surface superstructure rod
STM	scanning tunneling microscopy
STS	scanning tunneling spectroscopy

TED   transmission electron diffraction
TEM   transmission electron microscopy
TMO   transition-metal oxide
UHV   ultra high vacuum
XRMS   X-ray resonant magnetic scattering
YBCO $YBa_2Cu_3O_{7-\delta}$

7.1. Introduction

Understanding the complexity that develops in correlated electron materials (CEMs) is one of the grand challenges of the 21st century.[1-5] The exotic properties displayed by CEMs such as high-T_c superconductivity in cuprates[6,7] and iron-based superconductors,[8,9] "colossal" magnetoresistance (CMR) in manganites,[10] and heavy-fermion compounds,[11] are intimately related to the coexistence of competing nearly degenerate states which couple simultaneously active degrees of freedom—charge, lattice, orbital, and spin. In 2000, Birgeneau and Kastner wrote in the introduction to a special issue of *Science* dedicated to CEMs, "A remarkable variety of new materials have been discovered that cannot be understood at all with traditional ideas".[1] Little did they know that in 2008 superconductivity would be discovered in the iron-based materials.

Two examples of CEMs that are of relevance to this article are transition-metal oxides (TMOs)[12] and the newly discovered iron-based high-Tc superconductors.[8] Figure 7.1 presents a marble model of Ruddlesden-Popper (RP) series or ruthenates $Sr_{n+1}Ru_nO_{3n+1}$ ($n = 1, 2, 3, ...$) and two of the structures associated with the iron-based superconductors, iron pnictides ("122" type) and iron chalcogenides ("11" type). Structural changes are intimately related to the exotic behaviors in these materials. For example, in the layered Ruddlesden-Popper series the octahedra can tilt or rotate depending upon the doping. Rotation favors AFM ordering and Tilt favors FM.[13] A slight distortion in the octahedra breaks the symmetry and can drive different phases or physical properties. Subtle changes in structure have dramatic consequences on the physical properties. This article will address this issue by looking at the manifestations of broken symmetry at a surface.

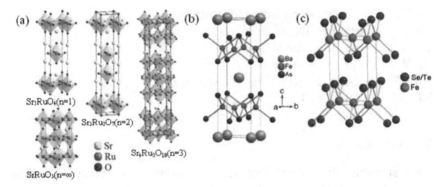

Fig. 7.1. (a) Artistic view of four members of the RP serials $(Sr_{n+1}Ru_nO_{3n+1})$ for $n = 1, 2,$ 3, and ∞. The RuO_6 octahedra display as the gray solid object. The structural properties of the $n = 1, 2,$ and 3 phases are clear low dimensional and are expected to lead to highly anisotropic physical properties, while the $n = \infty$ structure is 3D. Bulk crystal structure of (b) $BaFe_2As_2$ parent compound of "Ba-122" type, and (c) iron chalcogenide $FeTe_{1-x}Se_x$ "11" type iron-based superconductors. They share a common layered structure based on square planar sheets of iron in a tetrahedral environment with pnictogen or chalcogen anions. In $BaFe_2As_2$ the crystal structure is separated by Ba blocking layer, while $FeTe_{1-x}Se_x$ is composed of iron-chalcogenide slabs stacked together without any spacing layer.

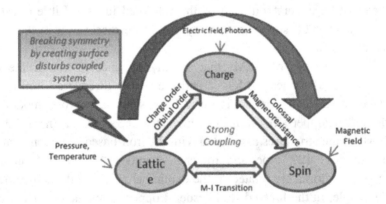

Fig. 7.2. Schematic illustration of close coupling between electronic, spin and lattice degrees of freedom in highly correlated materials. External probes such as temperature, pressure, photons, and electric or magnetic fields can drive remarkable change of physical properties in CEMs. For example, the temperature drives a MIT transition (electronic structure) and, in many cases, a lattice distortion. The application of magnetic field shifts the MIT transition temperature and induces orders of magnitude change in resistance in a CMR material. Conceptually, breaking symmetry by creating surface can be used to modify and probe this close coupling.

Figure 7.2 schematically illustrates the coupling in highly correlated systems. In these materials, remarkable changes in physical properties may happen with the application of an external stimulus due to the many competing "ground" states in the parent compound. For example, Fig. 7.3(a) illustrates a coupled magnetic-electronic transition in the doped perovskite $La_{0.65}Ca_{0.35}MnO_3$ as the temperature is rising.[14] The material changes from ferromagnetic metal at the low temperature to paramagnetic insulator at the high temperature. These coupled electronic and magnetic phase transitions are also associated with the static and dynamic lattice distortions.[15-17] The coupled phase transition temperature varies from ~ 60 K for $Pr_{0.7}Ca_{0.3}MnO_3$ to ~ 370 K for $La_{0.7}Sr_{0.3}MnO_3$ depending on the size and electronic properties of the divalent and trivalent components in the perovskites ($A_{1-x}B_xMnO_3$). This family (113) is the n = infinity, in the RP series, and is three dimensional in nature. Iron-based superconductors (layered materials) are other examples characterized by a strong coupling between structural, electronic, and magnetic properties leading to richness in their physical properties,

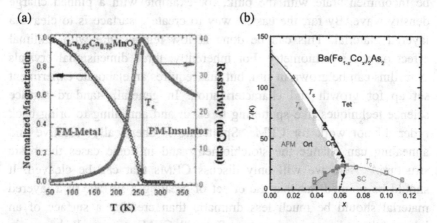

Fig. 7.3. Illustration of the macroscopic collective phenomena responding to external probes. (a) A coupled ferromagnetic to paramagnetic phase transition and a metal-insulator transition in $La_{0.65}Ca_{0.35}MnO_3$.[14] (b) The physical properties of $Ba(Fe_{1-x}Co_x)_2As_2$ are tuned by chemical substitution, which shown a superconducting dome for $0.03 < x < 0.12$. A structural transition from a high-temperature tetragonal to a low-temperature orthorhombic phase is always accompanied by a magnetic transition within a narrow temperature window.[18]

which can be tuned by chemical substitution. As shown in Fig. 7.3(b), in $Ba(Fe_{1-x}Co_x)_2As_2$ a structural transition from a high-temperature tetragonal (HTT) to a low-temperature orthorhombic (LTO) phase is always accompanied by a magnetic transition within a narrow temperature window.[18] In general, superconductivity is achieved by doping to kill the magnetic ordering and the accompanying orthorhombic phase. In the under doped regime, spin ordering, orthorhombicity and superconductivity coexist. It is also known that the application of pressure can drive some of the parent compounds (iron chalcogenides) into the superconducting state without chemical doping.[19]

Creating a surface is, or should be, a controlled way to investigate the consequences of broken symmetry. With half the sample missing (missing bonds) the surface structure will change compared to the bulk. The simplest case would be a surface relaxation where the symmetry of the two-dimensional unit cell is the same as the bulk, usually referred to as a (1×1) surface unit cell. But the stress created when the surface is formed can lead to a reconstruction with a surface unit cell larger but commensurate with the bulk. In extreme cases the surface unit cell could be incommensurate with the bulk, for example with a pinned charge density wave. By far, the easiest way to create a surface is to cleave a layered material. This can be done at low temperature with minimal effect on the stoichiometry. For inherently, three dimensional crystals thin films can be grown in situ, but this requires an elaborate experiment set up for growth and characterization. In general, standard surface science techniques like sputtering to clean and annealing to bring back order do not work for CEMs. Sputtering is chemically selective and annealing can change the stoichiometry and in some cases the basic structure.[20] Here we will only discuss CEMs that can be cleaved. It should be pointed out that the effect of creating a surface in a layered material should be much less dramatic than creating a surface of an inherent three dimensional system. But with CEMs very small changes in the structure of the surface can lead to dramatic property changes.

There is another way to use surface science to probe the functionality of CEMS. This is accomplished by modifying the surface after it is cleaved. Acceptor or donor adsorbates can be deposited in ordered or disordered configurations to shift the band structure, just like

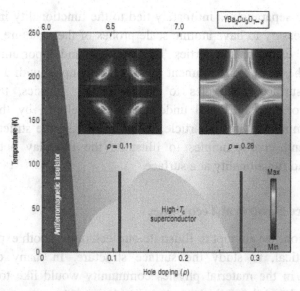

Fig. 7.4. Phase diagram of YBCO by ARPES. Fermi surface of as-cleaved YBCO exhibiting an effective hole doping $p = 0.28$ per planar Cu atom is marked by the right line. With potassium deposition, electrons are transferred to the topmost CuO_2 bilayer and the Fermi surfaces become progressively more hole-underdoped. With heavy potassium deposition, the effective hole doping becomes 0.11 as marked by the left line, and the E_F ARPES intensity reduces to the 1D CuO-chain Fermi surface and four disconnected nodal CuO_2 Fermi arcs. The created surface provides a controllable playground to tune the characterization of YBCO Fermi surface.[21]

in doping. Figure 7.4 indicates a successful control of the surface properties of cuperates $YBa_2Cu_3O_{7-\delta}$ (YBCO).[21] The polarity of the as-cleaved YBCO surface leads to overdoped-like Fermi surfaces due to reconstruction of the electronic states. Through *in situ* deposition of potassium atoms on cleaved YBCO surface, electrons are transferred to the topmost CuO_2 bilayer and the corresponding Fermi surfaces become progressively more hole-underdoped. Consequently, the surface doping level can be well controlled, and the evolution of the Fermi surface follows continuously from the overdoped to the underdoped regime.

In order to understand the surface physics of these complex compounds, comprehensive characterizations are necessary, starting with, in our view, the structural characterization but followed by determination of the electron, transport, magnetism and lattice dynamics.

Since phase separation is intimately tied to the functionality in CEMs it will be necessary to have atomic scale probes of the structure, chemical identity and electronic properties. The challenge and opportunity is quite clear: use the unique environment of a surface coupled with a variety of advanced surface techniques to do fascinating physics, taking full advantage of the skills and understanding developed by the surface science community. In this article, we will describe the structural tools available and give examples to illustrate the interplay between the structure and functionality at a surface.

7.2. Surface Structural Techniques

In this section, we attempt to illustrate our techniques, both experimental and theoretical, to study the surface structure. In many cases, the researchers in the material physics community would like to combine experimental and theoretical or several techniques together to determine the surface structure.

7.2.1. *Glazing Incidence X-ray Diffraction*

The grazing incidence X-ray scattering (GIXS)[22] is an experimental technique suitable for the determination of the surface structure of oxides. GIXS can be the method of choice in the case of insulating oxide materials, since it is not subject to charge effects like electron scattering techniques.

GIXS has many advantages when compared with the conventionally employed electron scattering techniques. The X-ray probing photons do not present a strong interaction with the atomic cores, contrasting with the multiple-scattering process in present in low energy electron diffraction (LEED). As a consequence of this weak interaction, usually just simple single scattering (kinematic) calculations are needed in order to yield the surface structure. Another advantage comes from the fact that X-ray can penetrate deep inside the crystal thus enabling GIXS to be also used to study the structure of buried interfaces. Atomic scattering cross sections are well known for all atoms as a consequence of the extensive volume of work already performed in 3D crystallography. Another point is that the X-ray scattering intensities can be experimentally obtained with high resolution in the perpendicular momentum and consequently enable the study of the diffraction line shape, usually a difficult task for other surface diffraction techniques. This line shape analysis can be used to obtain information of surface roughness as well as the lateral size of structural domains at the surface.

Advances in synchrotron radiation technology have created high intensity radiation sources that enable one to measure diffracted intensities from less than one atomic layer. By correctly choosing the grazing incidence angle it is possible to reduce the X-ray penetration depth (~ 25 Å) and thus eliminate most of the bulk inelastic and elastic contributions (Compton, Resonant Raman, Fluorescence and Thermal Diffuse) to the final yielded signal.[22] If the surface presents a reconstruction the specific diffraction rods related to the surface (non-integer reflections) can be measured and used to obtain structural information.

A schematic experimental set up for grazing incidence X-ray diffraction is presented in Fig. 7.5. The geometry is very similar to the 3-dimensional bulk case. The main difference is that the radiation (k_i) incidence angle is kept at a small value (α_i) in order to decrease the bulk

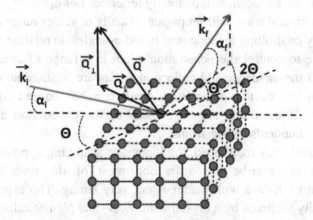

Fig. 7.5. Schematic presentation of the grazing incidence X-ray scattering experiment. In the GIXS experiment, the radiation (k_i) incidence angle is kept at a small value (α_i). The scattered intensity of a certain diffraction beam (k_f) is collected at an angle α_f with respect to the surface and at an angle 2Θ with respect to the transmitted beam. The transference of momentum Q is defined by $k_f - k_i$.

contribution to the scattering yield as previously discussed. Usually the detection process employs the following scheme. The scattered intensity of a certain diffraction beam is collected at an angle α_f with respect to the surface and at an angle 2Θ with respect to the transmitted beam. The transference of momentum $(Q = k_f - k_i)$ will present a component parallel $(Q_{\parallel})$ and one perpendicular to the surface $(Q_{\perp})$. The incidence and detection angles, α_i and α_f, will define the absolute value of the perpendicular component $Q_{\perp}$. For the case where α_i and α_f are both small Q will be approximately equal to $Q_{\parallel}$, so that the scattering plane is almost parallel to the surface and the diffraction planes perpendicular to it. Once the source incidence and detector angles are defined, the sample just need to be rotated about the surface normal to bring the diffraction net planes into diffraction condition. The condition is for the diffracted net planes to present an angle Θ with both the incident and the scattered beams. Usual procedure consists of collecting the scattered intensity for different $Q_{\perp}$ by increasing α_f while keeping α_i constant.

In the case of grazing incidence it is necessary to take into account the effects of refraction of the X-ray at the surface. In the case of X-ray the refractive index is a bit less than unity. There will be a critical incidence angle (α_c) and consequently for an incidence angle α_i smaller than this value the beam will be totally reflected. For α_i values larger than α_c the transmitted wave will propagate. Usually α_c values range from $0.1°$ to $0.5°$. By controlling the values of α_i and α_f angles, in relation with α_c, it is possible to control the penetration depth in a range of around 10 to 1000 Å. If the incidence and collection angles are adjusted so that $\alpha_i \ll \alpha_c$ and $\alpha_f \ll \alpha_c$ the penetration depth can be reduced to tens of angstroms. If α_i and α_f are larger than α_c the penetration angle will increase to hundreds of angstroms.

As previously mentioned, the kinematic scattering approach can be employed to describe the diffraction most of the time since the interaction of X-rays with matter is not very strong. The expression of the intensity scattered by a crystal with N_1, N_2 and N_3 unit cells along the three crystal axes a_1, a_2, a_3 is:[23]

$$I(Q) = AF^2(Q)S_{N1}^2(Q)S_{N2}^2(Q)S_{N3}^2(Q) \qquad (7.1)$$

with A being a constant and

$$S_N(Q.a_j) = \sum_{n=0}^{N-1} exp(iQ.a_j.n), j = 1,2,3. \tag{7.2}$$

$F(Q)$ is the so called structure factor and is expressed as a function of the atomic coordinates r_j within the unit cell:

$$F(Q) = \sum_{j\,unit\,cell} f_j \cdot \exp(iQ \cdot r_j) \tag{7.3}$$

where f_j is the atomic scattering factor and

$$S_{Nj}^2(Q.a_j) = \frac{sin^2(N_j Q.a_j/2)}{sin^2(Q.a_j/2)}\, j = 1,2,3, \tag{7.4}$$

the interference function of the N_j diffraction units.

S_N will tend to a periodic array of Dirac delta functions, spaced by $Q = 2\,p/a$, when N is very large. The intensity is not zero only if the $Q.a_1 = 2ph$, $Q.a_2 = 2pk$ and $Q.a_3 = 2pl$, with h, k and l integers. This condition means that Q is a vector of the reciprocal lattice, $Q = hb_1+kb_2+lb_3$ with b_1, b_2 and b_3 the basic vectors of the reciprocal lattice, satisfying the Laue condition. The intensity will be given by:

$$I_{hkl} = AF_{hkl}^2 N_1^2 N_2^2 N_3^2 \tag{7.5}$$

since now the structure form is written as:

$$F_{hkl} = \sum_{j\,unit\,cell} f_j \cdot \exp[2\pi i(hx_j + ky_j + lz_j) \cdot \exp(-M_j)] \tag{7.6}$$

with the summation extending over all atoms of the unit cell. The scattering factor, fractional coordinates and Debye-Waller factor for atom j are respectively represented by f_j, (x_j,y_j,z_j) and Mj respectively.

Figure 7.6 schematically presents the diffraction for a surface for four situations: a) an isolated monolayer; b) an ideal situation for monolayer plus bulk; c) a more realistic picture for surface plus bulk; and d) the surface layer presents a $c(2 \times 2)$ reconstruction.

As it can be seen for the case of an isolated layer there would be only rods (lines) in the reciprocal space. In this case of an isolated layer (2D crystal) $N_3 = 1$ in Eq. 7.1. The diffraction intensities will present sharp peaks in the two directions parallel to the layer, but the diffraction condition for $\mathbf{Q_3}$ (= $\mathbf{Q_\perp}$) is relaxed and the intensity will be continuous in the direction perpendicular to the layer.

For an ideal case for a monolayer plus bulk the result is a simple superposition of rods for the surface and points for the bulk (3-dimensional). Basically the picture is that the Bragg rods resulting from the surface diffraction pass through the Bragg points for the bulk. However this is a very simple picture. Figure 7.6(c) presents a more realistic picture for diffraction of surface plus bulk. The actual surface will involve more than one atomic layer and the rods will not present a flat profile for their intensities. All layers of the crystal will contribute. The intensity variation of the rods with $\mathbf{Q_\perp}$, called Crystal Truncation Rods (CTRs) will contain the structure information to be interpreted.

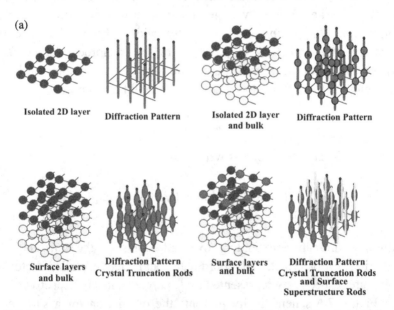

(a)

Isolated 2D layer   Diffraction Pattern   Isolated 2D layer and bulk   Diffraction Pattern

Surface layers and bulk   Diffraction Pattern Crystal Truncation Rods   Surface layers and bulk   Diffraction Pattern Crystal Truncation Rods and Surface Superstructure Rods

Fig. 7.6. Schematic presentation of diffraction patterns: (a) for one isolated atomic layer, (b) for the surface of a 3-dimensional crystal, (c) realistic representation of the diffraction pattern showing the intensity variation along the crystal truncation rods (CTRs). (d) c(2 × 2) reconstructed surface, with CTRs (red) and SSRs (yellow).

If the surface is reconstructed (presents a unit cell larger than the one for bulk) extra diffraction rods will be present and will be related to the surface structure. These rods are called Surface Superstructure Rods (SSRs). This situation is schematically presented in Fig. 7.6(d) for the case of a c(2×2) reconstruction. By employing both CTRs and SSRs data one is able to yield the structural information about the surface.

The structure determination of a solid surface, especially a reconstructed one, requires the measurement of as many diffracted beams as possible. With a larger unit cell consequently a lot of structural parameters will be involved in the analysis. As a consequence a wide range of data is necessary to assure reliability to the final structural results. The so-called atomic structure factors will basically describe the X-ray scattering by a certain type of atoms. These factors are well known and can be obtained either experimentally or theoretically.[24,25] Usually Fourier methods like the Patterson function method can be used at the beginning of the structural analysis. In order to obtain the experimental Patterson function one needs to calculate the Fourier transform of the experimental intensities, which is basically the electron density-density auto-correlation inside the unit cell. Any vector connecting the origin of coordinates to a peak in the Patterson function corresponds to an interatomic vector of the atomic structure. The Patterson function can be deconvoluted through a lot of different techniques.[26] With the knowledge of a set of interatomic vectors obtained from the Patterson function a trial structure can be obtained and calculations for this model can be compared with experimental data. Usually a least squares minimization of the difference between theory and experiment is performed in order to test and refine the structure.

A very good example to illustrate the surface structure determination of a complex oxide system is in a study of the $SrTiO_3(001)$ system.[27] $SrTiO_3(001)$ surface is of large scientific and technological importance since it consists of perhaps the most employed substrate for thin film growth of perovskites. Another interesting point about this surface is that several different reconstructions have been reported, depending on the surface preparation process, including (1×1), (2×1), (2×2), (6×2), c(4×2), c(6×2), c(4×4), ($\sqrt{5} \times \sqrt{5}$)R26.6° and ($\sqrt{13} \times \sqrt{13}$)R33.7°.[27-40]

The referred X-ray diffraction study had its focus on the surface of TiO$_2$-terminated SrTiO$_3$(001). A detailed structural analysis of this surface has been performed at room and high temperature (750 °C). Several structural models have been explored and the final structure obtained indicates the simultaneous coexistence of a relaxed (1×1) structure and (2×1) and (2×2) reconstructions. The structures disappeared upon heating, resulting in a (1×1) structure very similar to the one observed at room temperature. The final structural model indicates a weighted mix of the three phases. The three structures have a characteristic double TiO$_2$ top layer in common. The two reconstructed phases contain a zigzag repetition of a motif, in agreement with a previous study by Erdman et al.[35] Figure 7.7 presents a comparison between experimental data and the best fit results for two SSRs and two CTRs. This work is a good example of a surface structural determination of a very complex surface by GIXS. However the relation between structure and physical properties of the interesting TiO$_2$-terminated (001) surface of SrTiO$_3$ has not been explored.

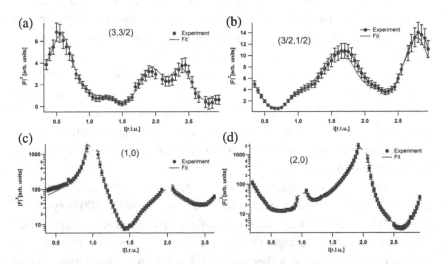

Fig. 7.7. Experimental and best fit surface X-ray diffraction data for the SrTiO$_3$(001) surface [75]. (a) and (b) are SSRs for (3, 3/2) and (3/2, 1/2) respectively. (c) and (d) are CTRs for (1, 0) and (2, 0) respectively.

7.2.2. *Electron Microscopy*

Electron microscopy uses electron beam to probe the material and produce magnified images. Although transmission electron microscopy/ transmission electron diffraction (TEM/TED) is used to probe bulk or interface lattice structure, it can be applied for determining the surface structure. This occurs due to their ability to allow for the capture of both real and reciprocal space crystallographic information from a single area of the specimen at atomic scale resolution. High resolution electron microscopy (HREM) achieves a resolution of less than 50 pm by using aberration-corrected optics,[41] which is sufficient to provide atomic scale information of surface structure by direct imaging. The experimental data clearly show the 47 pm spacing of Ge<114> with a sub-50-pm electron probe at 300 kV.

Generally speaking, the transmission electron microscope operates on the same basic principles as the light microscope but uses electromagnetic lenses to control the electrons instead of light. The focused electron beam is first diffracted by the selected area of the specimen in the transmission process, then the electromagnetic lenses refocus the beam into a Fourier-transformed image. After the post-experimental processing and simulation, the crystallographic information is formed in the magnified images. An excellent review of the historical development of TEM by Poppa can be accessed in Ref 42.

In a well-known example, TEM in combination with TED was used to determine the surface structure of Si(111)-7×7 by Takayanagi *et al.*, one of the most complicated and intriguing problems in the history of surface science,[43,44] in 1985. Since the first observation of the reconstruction by LEED in 1959,[45] various models had been proposed but without support of conclusive evidence because the analysis in LEED is not always without ambiguity and never without elaborate model-dependent calculation due to the strong multiple scattering, particularly in the case of a large number of atoms within the unit cell. By evaluating intensity distribution of the diffraction spots in the TED patterns, as shown in Fig. 7.8(a), Takayanagi and co-workers performed a comprehensive structural analysis of the surface and proposed the Dimer-Adatom-Stacking fault (DAS) model. The model basically consists of

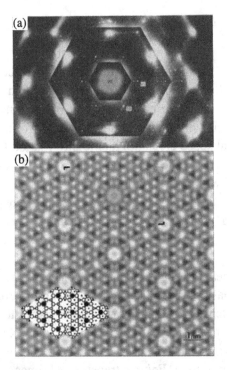

Fig. 7.8. The successful application of electron microscopy technique to solve the long-disputed surface structure of Si(111)-7×7. (a) Transmission electron diffraction of Si(111)-7×7 reconstructed surface taken by the almost $[111]_d$ incidence of the electron beam.[43] (b) Plan-view HRTEM image of Si(111)-7×7 surface after rotational and translationally averaging. The inset is an image simulation for a defocus of 236 nm. Atoms are black, with slightly darker features at the adatom locations where two atoms are superimposed.[52]

12 adatoms arranged locally in the 2×2 structure, nine dimers on the sides of the triangular subunits of the 7×7 unit cell and a stacking fault layer. In agreement with the results from GIXS,[46,47] LEED[45,48] and theoretical,[49,50] the DAS model was accepted widely and ended the long dispute which had existed among surface scientists.

In addition to successfully imaging the Si(111)-7×7 surface in real space with scanning tunneling microscopy (STM),[51] HREM is a powerful instrument able to probe many key features including not only the outermost surface but also the layers underneath which are too deep in the structure for STM. E. Bengu et al.[52] reported the results obtained using plan-view imaging coupled with noise reduction and numerical inversion to separate the top and bottom surfaces. The HREM images showed clearly not just the adatoms seen by STM but all the atoms in the top three layers, including the buried dimers in the third layer (Fig. 7.8(b)).

Through a combination of HREM and theoretical methods, Erdman *et al.* reported the determination of the $SrTiO_3$ (001) surface 2×1 structure as shown in Fig. 7.9.[35] In a diffraction experiment, the lost phase information usually prevents a direct Fourier inversion of the data. They

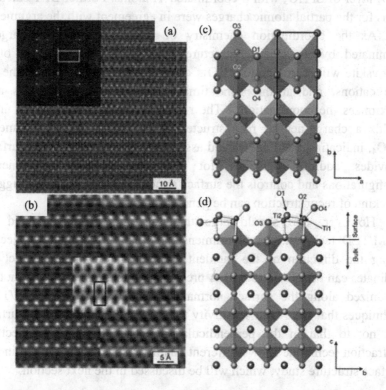

Fig. 7.9. The solution of the 2×1 $SrTiO_3$(001) surface structure. (a) High-resolution TEM image of the 2×1 $SrTiO_3$(001) surface with the inset showing the corresponding power spectrum. (b) Enlarged translation-averaged image of the 2×1 structure (left) and after filtering to separate translation/inversion domains on the top and bottom surface (right). The inset shows a multislice simulation of the structure with the unit cell marked. (c) and (d) Proposed model of the 2×1 $SrTiO_3$(001) surface structure ((c) Top view; (d) side view).[35]

solved the phase problem by exploiting probability relationships between the amplitudes and the phases of the diffracted beams. In HREM image (Fig. 7.9(b)) of a single domain after lattice averaging (left) and after filtering to separate translation/inversion domains of the structure (right), there is a zig-zag row of features with a local *p2mg* symmetry. A proposed structural model is shown in Fig. 7.9(c) and 7.9(d). The top layer has a TiO_2 stoichiometry, with half of the TiO_5 units displaced along the [110] direction. The TiO_5 units share edges between them as well as with the layer underneath, which is a part of the first full intact TiO_2 layer of $SrTiO_3$ with 6-coordinated Ti atoms. Further DFT calculations for the partial atomic charges were in agreement with the argument.

As the coordination chemistry of perovskite ABO_3 is largely dominated by its specific structure, they assume the surface of a perovskite will rearrange to a BO_{3-x} configuration, which screens the B-site cations, and at the same time minimizes surface dipoles and maximizes the coordination.[35] The retention of edge-shared TiO_5 units yields a charge-neutral reconstructed surface with the stoichiometry Ti_2O_4, indicating that edge-shared assembly of TiO_x units on the surface provides additional modes for rearrangement to lower-energy configurations and controls the surface structure of $SrTiO_3$. They suggest this kind of reconstruction can be general for perovskites.

However, one should note that the information obtained by TEM/TED techniques is two-dimensional (i.e. the structure projected along the direction of the incident beam). The in-plane atomic co-ordinates can be refined to 0.01 Å precision while the structure cannot be optimized along the surface normal.[53] In contrast to the TEM/TED techniques that have high sensitivity to movement parallel to the surface but not to that in the perpendicular direction, low energy electron diffraction technique shows different characteristics when applied in the surface structure study, which will be discussed in the next section.

7.2.3. Low Energy Electron Diffraction

The low energy electron diffraction technique[54,55] is perhaps the most reliable and suitable experimental approach for surface structure determination. The impinging low energy electrons strongly interact with the atoms in the surface. As a consequence their effective penetration depth is drastically reduced to a few atomic layers, turning LEED into a very surface sensitive method. The inelastic mean free path for pure copper,[56] electrons in the LEED energy range only penetrate 4 ~ 10 Å, which corresponds to a probing depth of only 2 to 5 atomic layers.

Figure 7.10(a) schematically shows the basics about the LEED experiment. Probing low energy electrons, typically in the 20 - 600 eV energy range, are emitted by the electron gun and reach the sample crystalline surface at defined polar and azimuthal incidence angles (Θ and Φ) and are scattered by the array of atoms at the top atomic

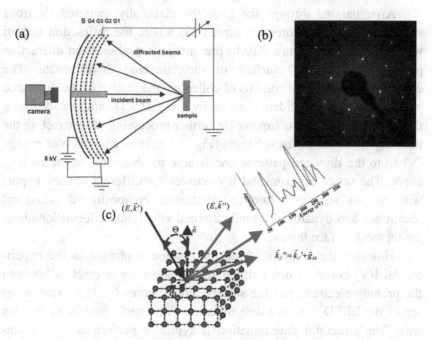

Fig. 7.10. (a) Schematic representation of a LEED experiment. Electrons emitted by the electron gun will reach the sample crystalline surface, then are scattered by the array of atoms at the top atomic layers. After passing through the grids the elastically scattered electrons will reach the phosphorescent screen. The diffracted patterns at different incident energies will be recorded by a camera. (b) A sample electron diffraction pattern for $Sr_3Ru_2O_7(001)$ surface. (c) The intensity of the diffracted spots collected as a function of energy [I(V) curve] will contain structural (atomic positions), compositional (chemical identities) and dynamical (atomic thermal vibrations) information about the topmost surface layers.

layers. The first metallic grid (G1) is grounded in order to provide a region free of electric field between itself and the sample (also grounded), i.e., along the electrons path. The two internal metallic grids (G2, G3) will act as a filter and select only the scattered electrons with the same energy as the incident beam (elastically scattered), which represent only 2 ~ 5% of all scattered electrons. G4 is also grounded in order to reduce the penetration depth of the electric field generated by G2 and G3. A positive high potential (~ 6 kV) is applied to the phosphorescent screen (S) in order to accelerate the electrons and produce a clear impression on the screen. The set of 4 metallic grids constitute the so-called Retarding Field Analyzer (RFA).[54]

After passing through the grids the elastically scattered electrons will reach the phosphorescent screen (S) where the diffraction pattern will be visualized. Figure 7.10(b) presents an actual electron diffraction pattern for the (001) surface of the bilayered Sr ruthenate. The experimental procedure consists of collecting the images of the diffracted patterns at different incident energies by using a CCD camera and storing this digitized data in a computer for further processing. By extracting the intensity (I) of the diffracted spots (Fig. 7.10(c)) as a function of energy (V) from the digitized patterns one is able to obtain the so-called I(V) curve. The set of experimental I(V) curves for different scattered spots will contain structural (atomic positions), compositional (chemical identities), and dynamical (atomic thermal vibrations) information about the topmost surface layers.

However the analysis of the information contained in the experimental I(V) curves is not a simple task. The strong interaction between the probing electrons and the surface atomic cores (multiple scattering) forces the LEED structure determination to be performed in an indirect way. The structural determination is typically performed by a quantitative comparison of experimentally collected curves of the intensity from diffracted spots as a function of incident electron energy [I(V)] with theoretically calculated ones (multiple scattering formalism)[54,55,57] for a proposed surface structural model. The electron scattering by the atomic core is modeled within the muffin-tin (MT) potential scenario.[57] In this scenario non-overlapping spheres (potential wells) are centered on the atomic positions (ion cores). In the interstitial region among spherical wells, the potential is assumed to be constant. Potential continuity at the atom-centered spheres and interstitial region interface is enforced. Exchange contribution to the potential is calculated by employing the Slater exchange approach in the calculations.

Due to the spherical nature of the MT approach the electron-atom scattering is fully described by the use of the phase shifts (partial wave method). Further electron multiple scattering inside one atomic plane and between atomic planes is the next step in the theoretical calculations leading to the final scattered intensities. Effects of atomic thermal vibrations are typically taken into account in the calculations using an isotropic Debye-Waller factor.[57]

The theory-experiment comparison of I(V) curves is made quantitative by the reliability factor (R-factor or Pendry R-factor). The lower final R-factor achieved, the more reliable is the structural determination. The surface structure determination by LEED turns into a search problem in which one needs to explore a parameters space (surface structural parameters, i.e., atomic positions) and find the global minimum of the R-factor. This minimum will ideally correspond to the actual surface structure. Figure 7.11 presents schematically a flowchart for the LEED structural search and optimization. As it can be inferred by the presented flowchart, the core part of the LEED structural analysis (green blocks) correspond to the optimization of search process, which involves the calculation of theoretical I(V) curves for different surface structural models and subsequent comparison with experimental data.

Historically LEED has mainly been applied to simple metal and semiconductor surfaces. It was challenging when LEED was used for the study of TMOs. The big surface unit cell, strong ionic character or surface polarization and strong electron correlation presented in TMOs in contrast with simple metals or semiconductors are the main obstacles for the surface structural determination of TMOs with LEED analysis. From the theoretical point of view the existing theory needs improvement in order to be more effectively applied to the complex TMOs. Usually the final theory-experiment for LEED on complex oxide is not as good as in the case of simple metallic surfaces. A direct consequence is a drastic reduction in the reliability and accuracy of the final structural results. One reason for this poor theory-experiment agreement is that the effects

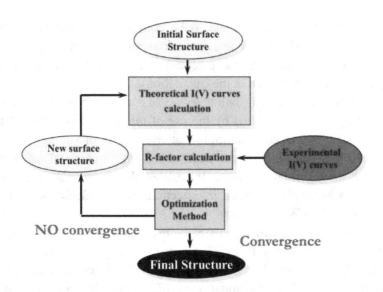

Fig. 7.11. Flowchart of the LEED surface structure determination process. From an initial surface structure, the theoretical I(V) curves will be calculated.Compare the theoretical I(V) curves and experimental ones, then the R-Factor of the initial structure can be calculated, which is used to judge the reliability of the initial surface structure. If a surface structure with reasonable R-Factor can be optimized, the final structure will be determined. Otherwise go back to the beginning with a new surface structure.

of the different ions must be taken into account in the theoretical calculations. A recent work has proposed a new effective method for LEED analysis of TMOs,[58] based on the optimized muffin-tin approach.[59] Within this approach the MT radii are energy dependent in order to assure potential continuity at interface between the spheres and interstitial potential. By adopting this procedure one is able to avoid undesired scattering resonance features in the phase shifts. Another difference from the typical MT approach is that the interstitial potential, related to the real part of inner potential (V_{OR}) as defined in LEED, is energy dependent due to the adoption of a Hedin-Lundquist or Sernelius[59] exchange-correlation approach. The optimized MT method offers a more realistic description of the electron scattering by the atomic species, and leads to an improved final theory-experiment agreement and consequently improved accuracy and reliability. This enhanced accuracy is extremely crucial in the structural investigation of surfaces of

TMOs, in which even a tiny change can lead to very dramatic changes. Some of the LEED structural investigations of TMOs surfaces described in this article have only been possible by adopting this new approach.

7.2.4. *Scanning Tunneling Microscopy*

Scanning tunneling microscopy is a well-developed technique for investigation of surface properties in the last three decades. As a direct imaging method in real space, the resolution can reach to 1 Å in-plane and 0.1 Å out-of plane fairly in vacuum.[60] In addition to imaging the surface topography with atomic-scale resolution, it allows one to probe the local density of states (LDOS) of a material on its surface. This experimental technique is called scanning tunneling spectroscopy (STS). Furthermore, the topographic and spectroscopic data can be recorded simultaneously in the experiments, which provide rich information of the relations between the surface crystallographic and local electronic structures.

The principle of scanning tunneling microscopy is based on quantum tunneling of electrons between two electrodes, a sharp metallic tip and a conducting sample, separated by a thin insulating barrier, usually vacuum. Figure 7.12 illustrates schematically how the tunneling

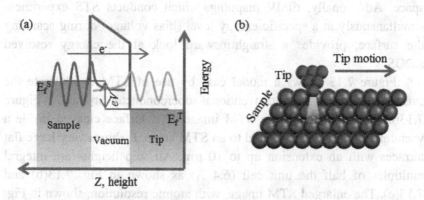

Fig. 7.12. Principle of STM. (a) Schematic energy level diagram illustrating the tunneling process between the tip and sample across a vacuum barrier. With a negative bias voltage *V* applied to the sample, electrons are attracted and tunnel preferentially from the sample into the tip. (b) Schematic illustration of the operation of STM.

process occurs during the operation of STM. An atomically sharp tip is located on the surface with several Å. Applying a bias voltage between the conducting sample and metallic tip results in a measureable tunneling current across the gap. The tunneling current is a function of tip position, applied voltage, and the local density of states of the sample. An electronic feedback loop is connected to maintain the tunneling current constant by adjusting the tip height. The tunneling current is extremely sensitive to tip-to-sample distance, which enable STM to achieve remarkable spatial resolution during scanning the surface, as shown in Fig. 7.12(b). When the metallic tip is scanning the surface, the variations in surface height and density of states causes changes in tunneling current, then the surface topography will be recorded and mapped out. The bias voltage determines the electronic levels involved in the tunneling process both filled and empty states depending on the voltage polarity. It should be noticed that STM is not a real structural tool. In principle what STM observes is the symmetry of the electron distribution but not the geometric structure.

Scanning tunneling spectroscopy, the information on the electronic structure in the local region on the surface, can be obtained by sweeping voltage and measuring current at a specific location with the feedback loop off. The advantage of STS over other surface measurements of density of the states lies in the extremely high resolution of STM in real space. Additionally, dI/dV mapping, which conducts STS experiment simultaneously at a specific energy level (bias voltage) during scanning the surface, provides a straightforward look at the energy resolved LDOS.

Figure 7.13 shows a model case by use of STM to illustrate the surface structure of the unconventional superconductor Sr_2RuO_4.[61] Figure 7.13(a) shows a large scale STM image of a surface cleaved inside a vacuum system and transferred to an STM stage. It shows very large flat terraces with an extension up to 10 μm. All step heights are integral multiples of half the unit cell (6.4 Å) as shown in Fig. 7.13(b) and 7.13(c). The enlarged STM image, with atomic resolution, shown in Fig. 7.13(d) indicates that the surface is not bulk truncated but reconstructed into a $(\sqrt{2} \times \sqrt{2})R45°$ structure. The measured and calculated surface structure is shown in Fig. 7.13(e). The octahedral in the surface plane are

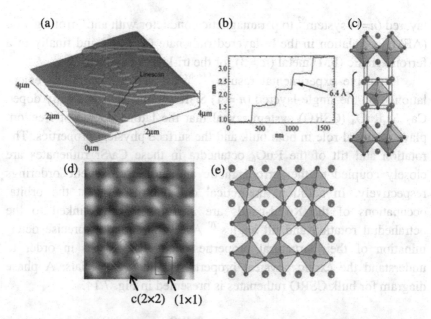

Fig. 7.13. (a) Large scale STM image of Sr_2RuO_4 cleaved in vacuum. It shows very large flat terraces with and extension up to 10 μm. (b) Height along the line scan shown in the STM image (a). (c) Ball model of the bulk unit cell of Sr_2RuO_4. Red, strontium; blue, oxygen; and green, ruthenium. (d) High resolution STM image of 26 Å × 26 Å containing 7 × 7 strontium sites. The bulk (1 × 1) (blue) and surface c(2 × 2) reconstructed unit cell (green) are indicated. (e) Ball model of the surface structure with rotated octahedra (top view).[61]

rotated alternating clock- and anticlockwise around the surface normal direction. The reconstructed surface has significant impact with the surface electronic structure in Sr_2RuO_4, which will be discussed in the next section.

7.3. Examples: The Interplay of Structure and Functionality

7.3.1. *Single-layered Ruthenate RP Series: $Ca_{2-x}Sr_xRuO_4$*

A myriad of electronic and magnetic properties are observed in the $Sr_{n+1}Ru_nO_{3n+1}$ ($n = 1, 2, 3,...$) ruthenate Ruddlesden-Popper (RP) series (shown in Fig. 7.1(a)). These materials present a wide range of exotic properties, ranging from a diamagnetic superconductor in the single-layered (n=1) system[62] to paramagnetic conductor with antiferromagnetic (AFM) correlation in the bi-layered ruthenate ($n = 2$)[63] and finally to a ferromagnetic (FM) metal ($n = 3$) for the tri-layered system.[64,65]

Extensive experimental results[61,66-69] as well as theoretical calculations[13] on the single-layered ($n = 1$) Sr_2RuO_4 and the isovalently doped $Ca_{2-x}Sr_xRuO_4$ (CSRO) system reveal that the lattice degree of freedom plays a critical role in both bulk and the surface physical properties. The rotation and tilt of the RuO_6 octahedra in these Ca/Sr ruthenates are closely coupled to the ferromagnetic and antiferromagnetic orderings respectively. In addition, theoretical calculations suggest the orbital occupations of the Ru d-bands are also intrinsically linked to the octrahedral rotation and tilt angles.[70] As a consequence, precise determination of their structural properties is a crucial step in order to understand the exotic physical properties of these materials. A phase diagram for bulk CSRO ruthenates is presented in Fig. 7.14.

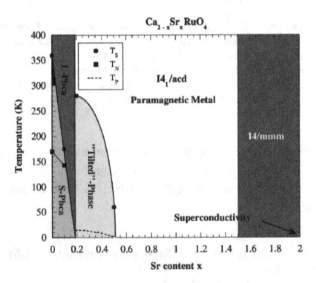

Fig. 7.14. Bulk phase diagram of $Ca_{2-x}Sr_xRuO_4$ including the different structural and magnetic phases. T_S is the structural transition temperatures determined by neutron diffraction. T_N is obtained from temperature dependence of ordered magnetic moment obtained by refinement of the B-centered antiferromagnetic structure of La_2NiO_4-type in $Ca_{1.9}Sr_{0.1}RuO_4$. Decrease of the Sr content below $x = 0.5$ causes a maximum in the temperature dependence of the susceptibility at T_p. All phases are metallic with exception made for S-$Pbca$.[67]

In the case of the RP layered ruthenates, as previously mentioned, a clean crystallographic surface can be obtained by cleaving bulk crystals. Intuitively one can expect that breaking the translational symmetry in a material with strong coupling between the lattice, electron, and spin can induce new functionality. The question arises whether the surface properties result from a simple shift of the bulk phase diagram, or a more severe distortion of bulk phases and properties. This section focuses on this question and shows the later scenario to be the case as the (001) surface of the single-layered $Ca_{2-x}Sr_xRuO_4$ ruthenate exhibits properties that are beyond any simple shift of the bulk phase diagram.

The bulk $Ca_{2-x}Sr_xRuO_4$ system presents an intricate coupling between lattice, electron, and spin degrees of freedom.[67,71,72] Substitutional doping of Ca^{2+} for Sr^{2+} in CSRO enhances both RuO_6 rotational and tilt distortions. Starting from a tetragonal structure (*I4/mmm*) for Sr_2RuO_4, a RuO_6 rotation is induced as Ca^{2+} is increased evolving to an *I4₁/acd* structure for $Ca_{1.5}Sr_{0.5}RuO_4$, and further increases in Ca^{2+} concentration induces an RuO_6 tilt resulting in an orthorhombic *S-Pbca* structure for Ca_2RuO_4.[67] A direct consequence of these structural changes is a gradual evolution of the ground state, starting from an unconventional superconducting state in Sr_2RuO_4[62] to a quantum critical point at $x_c = 0.5$ and to an antiferromagnetic Mott insulating phase for $x < 0.2$.[71,72]

To understand the implications of breaking translational symmetry at the surface, surface structural phases of $Ca_{2-x}Sr_xRuO_4$ were investigated using quantitative LEED-IV analysis.[69] Results obtained from this analysis show that the broken symmetry at the surface enhances the structural instability against the RuO_6 rotational distortion and diminishes the instability against the RuO_6 tilt distortion observed within the bulk crystal. For Sr_2RuO_4, an octahedral rotation is induced when the crystal is cleaved with a rotation angle similar to that observed in the bulk *I4₁/acd* phase for $x < 1.5$ as shown in Figs. 7.13 and 7.15. The surface structural analysis also reveals that the tilt instability is suppressed at $x = 0.2$ as it is not present on the surface at room temperature despite the fact that it is observed in the bulk. For $x < 0.2$, an octahedral tilt is observed on the surface at room temperature at a slightly enhanced angle from its bulk counterpart lying beneath the surface.

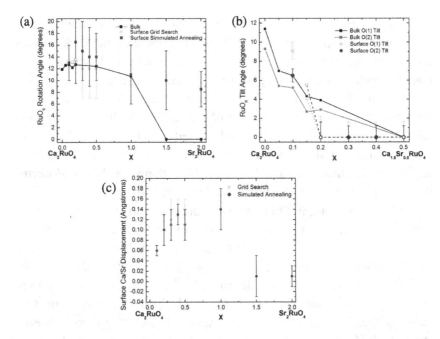

Fig. 7.15. LEED final structural results for $Ca_{2-x}Sr_xRuO_4$ at T = 300 K for different x. The closed squares and solid lines are from bulk powder data while the open squares and dashed/dotted lines are from bulk single crystal data. (a) RuO_6 rotation angle. (b) RuO_6 tilt angles. (c) Top Ca/Sr layer inward displacement.

In addition to the modifications of the RuO_6 rotation and tilt behavior, an inward motion of the topmost Ca/Sr layer is encountered when the smaller Ca^{2+} concentration is increased as shown in Fig. 7.16(b) and 7.16(c). Remarkably, the refined structure for the (001) surface of $Ca_{1.9}Sr_{0.1}RuO_4$ involves the inward motion of the top Ca/Sr ions (surface relaxation), which results in a static buckling in the surface Ca/Sr-O(2) layer of 0.23 Å. This buckling is schematically presented in Fig. 7.16(b). A straightforward and intuitive explanation for this compression of the surface layer is as follows. After cleaving the crystal the oxygen above the top Ca/Sr are missing due to the broken translational symmetry. Consequently the top Ca/Sr is pulled inward by the attractive forces from the oxygen below. This type of surface relaxation has been already observed for surfaces of perovskites like $SrTiO_3$,[73] $BaTiO_3$, and $PbTiO_3$.[74] At the surface, the broken translational symmetry causes a

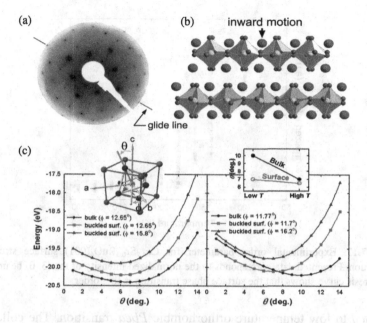

Fig. 7.16. (a) LEED image of $Ca_{1.9}Sr_{0.1}RuO_4(001)$ taken with an electron beam energy of 170 eV. The RuO_6 octahedra tilt distortion is evident by the existence of only one glide line (on which some spots are extinct) instead of two perpendicular glide lines as expected from a bulk truncated surface. (b) Schematic side view of the determined surface structure, presenting the large inward motion of Ca/Sr ions. (c) Calculated total energy as a function of RuO_6 tilt angle (Θ) for three different lattice structures: i) bulk room temperature structure (red symbols); ii) bulk low temperature structure (black symbols); iii) and surface structure (blue symbols). RuO_6 octahedron rotation and tilt distortions are schematically presented in the inset.

compression stress that increases the buckling of the Ca/Sr-O top surface plane as compared to the bulk.

While the modified RuO_6 angles and surface relaxation suggest a slight shift of the bulk phase diagram on the surface, temperature dependent data reveal more striking deviations from bulk behavior. Due to the octahedral rotation presenting in the *I4₁/acd*, a glide-plane symmetry is present and consequently the odd integer diffraction beams in the (0, n) and (n, 0) directions are extinguished along all energy range. When the RuO_6 tilts (*Pbca*) the glide symmetry is broken and the odd integer beams appear. As a consequence the intensity of the (3, 0) beam can be used as an order parameter for the high temperature tetragonal

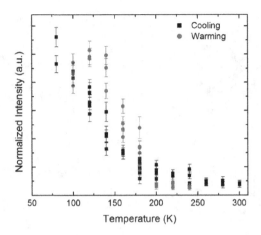

Fig. 7.17. Experimental order parameter for $Ca_{1.7}Sr_{0.3}RuO_4(001)$ surface structural transition. Presented data corresponds to the normalized intensity of the (3, 0) beam. The hysteresis curve shows that the surface phase transition is first order.

$I4_1/acd$ to low temperature orthorhombic *Pbca* transition. The collected intensity was normalized with another diffraction beam in order to eliminate the temperature effects (Debye-Waller). While the bulk phase transition is a smooth second order transition, the surface phase transition is first order as shown by the hysteresis curve in Fig. 7.17. While the surface phase transition occurs at a lower temperature than as observed in the bulk, it should also be noted that instabilities of the tilt are observed at temperatures well above the bulk transition temperature. It is postulated that the inward motion of the topmost Ca/Sr ions interferes with the tilting RuO_6 resulting in significant deviations from bulk behavior. While the general trend is to suppress the low temperature orthorhombic phase by suppressing the RuO_6 tilt, more profound modifications to the bulk phase diagram are observed at the bulk quantum critical point (QCP) at $x_c = 0.5$ where the tilt instability is observed on the surface. These results suggest a shift of the surface QCP to higher x, but more work is needed to determine the actual location of the surface QCP.

Figure 7.16(a) presents a typical LEED image for the (001) surface of $Ca_{1.9}Sr_{0.1}RuO_4$. The symmetry and the existence of a single glide line in the pattern (red line) indicates an unreconstructed p(1 × 1) [001] surface of a bulk terminated orthorhombic structure (*Pbca*). Static tilt and rotational distortions of the RuO_6 octahedra differentiate this orthorhombic structure from a simple cubic perovskite one. Structural analysis of the LEED data shows that surface structure remains unchanged across both the bulk and surface metal-insulator transitions, but for a gradual thermal relaxation. The final structures obtained by LEED analysis performed at 90 and 300 K show basically no difference.

Theoretical calculations have been performed in order to cast some light into these results. First principle calculations have reproduced the two stable bulk lattice structures as shown in Fig. 7.16(c): i) one with a RuO_6 tilt angle of about 6.5° for the metallic phase; ii) and the other with a tilt angle of ~ 10° for the insulating phase. Further calculations have also been performed to simulate the situation of the surface by adopting an enhanced buckling of the Ca/Sr-O plane, but still in a bulk calculation. As it can be inferred by the results presented in Fig. 7.16(c) this enhanced buckling pins the RuO_6 tilt angle at around 7°, and makes the increased tilt energetically unfavorable and thus inhibits the structural transition. The theoretically obtained tilt angle is in excellent agreement with LEED final determined results.

The original scenario proposed for a Mott metal-insulator transition is that it is completely driven by electron-electron interactions, without any structural change. However, due to the coupling between lattice and electronic and magnetic properties these transitions are usually accompanied in bulk by structural phase transitions. The vanadium oxide V_2O_3 is considered as the one known classical example of a Mott metal-insulator transition.[75-77] The LEED results for $Ca_{1.9}Ru_{0.1}RuO_4$ show no change in the surface structure across the metal-insulator transition temperature and thus other techniques are required to investigate the surface MIT.

One technique to be brought in for the study of surface properties of these TMOs is high resolution electron energy loss spectroscopy (HREELS). Probing the excitation spectra is critical for understanding the evolution of electron correlations, especially MIT in correlated electron materials.[78] For example, measuring the phonon and electron-hole pair excitations and determining the form of energy gap provide important information for revealing the nature of MIT. HREELS provides exactly the information of surface dielectric response function $\varepsilon(q,\omega)$, including collective excitations such as phonons and other quasi-particles,[78] thus probing electron-electron and electron-phonon interactions. In general, HREELS is similar to optical conductivity spectroscopy. However, the difference is that HREELS is quite surface sensitive, collecting information normally from the top 10 ~ 20 Å of a solid. Moreover, it has both energy and momentum resolution, thus providing valuable information of electronic band structure, quasi-particle and collective excitations of a material. HREELS has been proven to be very useful in studying surface lattice dynamics and MITs on both conventional surfaces.[79,80]

High-resolution electron energy loss spectroscopy measurements were performed with well-characterized single $Ca_{1.9}Sr_{0.1}RuO_4$ crystals over a temperature range from 300 K to 80 K. Incident electron energy is 20 eV and all data were taken in a specular arrangement with an incident angle of 45° and instrument resolution of 5 ~ 7 meV. Since the analyzer is fixed at the angle of specular reflection, only dipole active modes with atomic displacements perpendicular to the surface can be detected. This surface selection rule gives the mode assignment as shown in Fig. 7.18. As a result of the rough surface, instrument spectral count rate and resolution is reduced. Therefore a smaller sample ~ 3 mm is used for its smoother surfaces with fewer observable terraces created during the cleaving process.

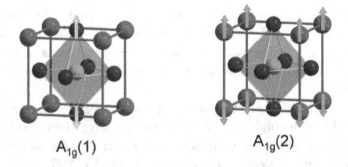

Fig. 7.18. Two surface dipole active modes along the direction perpendicular to the surface for K_2NiF_4 structure.

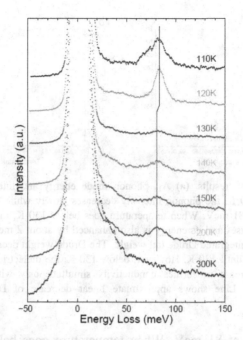

Fig. 7.19. Normalized HREELS data at different temperatures. The measurements are taken at specular angle, and the line indicates the phonon shifts with decreasing temperature. The change of phonon energy and intensity is clear at 130 K.

Typical HREELS data at different temperatures are shown in Fig. 7.19. Through the temperature-dependent spectra, the change of surface conductivity can be measured by tracking the spectral weight in the Drude tail. There are several models which can accurately describe the spectrum and are used to determine phonon dynamics and extract the spectral weight in the Drude tail.[78] Once the spectral peak and phonon are determined, they are subtracted from the spectrum data. The remaining inelastic data is an accurate representation of the spectral weight in the Drude tail and thus integrated for a qualitative description of the surface conductivity. Although quantitative conductivity measurements can be inferred from the data, we are only interested in the changes in conductivity over the temperature range of interest.

Figure 7.20(a) shows the change of phonon energy and intensity as a function of temperature. From room temperature to 130 K, it is found that the phonon intensity decreases slowly while the phonon energy

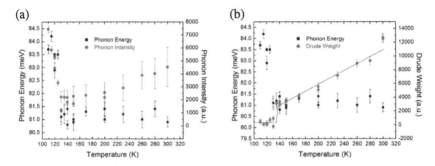

Fig. 7.20. HREELS results. (a) A_{1g} phonon mode energy and intensity. From room temperature to 130 K, the phonon intensity decreases slowly while the phonon energy keeps constant at 81 meV. When temperature goes below 130 K, the phonon intensity significantly increases and its energy is also enhanced by about 2 meV. (b) A_{1g} phonon mode energy and integrated Drude tail weight. The Drude weight decreases linearly with temperature down until 130 K. However, below 130 K, the dielectric response shows a discontinuous decrease in surface conductivity simultaneously with the changes in phonon character. Line shows approximate linear decrease of Drude weight with temperature.

keeps constant at 81 meV. When temperature goes below 130 K, the phonon intensity significantly increases and its energy is also enhanced by about 2 meV. Moreover, the phonon linewidth also has an obvious change from 20 meV to 7 meV at around 130 K. The evolution of surface conductivity as a function of temperature measured by HREELS is shown in Fig. 7.20(b). The Drude weight decreases linearly with temperature down until 130 K, as expected with a Drude metal. However, below 130 K, the dielectric response shows a discontinuous decrease in surface conductivity simultaneously with the changes in phonon character. All the results indicate that a distinct transition is observed on the $Ca_{1.9}Sr_{0.1}RuO_4$ surface.

The sudden decrease in surface conductivity below 130K is consistent with the Mott MIT observed in the bulk. Such similar behavior is also observed in Ca_2RuO_4 system for the A_{1g} phonon vibration mode across the bulk MIT.[81] Raman also shows the enhancement of phonon intensity and a shift of ~ 2 meV in the B_{1g} phonon across the MIT transition temperature.[82] However, the breaking of symmetry at the surface could distort the lattice stabilizing the metallic phase and creating a MIT which is ~ 20 K lower than the bulk MIT. In addition, an opening up of the insulating gap is also observed in scanning tunneling spectroscopy data. Thus a surface MIT with transition temperature ~ 20 K lower than the bulk is suggested by combination the HREELS and STS data.[68] The surface MIT on $Ca_{1.9}Sr_{0.1}RuO_4$, with T_c ~ 20 K lower than the bulk and without an accompanying structural transition presents a rare glimpse into the purely electronically driven MIT as first proposed by Mott.

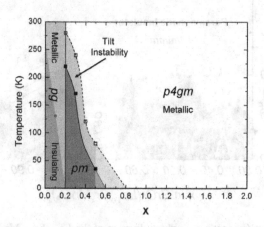

Fig. 7.21. Surface phase diagram for $Ca_{2-x}Sr_xRuO_4(001)$. The dashed line corresponds to the temperature where a tilt instability is observed and the solid line to the *p4gm-pm* structural phase boundary. No solid line is present between the metallic and insulating phases for $x < 0.2$ since no structural phase transition accompanies the surface MIT.

The surface phase diagram, presented in Fig. 7.21, shows that the surface structural and electronic properties of the CSRO family cannot be achieved by a simple shifting of the bulk phase diagram. Modification of the QCP near $x_c = 0.5$, realization of the first order phase transition for $x = 0.3$, and the creation of a purely electronic MIT without coupling to a lattice distortion demonstrates profound changes to physical properties and functionality at the surface of the Ruthenate compounds.

7.3.2. *Manganites $La_{2-2x}Sr_{1+2x}Mn_2O_7(001)$*

Varied magnetic and electronic states are presented by $La_{2-2x}Sr_{1+2x}Mn_2O_7$ system in its bulk form.[10,83] The Sr/La mixture results in a Mn mixed valence state (coexistence of Mn^{+3} and Mn^{+4}). Different magnetic orderings can be achieved by changing the Sr/La ratio, with the presence of ferromagnetic, paramagnetic, and antiferromagnetic phases. A bulk phase diagram for these bilayered manganites is presented in Fig. 7.22 as taken from reference.[83]

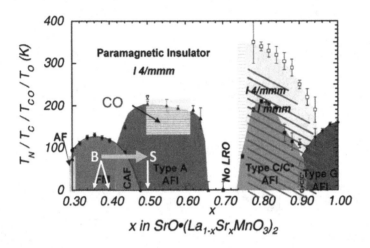

Fig. 7.22. Structural/magnetic bulk phase diagram of the $La_{2-2x}Sr_{1+2x}Mn_2O_7$ in the range $x = 0.3$ to 1.0 as determined by neutron powder diffraction. Solid markers are used to represent the magnetic transition temperature T_C or T_N. The structural transition from tetragonal to orthorhombic is delineated by open symbols. Ferromagnetic metal (FM), canted antiferromagnet (CAF), as well as A-, C-, and G-type antiferromagnetic insulators (AFI)magnetic phases are present in this phase diagram. No magnetic diffraction peaks are observed for temperatures down to 5 K in the "No LRO" region. "CO" stands for long-range charge ordering reflections observed by X-ray and/or electron diffraction. The change in the Jahn-Teller distortion value will act as a shift in the phase diagram to a higher concentration of Sr in the phase diagram, from $x = 0.36/0.40$ (FM) to 0.50 (AFM) and is schematically observed by the yellow arrow connecting B (bulk) to S (surface). In summary the surface layer would present an "effective" x equal to around 0.50.[83]

The (001) surface of $La_{2-2x}Sr_{1+2x}Mn_2O_7$ system, from $x = 0.36$ to 0.40 shows itself as a very interesting one. Characterization of the magnetic properties of the (001) surface ($x = 0.36$ to 0.40) indicates the existence of a non-FM surface bilayer on the top of a ferromagnetic bulk phase.[84] The surface magnetic state was probed by measuring the polarization-dependent magnetic reflectivity by using the X-ray

Resonant Magnetic Scattering (XRMS)[85] technique. The Mn L absorption edge presents chemical and magnetic scattering factors from the 3d electrons of similar magnitude.[86,87] Consequently scattering from an FM layer deeper down will interfere with scattering from the chemical interface and a corresponding change in the XRMS signal will be observed. By employing angle-dependent XRMS measurements [Fig. 7.23(a)], this interference was observed and the thickness of the nonmagnetic surface region was determined as corresponding to only the topmost bilayer.

Results obtained by point-contact tunneling indicate an insulating behavior for the top surface bilayer.[88] Both results seem to indicate an anti-ferromagnetic (and consequently insulating) phase at the surface. It is worthy to remark that in this system anti-ferromagnetism and insulating behavior are connected. This fact can be qualitatively explained within a simple double-exchange mechanism.[10,83] However, results from angle-resolved photoemission spectroscopy (ARPES) studies, although presenting conflicting results, clearly indicate a metallic state at the surface.[89-94]

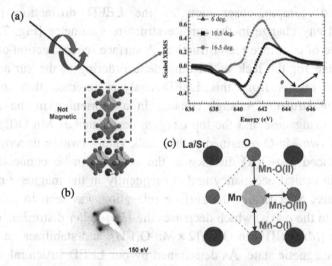

Fig. 7.23. (a) Schematic representation of the XRMS [75] experiment on the (001) surface of the La/Sr bilayered manganites. Obtained results indicate that the top bilayer presents no magnetic ordering. (b) LEED pattern collected at 150 eV, which shows no evidence for changes in symmetry at the surface. (c) Representation of the top half of the surface top bilayer, showing the three different Mn-O distances.

In order to understand the physics of the non-magnetic (001) surface of $La_{2-2x}Sr_{1+2x}Mn_2O_7$ a joint experimental and theoretical study has been performed.[95] As a first step a new surface magnetic property characterization was performed by using the XRMS technique. Contrasting to the first magnetic properties characterization this new investigation was performed under high-vacuum conditions in order to exclude any effect from surface contamination, which could explain the lack of magnetic ordering. In the newly obtained results a change of sign of the XRMS signal with incidence angle is clearly observed, as presented in the Fig. 7.23(a), and can be explained by the existence of a single non-magnetic surface bilayer on top of ferromagnetic bulk layers. No sign change would be observed if ferromagnetic ordering was present in the top surface bilayer.[84] These results are in agreement with the previous XRMS study performed *ex situ* and allowed to exclude surface contamination as a possible explanation.

The second step consisted on determining the structure of the (001) surface of the referred bilayered manganites by LEED.[54] Structural LEED analysis of (001) surface of $x = 0.36$ and 0.40 have been performed. Simple observation of the LEED diffraction patterns excluded any change in the surface structure symmetry [Fig. 7.23(b)], indicative of a surface reconstruction. A surface reconstruction could be responsible for the lack of ferromagnetic ordering at the surface. The structural results from this LEED analysis indicate that the only significant structural change consists in a shortening in the spacing between manganese and the top oxygen, designated as Mn-O(II), while the other two Mn-O remained at their bulk values (within uncertainties). This reduced Mn-O(II) distance at the surface can be connected to a change in orbital occupancy and consequently in the magnetic order at the surface. Specifically the surface relaxation was seen to produce a collapse in the c axis, which decreases the Jahn-Teller distortion, defined as $\Delta_{JT} = [\text{Mn-O(II)+Mn-O(I)}]/[2 \times \text{Mn-O(III)}]$, and stabilizes an A-type antiferromagnetic state. As determined by our LEED structural analysis the Mn-O(II) distance at the surface presents a value of (1.924 ± 0.06) Å, contrasting to a bulk value of 2.003 Å. According to the experimental results of the bulk structure[96,97] Δ_{JT} decreases from 1.034 at $x = 0.30$ (FM) to 1.011 at $x = 0.50$ (AFM). The change in magnetic order with increasing x results from the depopulation of the $3_z^2 - _r^2$ orbitals and is coupled to a decrease in Δ_{JT}[98]. This weakens the ferromagnetic double-exchange between the bilayers in favor of the t_{2g} AFM super-exchange, resulting in an A-type state with moments aligned ferromagnetically within the layer and antiparallel between the double layers. Using bulk values for Mn-O(I) and Mn-O(III) Δ_{JT} values equal to (1.0013 ± 0.0155) and (1.0044 ± 0.0181) are yielded for the (001) surface of $x = 0.36$ and 0.40 respectively. If a comparison with bulk is made it is possible to suggest an A-type AFM ordering on the surface. This change in the Jahn-Teller distortion at the surface plays the role of a shift to larger Sr concentration (~ 0.50) in the structural/functional phase diagram, as schematically presented in Fig. 7.22.

Theoretical calculations were performed on a bulk system as the third step of the investigation. The goal was to explore the structure-functionality relationship in the bilayered La/Sr manganite system and try to understand how the magnetic order correlates with Mn-O octahedron distortion at a fixed hole concentration (x). The calculations were performed within the FLAPW[99] density-functional approach to the electronic structure and properties of crystalline solids. The theoretical results for hole doping corresponding to $x = 0.40$, used as our initial structural parameters those of Mitchell *et al.*[96], are presented in Fig. 7.24. The total energy difference between AFM and FM configurations ($E_{AFM} - E_{FM}$) is presented as a function of the Mn-O(I)/Mn-O(II) ratio, which was varied by changing Mn-O(II) distance. Positive or negative values for this energy difference indicate a FM or AFM state, respectively. The LEED experimental values at the surface for the Mn-O(I)/Mn-O(II) ratio are presented in Fig. 7.24. As it can be seen in this figure the empirical Δ_{JT} surface values clearly indicate an AFM phase as the stable state.

This combined experimental and theoretical analysis has provided evidence for a nonmagnetic top surface bilayer in the $La_{2-2x}Sr_{1+2x}Mn_2O_7$ system, with x dopings of 0.36 and 0.40. Surface relaxation promotes changes in the Jahn-Teller distortion of the top half of the first bilayer. By comparing the surface Δ_{JT} with the bulk experimental results[96] and theoretical calculations, it was possible to explain the lack of magnetic order in the surface. This comparison may even suggest the existence of

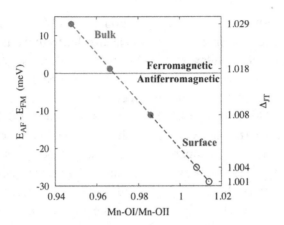

Fig. 7.24. Theoretical FLAPW results for bulk $x = 0.40$ system. The total energy difference between AF and FM configurations (E_{AF}-E_{FM}) is presented as function of the Mn-OI/Mn-OII ratio, which is varied by changing the Mn-O(II) distance. The associated Jahn-Teller distortion (Δ_{JT}) is also presented. The bulk experimental values for $x = 0.40$, as well as the surface values for $x = 0.36$ and 0.40 are presented in the graph.

an AFM (A-type) ordering which could not be confirmed experimentally. A very small contraction of the top Mn-O(II) distance is able to promote a dramatic change in the surface magnetic state.

7.3.3. *Iron-based Superconductors*

The discovery of high-temperature superconductivity in iron-based materials has triggered enormous excitement and activity reminiscent to the time when the first high-T_c superconductors were discovered 25 years ago in the material and physics community.[8] With the ferromagnetic element iron as potential building blocks, the first copper-free super-conductor with T_c reaching 55 K was completely unexpected.[100,101] So far there are six different families discovered in iron-based superconductors sharing a common layered structure based on square planar sheets of Fe in a tetrahedral environment with pnictogen or chalcogen anions.[9,102] Figure 7.1(b) and 7.1(c) show the bulk crystal structures of $BaFe_2As_2$ (Ba-"122" type) and $FeTe_{1-x}Se_x$ ("11" type). At the first glance, iron-based superconductors were advertised as the cousins of hight-T_c cuprates because of the similarities of the layered structure and proximity to magnetism. However, after three-years the investigations unveiled unique characteristics in the iron-based materials such as great chemical flexibility, small anisotropy in physical properties in spite of the layered structure, the metallic ground state, and multi-band electronic structures. Thus these new materials offer a new avenue to explore collective behavior in complex transition metal compounds.

In iron-based "122" and "1111" compounds, there is a clear indication of coupling between lattice and spin degrees of freedom: a structural transition from a high-temperature tetragonal to a low-temperature orthorhombic phase is accompanied by a magnetic transition with a narrow temperature window.[18] Further, the doping dependence of Fe ordered magnetic moment in $CeFeAs_{1-x}P_xO$ is tracking the bulk lattice orthorhombicity.[103] Superconductivity is intimately related to the angles and bonding of As-Fe-As tri-layer.[101] Application of pressure can drive some of the parent compounds into the superconducting state without chemical doping.[104]

Ba-based 122 system is the most studied of the six families so far with the use of surface techniques such as LEED, STM and ARPES,[105-107] and it is widely thought to capture the main traits of all iron-based superconductors. Layered structure makes these materials cleavable and produces an atomic flat and clean surface. Naively, the creation of a surface can be viewed as the application of uniaxial pressure. The single-crystal samples of $BaFe_2As_2$ can be easily cleaved and produce a clean (001) surface. Two types of ordered surface phases have been discovered. One is the 1×2 stripe phase and the other one is $(\sqrt{2} \times \sqrt{2})R45°$ phase. The LEED image shows a sharp 1×1 pattern (the inset of Fig. 7.27(a)) with reconstruction fractional spots missing probably due to the background created by the large area of disordered surface. STM observed both 1×2 and $(\sqrt{2} \times \sqrt{2})R45°$ (or orthorhombic (1×1)) ordered surface, as shown in Fig. 7.25. Interestingly, (1×1) orthorhombic unit cell domains and domain walls have been observed in Fig. 7.25(b). However, the imaged domain walls between different (1×1) regions display C_2 symmetry, namely the 45 degree boundary is not the same with the 135 degree boundary. It is in contrast to the geometric structure which has C_{2v} symmetry for the bulk-truncated surface. The lower symmetry actually reflects the symmetry of the spin

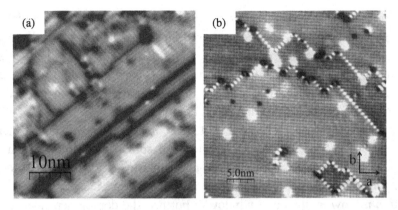

Fig. 7.25. Constant-Current STM topography of ordered surface of (a) 1 × 2 stripe phase, 50 nm × 50 nm, (V_{bias} = 1.0 V, I_{tip} = 200 pA) (b) ($\sqrt{2} \times \sqrt{2}$)R45° phase or orthorhombic 1 × 1 phase, 35.5 nm × 35.5 nm, (V_{bias} = 23 mV, I_{tip} = 200 pA) on (001) surface $BaFe_2As_2$ at 80 K. The blue and purple colors represent different domains with half-unit-cell shift in both a and b axis direction.

structure due to the strong orbital-spin coupling. The interesting question is how the surface structure has impacted or been impacted by the surface magnetism.

As stated before, the doping dependence of Fe ordered magnetic moment in bulk $CeFeAs_{1-x}P_xO$ is tracking the bulk lattice orthorhombicity,[103] suggesting a magnetic quantum critical point.[108] On the surface, there is a report that the surface orthorhombicity gets enhanced compared to the bulk.[109] S. H. Pan et al. also report that a surface rhombic distortion will take place of orthorhombic distortion when doping increases.[110] The surface phase diagram indicates a shifted surface QCP, as shown in Fig. 7.26. Does the enhanced surface orthorhombicity correspond to a large magnetic moment on the surface? Does the surface act as nuclei of bulk structural and magnetic transition? How does the quantum criticality affect the superconductivity? These questions still remain to be answered.

The HREELS measurements with the surface of $BaFe_2As_2$ were taken along ΓX direction of the Brillouin zone over a wide temperature range from 77 to 300 K across the Neel temperature. The Incident electron energy is 20 eV and the incident angle is 65°. The energy resolution is 2 - 4 meV. To avoid charging, the sample holder was coated

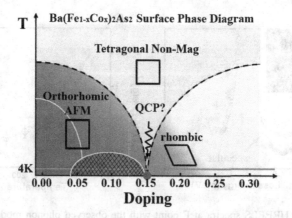

Fig. 7.26. Surface phase diagram for $Ba(Fe_{1-x}Co_x)_2As_2$. The y axis is the temperature axis but with no scale bar. 4 K is marked as a horizontal solid blue line. The yellow solid lines are the bulk phase boundaries. The crossed region is the bulk superconducting phase while the other yellow line region is the bulk orthorhombic antiferromagnetic phase. Dashed black lines define the three surface phase boundaries. The red region represents the surface orthorhombic antiferromagnetic phase (AFM). The yellow region represents the surface tetragonal non-magnetic phase. The blue region represents the surface rhombic phase.

with a solution of colloidal suspension of graphite. For the data fitting, as the Drude tail on the energy loss side causes the asymmetry of the elastic peak and also induces a Lorentzian part to the original Gaussian shape peak, which is supposed to be from the instrument, a Gaussian plus Lorentzian function (Voigt function) is used for the right side of the elastic peak. After subtracting the background, a multi-Lorentz function is used to fit the leftover part for the phonon loss peaks.

Among all the $BaFe_2As_2$ samples cleaved, there appear two categories of spectra with different loss peaks at the Brillouin zone center. On one category of spectrum, three resolved peaks are situated at 14 meV, 22 meV, and 32 meV respectively (Fig. 7.27(a)); on another category of spectrum, the peaks are resolved at 11 meV, 24 meV, and 37 meV correspondingly. Based on the results of the bulk $BaFe_2As_2$,[111] and also considering the HREELS selection rule, the loss peaks are assigned to the corresponding vibration modes. In the first spectrum shown in Fig. 7.27(a), peak I is the in-plane Ba stretching E_u mode. The two higher loss energy peaks are the dipole active modes along the

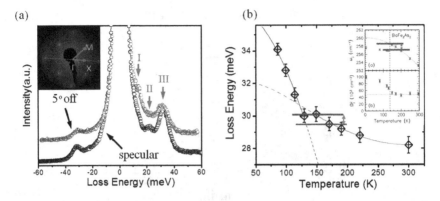

Fig. 7.27. (a) HREELS spectra at Γ point with the observed phonon modes at specular and 5° off on the Ba-dominated termination with peaks signed as I, II, and III. Inset: LEED image (energy: 113 eV, temperature: 77 K) showing 1 × 1 pattern. Peak I is the in-plane mode while peak II and III are the dipole-active mode with dipole moment normal to the surface. (b) Temperature dependent phonon frequency of Fe-Ba A_{2u} mode on the Ba dominated termination. As a guide to the eye, we fit the two different temperature regions using a mulit-phonon decay model. A phonon anomaly around the transition temperature T_N is highlighted by the red arrow. Inset: temperature dependence of frequency and intensity of the infrared-active mode in $BaFe_2As_2$ bulk observed at 253 cm^{-1}.[113] The dotted line represents the temperature dependence expected in the absence of a structural or magnetic transition.

surface normal. Peak II is identified as the A_{1g} mode, which is As breathing mode. Peak III is A_{2u} mode, a strongly mixed mode involving the out-of *ab*-plane vibration of both Fe and Ba atoms. For another category of spectrum, the two lower energy peaks are the dipole active modes while the higher energy peak is the in-plane mode of which the intensity increases as going out of specular. Peak I' is another A_{2u} mode, which involves the displacement of Ba and Fe atoms along z direction. Peak II' is the same A_{1g} mode as Peak II. Peak III' is assigned as the E_u mode with Fe and As atoms mixed vibrating in the *ab*-plane. Since $BaFe_2As_2$ is a layered compound, it may expose two terminations, Ba terminated and As terminated. The observed different phonon spectra relate to different surface structures and terminations. Comparing the two spectra, the first one shown here is determined to be the Ba-dominated termination, with a low energy pure Ba vibration and a very strong Ba-Fe vibration, while the second one should be the surface of the As-dominated termination, on which the A_{1g} mode of As vibration is much stronger with an exposed layer of As atoms in contrast.

Temperature dependent experiment is carried out for the out-of-plane Fe-Ba A_{2u} mode (peak III) on the Ba terminated surface. The main temperature effect is this Fe-Ba stretching mode shifts towards higher energy or hardening on cooling, which can be understood due to the contraction of the lattice with the decreasing temperature. The evolution of the phonon frequency as a function of temperature does not follow a simple anharmonic behavior.[112] Especially, upon closer inspection, we notice a phonon anomaly (a discontinuous shift in frequency) around the structural magnetic phase transition temperature T_N as marked by a red arrow in Fig. 7.27(b). This phonon signature clearly indicates the strong magneto-elastic coupling in this system. Compared with the IR result on $BaFe_2As_2$[113] presented as the inset of Fig. 7.27(b), the observed anomaly here is about 3% for the frequency change at T_N, much larger than the case of the bulk about 0.7%. This may suggest an enhanced ortho-rhombicity on the surface of the system. Now the conclusion for this part of phonon behavior is that both the change of phonon energy as well as the linewidth as functions of temperature cannot be simply explained by anharmonic effect. There is something behind: the strong effect of spin fluctuation with increasing temperature through spin-lattice coupling.

Furthermore, the temperature dependence of the linewidth change of the Fe-Ba A_{2u} mode was measured. The linewidth increases abruptly to the maximum value at T_N, which means a prolonged phonon lifetime at the phase transition point. But all the linewidths are larger than 10 meV, which may be due to the surface roughness. From the angle resolution plot, the intensity of the elastic peak decreases not too much but about a half with the incident angle changed as much as 3 degrees.

In contrast to the complexity on the surface of cleaved Ba-122 materials, the iron chalcogenide $FeTe_{1-x}Se_x$ is structurally and chemically the simplest in the all known iron-based superconductors.[114] As shown in Fig. 7.1(c), $FeTe_{1-x}Se_x$ is composed of iron-chalcogenide slabs stacked together without any spacing layer. The simplicity in the crystal structure of $FeTe_{1-x}Se_x$ provides a clear playground for surface techniques to perform on the exclusive termination after cleaving, the Te layer for FeTe or a mixed Te/Se layer for a doped compound. It was further affirmed by the observation of the large atomically flat terraces terminated by a step of rather single- or multiple-layer step height of bulk lattice constant c = 6.4 Å, as shown in Fig. 7.28(a).[115] What makes this family particularly interesting and unique is that the substitutional exchange of Te and Se is isovalent and the optimal superconducting transition occurs close to 50% mixture of Se and Te, while other compounds require only a small amount of doping for reaching the highest Tc.[116]

The high resolution STM images show the cleaved surfaces are 1 × 1 structure without any reconstruction in both cases of FeTe and FeTe$_{0.55}$Se$_{0.45}$. A very small corrugation of less than 8 pm is measured for

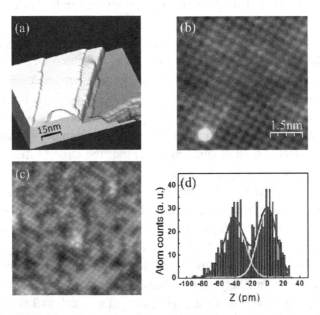

Fig. 7.28. (a) A large scale constant-current STM topographic image of the cleaved FeTe surface showing atomic steps of rather single- or multiple-layer step height of bulk lattice constant c = 6.4 Å. (b) A 59 Å × 59 Å high resolution filled-state STM topographic image of the FeTe surface (V_{bias} = -150 mV, I_{tip} = 500 pA). (c) A 111 Å × 111 Å high resolution constant-current STM topographic image of the FeTe$_{0.55}$Se$_{0.45}$ surface (V_{bias} = -150 mV, I_{tip} = 1.3 nA). (d) Histograms of atom heights on the FeTe$_{0.55}$Se$_{0.45}$ surface. The solid curves are the results by fitting to two Gaussian distributions. In each panel, the yellow (I) and green (II) peaks correspond to "bright" and "dark" atoms in the STM image (c), respectively, and the blue curve is the envelope sum of two peaks.[115]

the cleaved FeTe surface in Fig. 7.28(b). However, it is clear that there are two types of atoms forming irregular small patches on the $FeTe_{0.55}Se_{0.45}$ surface (Fig. 7.28(c)). A considerable height difference of ~ 47 pm between these two kinds of patches is observed from the STM topography. A statistical counting of the two kinds of atoms has the same ratio as that in the bulk (Fig. 7.28(d)). Thus, we are able to chemically identify the bright atoms as Te and the dark atoms as Se. The high-resolution STM images clearly reveal a nanophase separation between Te and Se atoms in $FeTe_{0.55}Se_{0.45}$, in sharp contrast with normally expected picture for a random alloy. From STM topographies, the average patch size is estimated to be ~ 1 nm^2 which contains 9-10 atoms on the surface. The local chemical inhomogeneity in the Fe-Te/Se layer of $FeTe_{0.55}Se_{0.45}$ induces a significant height difference between Te and Se atoms, leading to larger surface corrugation compared to undoped FeTe surface.

Remarkably, there is no electronic phase separation seen in the tunneling spectroscopy on $FeTe_{0.55}Se_{0.45}$ surface as shown in Fig. 7.29, which reveals a surprising feature that inhomogeneous chemical distribution gives rise to homogeneous electronic behavior. What has been expected for many doped CEMs is just the opposite: chemical homogeneity but electronic and magnetic phase separation.[117] The homogeneous electronic properties in normal and superconducting states[118,119] in this compound with nanoscale chemical phase separation

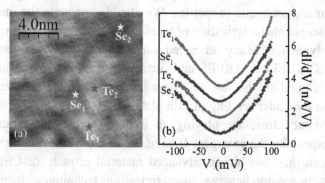

Fig. 7.29. (a) A 128 Å × 128 Å constant-current STM topographic image of the $FeTe_{0.55}Se_{0.45}$ surface. (b) Tunneling spectra taken at several Te and Se sites indicated in (a) show homogeneous electronic behavior on the chemically inhomogeneous surface.[115]

must be a consequence of the fact that the sizes of chemical patches are smaller than the superconducting coherence length in the iron chalcogenide materials.

7.4. Conclusion

The complexity of CEMs, owning to their fundamental character of the subtle coupling between charge, lattice, orbital and spin degrees of freedom, is responsible for the tenability of their functionality. Creating a surface is a controlled approach to disturb these coupled systems and possible generating unexpected properties. In the past few years, it is becoming increasingly clear that surfaces, interfaces, thin films, and heterostructures of CEMs display a rich diversity of fascinating properties that are related to, but not identical to, the bulk phenomena. Correlated electron devices will involve the fabrication of thin films, superstructures and junctions. Virtually all electronic devices began with an understanding of interface barrier formation, electronic/magnetic structure, and control — "the interface is the device".[4] Knowledge of the surface/interface properties as well as the effects derived from broken symmetry and spatial confinement is essential if these devices are to be made to work at optimized functionality.

Understanding the surface properties begins with the understanding of surface structure. In this review, we have shown that the surface structural study is an essential step toward the understanding of surface electronic and magnetic properties. We also demonstrate the importance of microscopic characterization of local lattice and electronic structure as well as dynamics. Many advanced surface techniques such as LEED, STM/STS, HREELS, ARPES and surface X-ray scattering, which have been developed in the last few decades and provided an tremendous impact on the understanding of "simple" materials such as simple metals and semiconductors, are playing an important role in the studies of CEMs, especially the emergent filed of the oxide heterostructures. It can be foreseen that, combining advanced material growth for CEM architecture with comprehensive characterization techniques, many unexpected functionalities in the proximity of surface/interface of CEMs will be expected to be discovered.

Acknowledgments

This work was founded by the National Science Foundation through grant NSF DMR-1002622 and DMR-0451163. X.H. received the support from DOE DE-SC0002136. We would like to thank Dr. P. Willmott (Paul Scherrer Institut, Switzerland) for allowing us to use the surface x-ray data (SSR and CTR) for $SrTiO_3(001)$ surface.

References

1. R. J. Birgeneau and M. A. Kastner, *Science* **288**, 437 (2000).
2. Y. Chu, L. W. Martin, M. B. Holcomb, and R. Ramesh, *Mater. Today* **10**, 16 (2007).
3. C. Israel, M. J. Calderon, and N. D. Mathur, *Mater. Today* **10**, 24 (2007).
4. H. Habermeier, *Mater. Today* **10**, 34 (2007).
5. H. Ohta, *Mater. Today* **10**, 44 (2007).
6. P.W. Anderson, The Theory of Superconductivity in the High-T_c Cuprate Superconductors (Princeton University Press, New Jersey, 1997)
7. J. G. Bednorz and K. A. Muller, *Z. Phys. B: Condens. Matter* **64**, 189 (1986).
8. Y. Kamihara, T. Watanabe, M. Hirano, and H. Hosono, *J. Am. Chem. Soc.* **130**, 3296 (2008).
9. J. Paglione and R. L. Greene, *Nature Phys.* **6**, 645 (2010).
10. T. Kimura, Y. Tomioka, H. Kuwahara, A. Asamitsu, M. Tamura, and Y. Tokura, *Science* **274**, 1698 (1996).
11. Q. Si and F. Steglich, *Science* **329**, 1161 (2010).
12. P. A. Cox, Transition Metal Oxides: An Introduction to their Electronic Structure and Properties (International Series of Monographs on Chemistry) (Oxford University Press, New York, 1995).
13. Z. Fang and K. Terakura, *Phys. Rev.* **B64**, 020509(R) (2001).
14. E. W. Plummer, Ismail, R. Matzdorf, A. V. Melechko, J. P. Pierce, and J. Zhang, *Surf. Sci.* **500**, 1 (2002).
15. P. Dai, J. Zhang, H. A. Mook, S.-H. Liou, P. A. Dowben and E. W. Plummer, *Phys. Rev.* **B54**, R3694 (1996).
16. S. J. L. Billinge, R. G. DiFrancesco, G. H. Kwei, J. J. Neumeier and J. D. Thompson, *Phys. Rev. Lett.* **77**, 715 (1996).
17. P. G. Radaelli, D. E. Cox, M. Marezio, S-W. Cheong, P. E. Schiffer and A. P. Ramirez, *Phys. Rev. Lett.* **75**, 4488 (1995).
18. S. Nandi, M. G. Kim, A. Kreyssig, R. M. Fernandes, D. K. Pratt, A. Thaler, N. Ni, S. L. Bud'ko, P. C. Canfield, J. Schmalian, R. J. McQueeney and A. I. Goldman, *Phys. Rev. Lett.* **104**, 057006 (2010).
19. S. Medvedev, T. M. McQueen, I. A. Troyan, T. Palasyuk, M. I. Eremets, R. J. Cava, S. Naghavi, F. Casper, V. Ksenofontov, G. Wortmann and C. Felser, *Nature Mater.* **8**, 630 (2009).

20. H. Dulli, P. A. Dowben, S.-H. Liou and E. W. Plummer, *Phys. Rev.* **B62**, R14629 (2000).
21. M. A. Hossain, J. D. F. Mottershead, D. Fournier, A. Bostwick, J. L. McChesney, E. Rotenberg, R. Liang, W. N. Hardy, G. A. Sawatzky, I. S. Elfimov, D. A. Bonn and A. Damascelli, *Nature Phys.* **4**, 527 (2008).
22. G. Renaud, *Surf. Sci. Rep.* **32**, 1 (1998).
23. B. E. Warren, *X-ray Diffraction* (Dover Publications, New York, 1990).
24. I. R. Entin, V. I. Khrupa, and O. V. Petrosyan, *Phys. Status Solidi* **A127**, 321 (1991).
25. G. S. Chandler, B. N. Figgis, P. A. Reynolds, and S. K. Wolff, *Chem. Phys. Lett.* **225**, 421 (1994).
26. M. F. C. Ladd and R. A. Palmer, *Structure Determination by X-Ray Crystallography* (Springer, Berlin, Germany, 2003).
27. J. E. T. Andersen and P. J. Moller, *Appl. Phys. Lett.* **56**, 1847 (1990).
28. T. Matsumoto, H. Tanaka, T. Kawai and S. Kawai, *Surf. Sci.* **278**, L153 (1992).
29. M. Naito and H. Sato, *Physica* **C229**, 1 (1994).
30. Q. D. Jiang and J. Zegenhagen, *Surf. Sci.* **338**, L882 (1995).
31. Q. D. Jiang and J. Zegenhagen, *Surf. Sci.* **425**, 343 (1999).
32. G. Charlton, S. Brennan, C. A. Muryn, R. McGrath, D. Norman, T. S. Turner and G. Thornton, *Surf. Sci.* **457**, L376 (2000).
33. T. Kubo and H. Nozoye, *Phys. Rev. Lett.* **86**, 1801 (2001).
34. A. Kazimirov, D. M. Goodner, M. J. Bedzyk, J. Bai and C. R. Hubbard, *Surf. Sci.* **492**, L711 (2001).
35. N. Erdman, K. R. Poeppelmeier, M. Asta, O. Warschkow, D. E. Ellis and L. D. Marks, *Nature* **419**, 55 (2002).
36. M. R. Castell, *Surf. Sci.* **505**, 1 (2002).
37. M. R. Castell, *Surf. Sci.* **516**, 33 (2002).
38. K. Johnston, M. R. Castell, A. T. Paxton and M. W. Finnis, *Phys. Rev.* **B70**, 085415 (2004).
39. V. Vonk, S. Konings, G. J. V. Hummel, S. Harkema and H. Graafsma, *Surf. Sci.* **595**, 183 (2005).
40. F. Silly, D. T. Newell and M. R. Castell, *Surf. Sci.* **600**, L219 (2006).
41. R. Erni, M. D. Rossell, C. Kisielowski, and U. Dahmen, *Phys. Rev. Lett.* **102**, 096101 (2009).
42. H. Poppa, *J. Vac. Sci. Technol.* **A22**, 1931 (2004).
43. K. Takayanagi, Y. Tanishiro, M. Takahashi and S. Takahashi, *J. Vac. Sci. Technol.* **A3**, 1502 (1985).
44. K. Takayahagi, Y. Tanishiro, S. Takahashi and M. Takahashi, *Surf. Sci.* **164**, 367 (1985).
45. R. E. Schlier and H. E. Farnsworth, *J. Chem. Phys.* **30**, 917 (1959).
46. I. K. Robinson, W. K. Waskiewicz, P. H. Fuoss, J. B. Stark and P. A. Bennett, *Phys. Rev.* **B33**, 7013 (1986).
47. I. K. Robinson, W. K. Waskiewicz, P. H. Fuoss and L. J. Norton, *Phys. Rev.* **B37**, 4325 (1988).

48. H. Huang, S. Y. Tong, W. E. Packard and M. B. Webb, *Phys. Lett.* **A130**, 166 (1988).
49. K. Takayanagi, Y. Tanishiro and K. Kajiyama, *J. Vac. Sci. Technol.* **B4**, 1074 (1986).
50. G. Qian and D. J. Chadi, *J. Vac. Sci. Technol.* **B4**, 1079 (1986).
51. G. Binnig, H. Rohrer, Ch. Gerber and E. Weibel, *Phys. Rev. Lett.* **50**, 120 (1983).
52. E. Bengu, R. Plass, L. D. Marks, T. Ichihashi, P. M. Ajayan and S. Iijima, *Phys. Rev. Lett.* **77**, 4226 (1996).
53. A. Subramanian and L. D. Marks, *Ultramicroscopy* **98**, 151 (2004).
54. M. A. Van Hove, W. H. Weinberg and C.-M Chan, Low-Energy Electron Diffraction Experiment, Theory, and Surface Structure Determination (Springer, Berlin, Germany, 1986).
55. L. J. Clarke, Surface Crystallography: An Introduction to Low Energy Electron Diffraction (John Wiley & Sons Inc, New Jersey,1985).
56. C. J. Powell and A. Jablonski, *J. Phys. Chem. Ref. Data* **28**, 19 (1999).
57. S. Y. Tong, *Prog. Surf. Sci.* **7**, 1 (1975).
58. V. B. Nascimento, R. G. Moore, J. Rundgren, J. Zhang, L. Cai, R. Jin, D. G. Mandrus and E. W. Plummer, *Phys. Rev.* **B75**, 035408 (2007).
59. J. Rundgren, *Phys. Rev.* **B68**, 125405 (2003).
60. C. Bai, *Scanning tunneling microscopy and its applications* (Springer, Berlin, Germany, 2000).
61. R. Matzdorf, Z. Fang, Ismail, J. Zhang, T. Kimura, Y. Tokura, K. Terakura and E. W. Plummer, *Science* **289**, 746 (2000).
62. Y. Maeno, H. Hashimoto, K. Yoshida, S. Nishizaki, T. Fujita, J. G. Bednorz and F. Lichtenberg, *Nature* **372**, 532 (1994).
63. Q. Huang, J. W. Lynn, R. W. Erwin, J. Jarupatrakorn and R. J. Cava, *Phys. Rev.* **B58**, 8515 (1998).
64. G. Cao and P. Schlottmann, *Mod. Phys. Lett.* **B22**, 1785 (2008).
65. D. Fobes, M. H. Yu, M. Zhou, J. Hooper, C. J. O'Connor, M. Rosario and Z. Q. Mao, *Phys. Rev.* **B75**, 094429 (2007).
66. S. Nakatsuji and Y. Maeno, *Phys. Rev.* **B62**, 6458 (2000).
67. O. Friedt, M. Braden, G. Andre, P. Adelmann, S. Nakatsuji and Y. Maeno, *Phys. Rev.* **B63**, 174432 (2001).
68. R. G. Moore, J. Zhang, V. B. Nascimento, R. Jin, J. Guo, G. T. Wang, Z. Fang, D. Mandrus and E. W. Plummer, *Science* **318**, 615 (2007).
69. R. G. Moore, V. B. Nascimento, J. Zhang, J. Rundgren, R. Jin, D. Mandrus and E. W. Plummer, *Phys. Rev. Lett.* **100**, 066102 (2008).
70. Z. Fang, K. Terakura and N. Nagaosa, *New J. Phys.* **7**, 66 (2005).
71. S. Nakatsuji and Y. Maeno, *Phys. Rev. Lett.* **84**, 2666 (2000).
72. S. Nakatsuji, D. Hall, L. Balicas, Z. Fisk, K. Suhahara, M. Yoshioka and Y. Maeno, *Phys. Rev. Lett.* **90**, 137202 (2003).
73. E. Heifets, R. I. Eglitis, E. A. Kotomin, J. Maier and G. Borstel, *Phys. Rev.* **B64**, 235417 (2001).
74. S. Piskunov, E. A. Kotomin, E. Heifets, J. Maier, R. I. Eglitis and G. Borstel, *Surf. Sci.* **575**, 75 (2005).

75. N. F. Mott, *Rev. Mod. Phys.* 40, 677 (1968).
76. N. F. Mott, *Metal-Insulator Transitions* (Taylor & Francis, London, UK, 1990).
77. F. Gebhard, *The Mott Metal-Insulator Transition* (Springer, Berlin, Germany, 1997).
78. H. Ibach and D. L. Mills, Electron Energy Loss Spectroscopy and Surface Vibrations (Academic Press, New York, 1982).
79. J. E. Demuth and B. N. J. Persson, *Phys. Rev. Lett.* 54, 584 (1985).
80. S. Modesti, V. R. Dhanak, M. Sancrotti, A. Santoni, B. N. J. Persson and E. Tosatti, *Phys. Rev. Lett.* 73, 1951 (1994).
81. J. H. Jung, Z. Fang, J. P. He, Y. Kaneko, Y. Okimoto and Y. Tokura, *Phys. Rev. Lett.* 91, 056403 (2003).
82. H. Rho, S. L. Cooper, S. Nakatsuji, H. Fukazawa and Y. Maeno, *Phys. Rev.* B68, 100404(R) (2003).
83. J. F. Mitchell, D. N. Argyriou, A. Berger, K. E. Gray, R. Osborn and U. Welp, *J. Phys. Chem.* B105, 10731 (2001).
84. J. W. Freeland, J. J. Kavich, K. E. Gray, L. Ozyuzer, H. Zhang, J. F. Mitchell, M. P. Warusawithana, P. Ryan, X. Zhai, R. H. Kodama and J. N. Eckstein, *J. Phys.: Condens. Matter* 19, 315210 (2007).
85. D. R. Lee, S. K. Sinha, D. Haskel, Y. Choi, J. C. Lang, S. A. Stepanov and G. Srajer, *Phys. Rev.* B68, 224409 (2003).
86. C. Kao, J. B. Hastings, E. D. Johnson, D. P. Siddons, G. C. Smith and G. A. Prinz, *Phys. Rev. Lett.* 65, 373 (1990).
87. J. B. Kortright and S. Kim, *Phys. Rev.* B62, 12216 (2000).
88. J. W. Freeland, K. E. Gray, L. Ozyuzer, P. Berghuis, E. Badica, J. Kavich, H. Zhang and J. F. Mitchell, *Nature Mater.* 4, 62 (2005).
89. Y.-D. Chuang, A. D. Gromko, D. S. Dessau, T. Kimura and Y. Tokura, *Science* 292, 1509 (2001).
90. N. Mannella, W. L. Yang, X. J. Zhou, H. Zhang, J. F. Mitchell, J. Zaanen, T. P. Devereaux, N. Nagaosa, Z. Hussain and Z.-X. Shen, *Nature* 438, 474 (200).
91. N. Mannella, W. L. Yang, K. Tanaka, X. J. Zhou, H. Zhang, J. F. Mitchell, J. Zaanen, T. P. Devereaux, N. Nagaosa, Z. Hussain and Z.-X. Shen, *Phys. Rev.* B76, 233102 (2007).
92. Z. Sun, Y.-D. Chuang, A. V. Fedorov, J. F. Douglas, D. Reznik, F. Weber, N. Aliouane, D. N. Argyriou, H. Zheng, J. F. Mitchell, T. Kimura, Y. Tokura, A. Revcolevschi and D. S. Dessau, *Phys. Rev. Lett.* 97, 056401 (2006).
93. S. de Jong, Y. Huang, I. Santoso, F. Massee, R. Follath, O. Schwarzkopf, L. Patthey, M. Shi and M. S. Golden, *Phys. Rev.* B76, 235117 (2007).
94. Z. Sun, J. F. Douglas, A. V. Fedorov, Y.-D. Chuang, H. Zheng, J. F. Mitchell and D. S. Dessau, *Nature Phys.* 3, 248 (2007).
95. V. B. Nascimento, J. W. Freeland, R. Saniz, R. G. Moore, D. Mazur, H. Liu, M. H. Pan, J. Rundgren, K. E. Gray, R. A. Rosenberg, H. Zheng, J. F. Mitchell, A. J. Freeman, K. Veltruska and E. W. Plummer, *Phys. Rev. Lett.* 103, 227201 (2009).
96. J. F. Mitchell, D. N. Argyriou, J. D. Jorgensen, D. G. Hinks, C. D. Potter and S. D. Bader, *Phys. Rev.* B55, 63 (1997).

97. M. Kubota, H. Fujioka, K. Hirota, K. Ohoyama, Y. Moritomo, H. Yoshizawa and Y. Endoh, *J. Phys. Soc. Jpn.* **69**, 1606 (2000).

98. K. Hirota, S. Ishihara, H. Fujioka, M. Kubota, H. Yoshizawa, Y. Moritomo, Y. Endoh and S. Maekawa, *Phys. Rev.* **B65**, 064414 (2002).

99. E. Wimmer, H. Krakauer, M. Weinert and A. J. Freeman, *Phys. Rev.* **B24**, 864 (1981).

100. Z. Ren, W. Lu, J. Yang, W. Yi, X. Shen, Z. Li, G. Che, X. Dong, L. Sun, F. Zhou and Z. Zhao, *Chin. Phys. Lett.* **25**, 2215 (2008).

101. David C. Johnston, *Adv. Phys.* **59**, 803 (2010).

102. J. Guo, S. Jin, G. Wang, S. Wang, K. Zhu, T. Zhou, M. He and X. Chen, *Phys. Rev.* **B82**, 180520(R) (2010).

103. C. de la Cruz, W. Z. Hu, S. Li, Q. Huang, J. W. Lynn, M. A. Green, G. F. Chen, N. L. Wang, H. A. Mook, Q. Si and P. Dai, *Phys. Rev. Lett.* **104**, 017204(R) (2010).

104. S. A. J. Kimber, A. Kreyssig, Y. Zhang, H. O. Jeschke, R. Valenti, F. Yokaichiya, E. Colombier, J. Yan, T. C. Hansen, T. Chatterji, R. J. McQueeney, P. C. Canfield, A. I. Goldman and D. N. Argyriou, *Nature Mater.* **8**, 471 (2009).

105. V. B. Nascimento, A. Li, D. R. Jayasundara, Y. Xuan, J. O'Neal, S. Pan, T. Y. Chien, B. Hu, X. B. He, G. Li, A. S. Sefat, M. A. McGuire, B. C. Sales, D. Mandrus, M. H. Pan, J. Zhang, R. Jin and E. W. Plummer, *Phys. Rev. Lett.* **103**, 076104 (2009).

106. G. Li, X. He, A. Li, S. Pan, J. Zhang, R. Jin, A. S. Sefat, M. A. McGuire, D. Mandrus, B. C. Sales and E. W. Plummer, *arXiv*: 1006.5907v1 (2010).

107. T. Shimojima, K. Ishizaka, Y. Ishida, N. Katayama, K. Ohgushi, T. Kiss, M. Okawa, T. Togashi, X.-Y. Wang, C.-T. Chen, S. Watanabe, R. Kadota, T. Oguchi, A. Chainani and S. Shin, *Phys. Rev. Lett.* **104**, 057002 (2010).

108. E. Abrahams and Q. Si, *arXiv*: 1101.4701v1 (2011).

109. A. Li, J. Ma, A. Sefat, M. McGuire, B. Sales, D. Mandrus, R. Jin, C. Zhang, P. Dai and S. Pan, *APS March Meeting Bulletin, Dallas, USA* (2011).

110. S. Pan, A. Li, J. Ma, A. Sefat, M. McGuire, B. Sales, D. Mandrus, R. Jin and E. Plummer, *APS March Meeting Bulletin, Dallas, USA* (2011).

111. E. Akturk and S. Ciraci, *Phys. Rev.* **B79**, 184523 (2009).

112. J. Menendez and M. Cardona, *Phys. Rev.* **B29**, 2051 (1984).

113. A. Akrap, J. J. Tu, L. J. Li, G. H. Cao, Z. A. Xu and C. C. Homes, *Phys. Rev.* **B80**, 180502(R) (2009).

114. F. Hsu, J. Luo, K. Yeh, T. Chen, T. Huang, P. M. Wu, Y. Lee, Y. Huang, Y. Chu, D. Yan and M. Wu, *Proc. Natl. Acad. Sci. USA* **105**, 14262 (2008).

115. X. He, G. Li, J. Zhang, A. B. Karki, R. Jin, B. C. Sales, A. S. Sefat, M. A. McGuire, D. Mandrus and E. W. Plummer, *Phys. Rev.* **B83**, 220502(R) (2011).

116. N. Ni, M. E. Tillman, J.-Q. Yan, A. Kracher, S. T. Hannahs, S. L. Bud'ko and P. C. Canfield, *Phys. Rev.* **B78**, 214515 (2008).

117. E. Dagotto, *Science* **309**, 257 (2005).

118. T. Kato, Y. Mizuguchi, H. Nakamura, T. Machida, H. Sakata and Y. Takano, *Phys. Rev.* **B80**, 180507(R) (2009).

119. T. Hanaguri, S. Niitaka, K. Kuroki and H. Takagi, *Science* **328**, 474 (2010).

Chapter 8

INSTABILITY OF THE $J_{eff} = 1/2$ INSULATING STATE IN $Sr_{n+1}Ir_nO_{3n+1}$ (n = 1 AND 2)

G. Cao and L. E. DeLong

Center for Advanced Materials
Department of Physics and Astronomy,
University of Kentucky,
Lexington, KY 40506, USA
E-mail: cao@uky.edu; delong@pa.uky.edu

The most profound consequence of the spin-orbit interaction in layered iridates is the realization of a recently proposed $J_{eff} = 1/2$ insulating state, which explains novel insulating behavior observed in the $Sr_{n+1}Ir_nO_{3n+1}$ materials. We review the basic physical properties of layered iridates, including our recent transport and thermodynamic data for $Sr_{n+1}Ir_nO_{3n+1}$ (n = 1, 2). These results indicate the spin-orbit interaction vigorously competes with Coulomb interactions, non-cubic crystal electric field interactions, and the Hund's rule coupling, which leads to states with exotic properties that are highly susceptible to small perturbations. The effects of chemical doping, application of pressure and magnetic field are emphasized; these perturbations are relatively weak, but capable of influencing the spin-orbit interaction and generating a rich phase diagram of strongly competing ground states.

Contents

8.1. Introduction

Traditional arguments suggest that iridium oxides should be more metallic and less magnetic than materials based upon 3*d* and 4*f* elements, because 5d-electron orbitals are more extended in space, which increases their electronic bandwidth. This conventional wisdom conflicts with two trends observed among layered iridates such as the Ruddlesden-Popper phases, $Sr_{n+1}Ir_nO_{3n+1}$ (n = 1 and 2; n defines the number of Ir-O layers in a unit cell) [1-8] and hexagonal $BaIrO_3$ [9]. First, complex magnetic states occur with high critical temperatures but unusually low ordered moments. Second, "exotic" insulating states are observed rather than metallic states [7-9] (see **Table 8.1**).

Table 8.1. *Examples of Layered Iridates*

System	Néel/Curie Temperature (K)	Ground State
Sr_2IrO_4 (n = 1)	240	*Canted AFM insulator*
$Sr_3Ir_2O_7$ (n = 2)	285	*AFM insulator*
$BaIrO_3$	183	*Magnetic insulator*

A critical underlying mechanism for these unanticipated states is a strong spin-orbit interaction (SOI) that vigorously competes with Coulomb interactions, non-cubic crystalline electric fields, and Hund's rule coupling. The net result of this competition is to stabilize ground states with exotic behavior. The most profound effect of the spin-orbit interaction (SOI) on the iridates is the $J_{eff} = 1/2$ insulating state [1-3], a new quantum state that embodies the novel physics in 5d-based systems.

It is now recognized that strong SOI can drive novel narrow-gap insulating states in iridates [1-3]. The SOI is a relativistic effect that is proportional to Z^4 (Z is the atomic number), and is approximately 0.4 eV in the iridates (compared to ~ 20 meV in 3d materials), and splits the t_{2g} bands into states with $J_{eff} = 1/2$ and $J_{eff} = 3/2$, the latter having lower energy [1-2] (see **Table 8.2**). Since Ir^{4+} ($5d^5$) ions provide five 5d-electrons to bonding states, four of them fill the lower $J_{eff} = 3/2$ bands, and one electron partially fills the $J_{eff} = 1/2$ band where the Fermi level E_F resides. The $J_{eff} = 1/2$ band is so narrow that even a reduced on-site Coulomb repulsion (U ~ 0.5 eV, due to the extended nature of 5d-electron orbitals), is sufficient to open a small gap Δ that underpins the insulating state in the iridates [1, 2, 11]. The splitting between the $J_{eff} = 1/2$ and $J_{eff} = 3/2$ bands narrows as the dimensionality (i.e., n) increases in $Sr_{n+1}Ir_nO_{3n+1}$, and the two bands progressively broaden and contribute to the density of states (DOS) near the Fermi surface. In particular, the bandwidth W of $J_{eff} = 1/2$ band increases from 0.48 eV for n = 1 to 0.56 eV for n = 2 and 1.01 eV for n = ∞ [2, 11]. The ground state evolves with decreasing Δ, from a robust insulating state for Sr_2IrO_4 (n = 1) to a metallic state for $SrIrO_3$ (n = ∞) as n increases. A well-defined, yet weak, insulating state lies between them at $Sr_3Ir_2O_7$ (n = 2). Given the delicate balance between relevant interactions, a recent theoretical proposal predicts $Sr_3Ir_2O_7$ to be a spin-orbit, band insulator [12].

Table 8.2. *Comparison between 3d and 4d/5d Electrons*

Electron Type	U(eV)	λ_{os}(eV)	Spin State	Interactions	Phenomena
3d	5-7	0.01-0.1	High	$U > CF > \lambda_{so}$	Magnetism/HTSC
4d/5d	0.4-2	0.1-1	Low	$U \sim CF \sim \lambda_{so}$	J_{eff}=1/2 State

The onset of weak ferromagnetic order is observed at T_C = 240 K [4-7] and 285 K [8] in the case of Sr_2IrO_4 and $Sr_3Ir_2O_7$, respectively (see **Fig. 8.1**). It is generally recognized that the magnetic ground state for Sr_2IrO_4 and $Sr_3Ir_2O_7$ (i.e., for both n = 1, 2) is antiferromagnetic (AFM) and is closely associated with the rotation of the IrO_6 octahedra about the **c**-axis, which dictates the crystal structure of both Sr_2IrO_4 and $Sr_3Ir_2O_7$ [12, 13]. Indeed, the temperature dependence of the magnetization M(T) closely tracks the rotation of the octahedra, as characterized by the Ir-O-Ir bond angle θ, for both n = 1 and 2, as shown in **Figs. 8.1a** and **8.1b**. A recent neutron study of single-crystal Sr_2IrO_4 reveals a canted AFM structure in the basal plane with spins primarily aligned along the

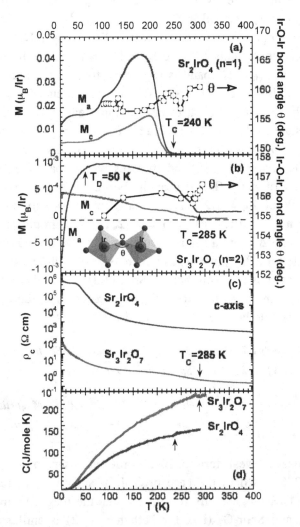

Fig. 8.1. Temperature dependence of: **(a)** the magnetization and Ir-O-Ir bond angle θ (right scale) for Sr_2IrO_4; **(b)** the magnetization M and Ir-O-Ir bond angle θ (right scale) for $Sr_3Ir_2O_7$; **(c)** the c-axis resistivity ρ_c for Sr_2IrO_4 and $Sr_3Ir_2O_7$; and **(d)** the specific heat C for Sr_2IrO_4 and $Sr_3Ir_2O_7$.

a-axis [21]. Unlike Sr_2IrO_4, $Sr_3Ir_2O_7$ exhibits an intriguing magnetization reversal in M(T) below T_D = 50 K; both T_C and T_D can be observed only when the system is field-cooled (FC) from above T_C (**Fig. 8.1b**) [8]. Such field-induced magnetic behavior suggests that any perturbation may tip the balance between nearly degenerate, competing ground states in $Sr_3Ir_2O_7$ [12, 14].

In spite of the extremely small magnitude of M(T,H), corresponding anomalies in the electrical resistivity ρ and specific heat C(T) are observed at T_C = 285 K in $Sr_3Ir_2O_7$. In sharp contrast, correlated anomalies in M(T), C(T) and ρ(T) at T_C are either weak, or conspicuously absent in Sr_2IrO_4 [7, 15], as illustrated in **Fig. 8.1**. For example, the observed specific heat anomaly $|\Delta C|$ ~ 10 J/mole K at T_C for $Sr_3Ir_2O_7$, indicating a sizable entropy change; this change is tiny (~ 4 mJ/mole K) for Sr_2IrO_4, in spite of its robust, long-range magnetic order below T_C = 240 K [7, 10, 15-17]. The weak phase transition signatures suggest that thermal and transport properties may not be driven by the same interactions that dictate the magnetic behavior; and more generally, they point to a novel ground state for Sr_2IrO_4. Recently, a time-resolved optical study indicates that Sr_2IrO_4 is a unique system in which Slater- and Mott-Hubbard-type behaviors coexist, which might explain the absence of anomalies at T_C in transport and thermodynamic measurements [18].

The nature of magnetism and its implications in the iridates are currently open to debate, in part because the spin degree of freedom is no longer an independent parameter, owing to the strong SOI in the iridates; and it is clear that the ground state of the these materials is strongly influenced by the lattice degrees of freedom [15-17]. This unusual state of affairs will be documented in this Chapter, where we review the physical properties of the Ruddelsen-Popper iridates as a function of chemical doping, magnetic field and pressure.

8.2. Sr_2IrO_4

Sr_2IrO_4, with a crystal structure similar to that of La_2CuO_4 and the p-wave superconductor Sr_2RuO_4, is a weak ferromagnet with a Curie temperature T_C = 240 K [4-7]. A unique and important structural feature of Sr_2IrO_4 is that it crystallizes in a tetragonal structure (space-group *I41/acd*) due to a rotation of the IrO_6-octahedra about the **c**-axis by ~11°, which results in a larger unit cell volume by a factor $\sqrt{2}$ x $\sqrt{2}$ x 2, as shown in **Fig. 8.2** [4-7]. This rotation corresponds to a distorted in-plane Ir1-O2-Ir1 bond angle θ that is critical to the electronic structure. A reduced crystal structure symmetry is also consistent with a recent neutron study [21] of single-crystal Sr_2IrO_4, which revealed forbidden nuclear reflections of space group *I41/acd* over a wide temperature interval, 4 K < T< 600 K. In essence, strong crystal fields split off 5d band states with e_g symmetry, and t_{2g} bands arise from $J_{eff} = 1/2$ and $J_{eff} = 3/2$ multiplets via a strong SOI (~0.4 eV), as described above. A similar mechanism also describes insulating states observed in other iridates, such as $Sr_3Ir_2O_7$ [2, 8] and $BaIrO_3$ [9, 22].

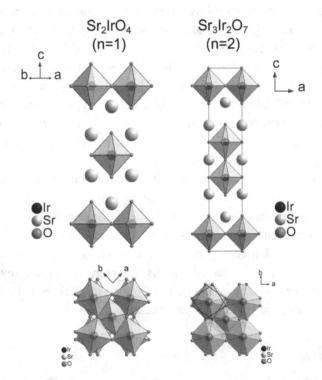

Fig. 8.2. The crystal structure of Sr_2IrO_4 (left) and $Sr_3Ir_2O_7$ (right) generated from our single-crystal X-ray diffraction data. Note that the lower two figures illustrate the rotation of IrO_6 octahedra about the **c**-axis for Sr_2IrO_4 and $Sr_3Ir_2O_7$, respectively.

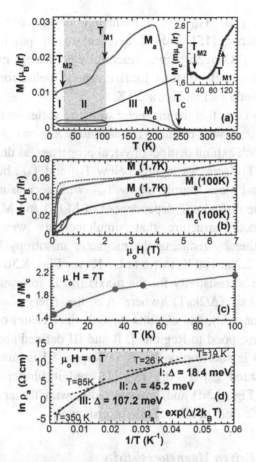

Fig. 8.3. The field-cooled magnetization for the a-axis and the c-axis, M_a and M_c, respectively, as functions of (a) temperature at $\mu_oH = 0.1$ T, and (b) magnetic field at T = 1.7 and 100 K. (c) The magnetic anisotropy M_a/M_c as a function of temperature. (d) The logarithm of resistivity for the a-axis, ln ρ_a, as a function of 1/T. Inset in (a): Enlarged low-T M_c. Note that the data in (a) and (d) define Regions I, II, and III [Ref. 16].

A few critical magnetic features of Sr_2IrO_4 need to be addressed at the outset. Both the a-axis $M_a(T)$ and the c-axis $M_c(T)$ exhibit ferromagnetic (FM) order below $T_C = 240$ K, and a positive Curie-Weiss temperature, $\theta_{CW} = +236$ K, confirming a FM exchange coupling at high temperatures [4-7, 15-17]. A close examination of the low-field M(T) reveals two additional anomalies at $T_{M1} \approx 100$ K and $T_{M2} \approx 25$ K in $M_a(T)$ and $M_c(T)$ (see **Fig. 8.3a**). Our ac magnetic susceptibility data also exhibit a peak near T_{M1} and a frequency dependence indicative of magnetic frustration [15]. Indeed, a recent muon-spin rotation (μSR) study [23] of Sr_2IrO_4 reports two structurally equivalent muon sites that experience increasingly distinct local magnetic fields for $T < 100$ K, which subsequently lock in below 20 K.

It becomes clear that the magnetic structure varies with temperature, resulting in three well-defined *temperature Regions I, II, and III* (**Fig. 8.3a**), which exhibit distinct physical properties, as discussed below. Moreover, $M_a(T)$ decreases rapidly below T_{M1} and T_{M2}, but $M_c(T)$ rises below 50 K, and more sharply below T_{M2}, with decreasing T (see **Fig. 8.3a inset**). The different T-dependences of $M_a(T)$ and $M_c(T)$ signal an evolving magnetic structure that simultaneously weakens M_a but enhances M_c, thereby reducing the magnetic anisotropy M_a/M_c, which decreases from 2.2 at 100 K, to 1.5 at 1.7 K (see **Figs. 8.3b** and **8.3c**).

The electrical resistivity for the **a**-axis ρ_a (T) follows an activation law, $\rho_a(T) \sim \exp(\Delta/2k_BT)$ (where Δ is the energy gap and k_B is Boltzmann's constant), and exhibits three distinct values of Δ in regions that closely correspond to Regions I, II and III defined above, as shown in **Fig. 8.3d**. It is noteworthy that Δ in Region III is quite close to the optically measured gap (~ 0.1 eV) [1], and it further narrows with decreasing T (**Fig. 8.3d**) and, unexpectedly, with the application of a modest magnetic field of a few Tesla (not shown).

8.2.1. *Lattice-Driven Magnetoresistivity*

The transport properties of Sr_2IrO_4 are coupled to magnetic field in such a peculiar fashion that no available model can describe the observed magnetoresistivity shown in **Figs. 8.4** and **8.5**. We focus on a representative temperature $T = 35$ K that is within Region II. Both the **a**-axis resistivity $\rho_a(H\|a)$ (**Fig. 8.4b**) and the **c**-axis resistivity $\rho_c(H\|a)$ (**Fig. 8.4c**) exhibit an abrupt drop by $\sim 60\%$ near $\mu_oH = 0.3$ T applied along the **a**-axis, where a metamagnetic transition (spin-reorientation) occurs, consistent with earlier studies [5, 16]. These data partially track the field dependences of $M_a(H)$ and $M_c(H)$ shown in **Fig. 8.4a**, and suggest a reduction of spin scattering [23]. However, given the small ordered moment $m_s < 0.07$ μ_B/Ir, a reduction of spin scattering alone certainly cannot account for such a drastic reduction in ρ(H). Even more strikingly, for **H ∥ c**-axis, both ρ_a(**H∥c**) and ρ_c(**H∥c**) exhibit multiple anomalies at $\mu_oH = 2$ T and 3 T, which leads to a large overall reduction

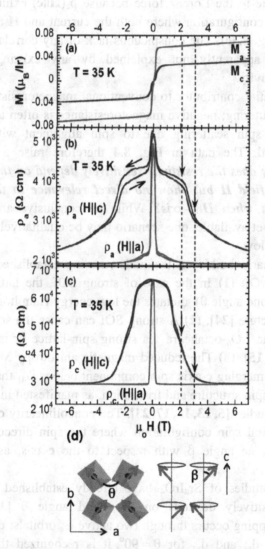

Fig. 8.4. The field dependences at T = 35 K of: (a) the magnetizations M_a and M_c. (b) The **a**-axis resistivity ρ_a for **H∥a** and **H∥c**. (c) The **c**-axis resistivity ρ_c for **H∥a** and **H∥c**. (d) The schematics of the spin configuration for the basal plane (left) and the **ac**-plane (right) [Ref. 16].

of resistivity by more than 50%. However, no anomalies corresponding
to these transitions in $M_a(H)$ and $M_c(H)$ are discerned! In addition,
dM/dH shows no slope change near $\mu_o H = 2$ and 3 T. Such behavior is
clearly not due to the Lorenz force because $\rho_c(H\|c)$ exhibits the same
behavior in a configuration where both the current and **H** are parallel to
the c-axis (**Fig. 8.4c**). The conspicuous lack of any correlation between
ρ and M is apparently not explained by any existing model for
magnetoresistivity.

An essential contributor to conventional magnetoresistance is spin-
dependent scattering; negative magnetoresistance is often a result of the
reduction of spin scattering due to spin alignment with increasing
magnetic field. The data in **Fig. 8.4** therefore raise a fundamental
question: ***Why does the resistivity sensitively depend on the orientation
of magnetic field H but show no direct relevance to the measured
magnetization when H∥c-axis?*** While no conclusive answers to the
question are yet available, one scenario may be qualitatively relevant, as
we explain below.

This scenario builds on the understanding established by existing
work on Sr_2IrO_4: **(1)** In the case of strong SOI, the lattice distortion
(Ir1-O2-Ir1 bond angle θ) dictates the low-energy Hamiltonian [13], and
the band structure [24]. **(2)** A strong SOI can cause the spins to rigidly
rotate with the IrO_6-octahedra via strong spin-lattice or magnetoelastic
coupling [13, 15]. **(3)** The reduced magnetic anisotropy M_a/M_c strongly
indicates an emerging c-axis spin component below T_{M1} that generates a
noncollinear spin structure and frustration, as manifested in **Fig. 8.3**, and
in previous studies [5, 13, 15, 17, 23]. The noncollinearity could take the
form of a spiral spin configuration where the spin direction is rigidly
maintained at an angle β with respect to the c-axis, as sketched in
Fig. 8.4d.

Recent studies of Sr_2IrO_4 have already established that electron
hopping sensitively depends on the bond angle θ [15-17, 24]. In
particular, hopping occurs through two active t_{2g} orbitals: d_{xy} and d_{xz} for
$\theta = 180°$, and d_{xz} and d_{yz} for $\theta = 90°$. It is recognized that the larger
θ, the more energetically favorable it is for electron hopping and
stronger superexchange interactions. Since the IrO_6-octahdra rotate with
the spins, we conclude the application of **H** ∥ c-axis must at least slightly

rotate the IrO_6-octahdra about the **c**-axis, which, in turn, changes θ. It is important to realize that even a small increase in θ due to increasing H can be sufficient to drastically enhance the hopping, which could explain the multiple downturns in $\rho(H)$. The clear hysteresis exhibited in **Fig. 8.4b** reinforces the notion that the magnetoresistivity is primarily driven by field-induced lattice distortions when **H ‖ c**. The absence of anomalies in $M_a(H)$ and $M_c(H)$ corresponding to the transitions in $\rho_a(H)$ and $\rho_c(H)$ can be attributed to a spiral spin configuration: the spins respond to **H** only by rotating about the **c**-axis, and this rotation changes θ but does not affect β or the **c**-axis and **a**-axis projections of the magnetic moment, as schematically illustrated in **Fig. 8.4d**; therefore, $M_a(H)$ and $M_c(H)$ remain unchanged, at least, up to $\mu_0H = 14$ T. Such behavior may be extend to even higher fields, given the unsaturated $M_a(H)$ and $M_c(H)$ we have observed at 14 T (not shown).

The delicate nature of the coupling of the magneto-transport behavior to the lattice and magnetic structures is apparent from several points of view, as illustrated in **Fig. 8.5**. The transport behavior seen in Region II is no longer observable in Region I, where the magneto-resistivity is remarkably weak; this is evident in $\rho_a(H)$ and $\rho_c(H)$ at T = 10 K, as shown in **Fig. 8.5a**. Moreover, the application of **H ‖ a**-axis causes a pronounced rise in $\rho_a(H)$ rather than the sharp drop observed in Region II at low H (**Figs. 8.4, 8.5b** and **8.5c**), and a reversal of the resistivity anisotropy (**Fig. 8.5a**). On the other hand, as T approaches Region III, the field dependences of $\rho_a(H)$ and $\rho_c(H)$ retain some resemblance to that in the Region II, but become far weaker.

8.2.2. Rh Doping to Tune the Spin-Orbit Interaction

Although Ir^{4+} ($5d^5$) ions are isoelectronic to Rh^{4+} ($4d^5$) ions, Sr_2RhO_4 has a weaker SOI (~ 0.16 eV), and therefore a smaller splitting between the $J_{eff} = 1/2$ and $J_{eff} = 3/2$ bands that are more evenly filled by the five 4d-electrons [25-29]. The weaker SOI, combined with more effectively screened Coulomb interactions between O-2p and Rh-4d electrons, favors a metallic state [26]. Indeed, Sr_2RhO_4 is a paramagnetic, correlated metal [27-29], in sharp contrast to the magnetic insulator Sr_2IrO_4 that orders at $T_C = 240$ K [4-7]. In addition, comparisons of Sr_2RhO_4

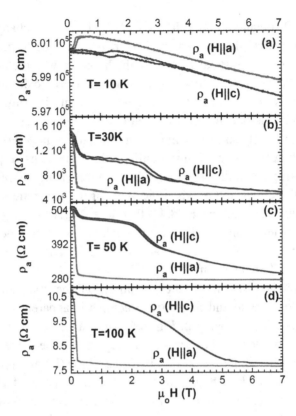

Fig. 8.5. The field dependence of the **a**-axis resistivity ρ_a for H‖a and H‖c at the following representative temperatures: **(a)** T = 10 K (Region I), **(b)** T = 30 K (Region II), **(c)** T = 50 K (Region II), **(d)** T = 100 K (approaching Region III) [Ref. 16].

with another 4d-based compound, Sr_2RuO_4 (a p-wave superconductor [30]), suggest that the impact of the SOI strongly depends on details of the electronic band structure near the Fermi surface E_F, Coulomb interactions, and lattice distortions [13, 26, 31, 32]. Despite the similar strength of the SOI in both materials [23], the t_{2g} bands in Sr_2RhO_4 are less dispersive than those in Sr_2RuO_4, and are therefore more susceptible to the SOI-induced band shifts near E_F than in Sr_2RuO_4. This is in part because the Ru^{4+} ($4d^4$) ion has *four* 4d electrons instead of five; and Ru doping therefore adds holes to the d-bands.

Sr_2IrO_4 and Sr_2RhO_4 are not only isoelectronic, but also isostructural. An important structural feature shared by both Sr_2IrO_4 and Sr_2RhO_4 is that they crystallize in a reduced tetragonal structure with space-group $I4_1/acd$, due to a rotation of the IrO_6- or RhO_6-octahedra about the **c**-axis by $\sim 12°$ or $\sim 9.7°$, respectively. That the two isostructural and isoelectronic compounds exhibit the sharply contrasting physical properties underscores the critical role SOI plays in determining the ground state of the iridate.

The ground state can be tuned by reducing the SOI, via substitution of $Rh^{4+}(4d^5)$ for $Ir^{4+}(5d^5)$ in single-crystal $Sr_2Ir_{1-x}Rh_xO_4$ ($0 \leq x \leq 1$). As schematically illustrated in **Fig. 8.6a**, Rh substitution, unlike other chemical substitutions, directly reduces the SOI and consequently, the

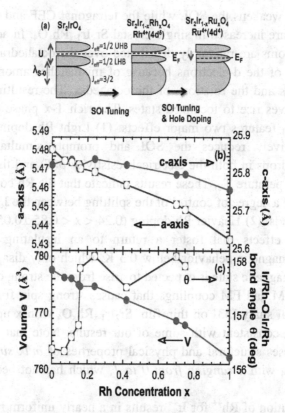

Fig. 8.6. (a) The schematics for the effects of Rh and Ru doping on the $J_{eff} = 1/2$ and $J_{eff} = 3/2$ bands. The Rh concentration (x) dependence at 90 K of: (b) the **a**-, and **c**-axis lattice parameters (right scale), and (c) the unit cell volume V and the Ir-O-Ir angle θ (right scale) [Ref. 54].

splitting between the $J_{eff} = 1/2$ and $J_{eff} = 3/2$ bands, but without signif-icant alteration of the band filling. Hence, the system remains positioned at the Mott instability and is very susceptible to electron scattering by disorder, which gives rise to localization. For comparison, one may also substitute $Ru^{4+}(4d^4)$ for $Ir^{4+}(5d^5)$ in $Sr_2Ir_{1-x}Ru_xO_4$ ($0 \leq x \leq 1$), where Ru not only reduces the SOI, but also fills the t_{2g} bands with holes, which lowers E_F and moves the system away from the Mott instability. Disorder scattering is then less relevant, and Ru doping systematically drives the system into a robust metallic state. The underlying effects of Ru doping on the $J_{eff} = 1/2$ and $J_{eff} = 3/2$ bands are also schematically illustrated in **Fig. 8.6a**.

Doping weakens the SOI, while the tetragonal CEF and the Hund's coupling J_H are increased in single-crystal $Sr_2Ir_{1-x}Rh_xO_4$. In addition, the Rh and Ir atoms are randomly distributed over the octahedra, hindering the hopping of the d-electrons because of mismatches among both the energy levels and the rotations of the octahedra. The resulting disorder scattering gives rise to localized states. The rich T-x phase diagram of $Sr_2Ir_{1-x}Rh_xO_4$ features two major effects: **(1)** Light Rh doping ($0 \leq x \leq 0.16$) effectively reduces the SOI and prompts simultaneous and precipitous drops in both the electrical resistivity $\rho(T)$ and the magnetic ordering temperature T_C. These results indicate that the Rh concentration does provide a degree of control of the splitting between the $J_{eff} = 1/2$ and $J_{eff} = 3/2$ bands. **(2)** Heavier Rh doping ($0.24 < x < 0.85 \pm 0.05$) increases localization effects that foster a return to an insulating state with anomalous magnetic behavior below 0.3 K, which only disappears near $x = 1$. The magnetic state is expected to arise from the strong competition between AFM and FM couplings that causes strong spin frustration. A recent optical study [33] on thin-film $Sr_2Ir_{1-x}Rh_xO_4$ with x up to 0.26 is qualitatively consistent with some of our results. Note that the present work addresses structural and physical properties of *bulk single-crystal* $Sr_2Ir_{1-x}Rh_xO_4$ with *x ranging from 0 to 1*, which has not been reported before.

Substitution of Rh^{4+} for Ir^{4+} results in a nearly uniform reduction in the **a**- and **c**-axis lattice parameters and a unit cell volume V that shrinks by ~ 2%, as shown in **Figs. 8.6b** and **8.6c**. This behavior is expected for Rh^{4+} doping because the ionic radius of Rh^{4+} (0.600 Å) is smaller than that of Ir^{4+} (0.625 Å). (An increase in the lattice parameters would be anticipated for Rh^{3+} ($4d^6$) doping because of the larger ionic radius = 0.670 Å for Rh^{3+}.) The **a**-axis is compressed by 0.87%, whereas the **c**-axis is compressed by only 0.26%, which enhances the tetragonal CEF. In addition, the Ir-O-Ir bond angle θ increases significantly near x = 0.16, indicating a less distorted lattice for x > 0.16. It was already established that θ is critical to the electronic and magnetic structure of Sr_2IrO_4 [15-17, 21, 24].

Rh doping effectively suppresses the magnetic transition T_C from 240 K at x = 0 to zero at x = 0.16, as shown in **Figs. 8.7a, 8.7b** and **8.7c**. With increasing x, the **c**-axis magnetization M_c stengthens at $\mu_oH = 0.1$ T. This is consistent with the reduced magnetic anisotropy in isothermal magnetization M(H) for x = 0.11 at T = 1.7 K and $\mu_oH_c = 1$ T, as shown in **Fig. 8.7b**. The ratio M_a/M_c at lower fields (< 2T) is significantly weaker for x = 0.11 than for x = 0 [16]. This change could be due to a change in the relative strength of the SOI and tetragonal CEF, as the tetragonal CEF is enhanced due to the increased c/a ratio (**Fig. 8.6b**), and encourages a spin alignment along the **c**-axis [13]. The magnetic data in **Fig. 8.7** were first fit to a Curie-Weiss law $\chi = \chi_o + C/(T+\theta_{CW})$ over the temperature range of 50-350 K for x > 0.16 (χ_o is a temperature-independent constant, θ_{CW} the Curie-Weiss temperature and C the Curie constant). Then χ_o could be used to obtain $\Delta\chi = C/(T+\theta_{CW})$, and the data plotted in terms of $\Delta\chi^{-1}$ vs. T, as shown in **Fig. 8.7c** (right scale). θ_{CW} tracks the rapidly decreasing T_C for $0 \leq x \leq 0.16$, and becomes nearly zero at x = 0.16, and then changes its sign from positive to negative as x increases further. It is remarkable that θ_{CW} is -72 K at x = 0.42, and then attains -2 K at x = 1, as shown in **Fig. 8.7d**.

Since θ_{CW} measures the strength of the magnetic interaction, a large absolute value $|\theta_{CW}| = 72$ K for a system without magnetic ordering above 0.3 K (magnetic order is observed below 0.3 K, as discussed below) implies a strong depression of the ordering temperature by magnetic frustration. It becomes conceivable that both the disappearance of magnetic order at x = 0.16 and the appearance of frustration at higher x are a consequence of atomic disorder among the Rh and Ir sites. Furthermore, changes in local atomic energies with x, such as the SOI, the non-cubic CEF, and the Hund's rule coupling, could influence the

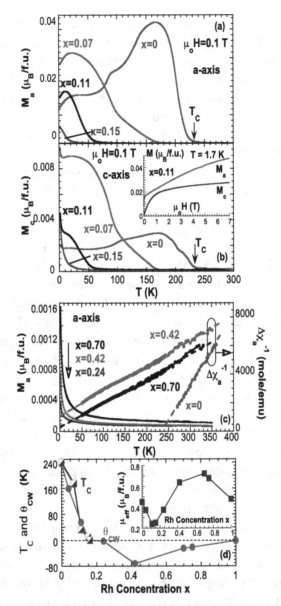

Fig. 8.7. (a) The temperature dependence at $\mu_o H = 0.1$ T of the magnetization M_a, and (b) M_c for $0 \leq x \leq 0.15$ of single-crystal $Sr_2Ir_{1-x}Rh_xO_4$. Inset in (b) Isothermal magnetization M_a and M_c for $x = 0.11$ at $T = 1.7$ K. (c) M_a for $0.24 \leq x \leq 0.75$, and $\Delta\chi_a^{-1}$ (right scale) for $x = 0, 0.42,$ and 0.70. (d) The Rh concentration (x) dependence of T_C and θ_{CW}. Inset in (d) The magnetic effective moment μ_{eff} [Ref. 54].

competition between AFM and FM couplings. In addition, the magnetic susceptibility $\chi(T)$ follows a power law, $\chi(T) \sim T^{-\alpha}$ for T < 40 K and $0.24 \leq x \leq 0.75$. The exponent α increases from 0.35 to 0.57 with increasing x, suggesting strong spin interactions exist among unscreened spins even at low temperatures.

Rh substitution for Ir unexpectedly generates three doping regions having distinct electrical transport behavior: Region I spans $0 \leq x \leq 0.24$, Region II spans $0.24 < x < 0.85$ (± 0.05), and Region III spans 0.85 (± 0.05) $< x \leq 1$. Region III represents a metallic state that exists in a very narrow region close to x = 1 (i.e., Sr_2RhO_4), and is thoroughly discussed in Ref. 17. Here we focus on Regions I and II, as discussed below.

Region I, $0 \leq x \leq 0.24$: Rh substitutions reduce the electrical resistivity $\rho(T)$ along the **a**- and **c**-axis by nearly six orders of magnitude at low temperatures (e.g., from $\sim 10^6$ Ω cm at x = 0 to ~ 1 Ω cm at x = 0.07), as shown in **Fig. 8.8a**. For $0.07 < x \leq 0.24$, the **a**-axis resistivity $\rho_a(T > 50$ K) exhibits metallic behavior with $d\rho_a/dT > 0$ (which is most obvious at x = 0.11), but with a moderately high magnitude, ranging from 10^{-3} to 10^{-1} Ω cm (see **Fig. 8.8b**). The corresponding **c**-axis resistivity $\rho_c(T)$ is slightly larger, but with negative $d\rho_c/dT$, as shown in **Fig. 8.8d**. Both $\rho_a(T)$ and $\rho_c(T)$ exhibit a noticeable upturn below 50 K, indicating that a low-temperature metallic state is not fully realized although $\rho_a(T)$ and $\rho_c(T)$ are radically reduced by six orders of magnitude.

We note the bond angle θ, which is critical to electron hopping, remains essentially unchanged until x > 0.16 (**Fig. 8.6c**), and Rh doping adds no holes or electrons to the bands. Therefore, the drastic reductions in $\rho_a(T)$ and $\rho_c(T)$ may symptomatic of a weakened SOI. In addition, the vanishing magnetic state in this doping range may also help reduce the band gap, since the internal magnetic field lifts the degeneracy along the edge of the AF Brillouin zone, which facilitates the opening of a larger gap by the SOI in the presence of U [25].

There is also evidence that Anderson localization comes into play at x = 0.24 and T < 50 K, where $\rho_a(T)$ follows the variable range hopping (VRH) model, $\rho \sim \exp (1/T)^{1/2}$. The persistent nonmetallic state below 50 K suggests that the band gap is not fully closed, despite the weakened SOI and reduced internal magnetic field. It is interesting that 14% Ru

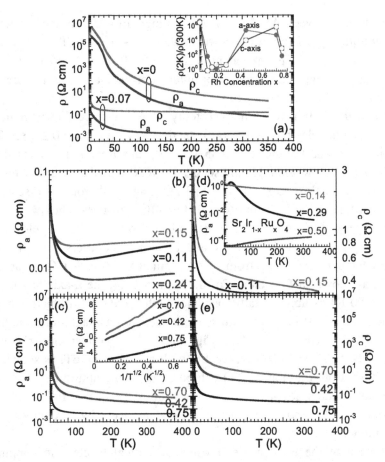

Fig. 8.8. Temperature dependence of electrical resistivity for $Sr_2Ir_{1-x}Rh_xO_4$. **(a)** The resistivity ρ, and the **a**-axis and **c**-axis resistivities ρ_a and ρ_c, respectively, for x = 0 and 0.07. Inset in **(a)**: The ratio $\rho(2K)/\rho(300K)$ vs x. **(b)** ρ_a for x = 0.11, 0.15 and 0.24 **(c)** ρ_a for x = 0.42, 0.70 and 0.75. Inset in **(c)**: $\ln \rho_a$ vs $T^{-1/2}$. **(d)** ρ_c for x = 0.11 and 0.15. Inset in **(d)**: ρ_a vs. T for $Sr_2Ir_{1-x}Ru_xO_4$, where a robust metallic state occurs at x = 0.50. **(e)** ρ_c for x = 0.42, 0.70 and 0.75 [Ref. 54].

substitution, which not only reduces SOI, but also adds holes to the bands, also fails to induce a metallic state (see Inset in **Fig. 8.8d**).

　　Region II, 0.24 < x < 0.85 (±0.05): If the reduction of SOI were the only important mechanism behind the electrical transport behaviors discussed above, an enhanced metallicity would be expected with increasing x. However, both $\rho_a(T)$ and $\rho_c(T)$ increase significantly, reaching 10^5 and 10^7 Ω cm, respectively, at low temperatures for x = 0.70, before dropping back to 10^{-1} Ω cm for x = 0.75, as shown in **Figs. 8.8c** and **8.3e**. No metallic behavior (dρ/dT > 0) is observed over the entire temperature range measured for x = 0.42, 0.70 and 0.75. The insulating state extant in this region is evidently the consequence of Anderson localization due to disorder on the Rh/Ir sites, since the resistivity at these Rh concentrations fits the VRH form, $\rho \sim \exp (1/T)^{1/2}$ for 2 < T < 100 K **[34]**. However, the existence of Rh clusters cannot be ruled out, although the single crystals studied are highly ordered (**Fig. 8.8e**). It sharply contrasts the well-established metallic state in $Sr_2Ir_{1-x}Ru_xO_4$ with x = 0.50 (inset of **Fig. 8.8d**). It is important to note that $\rho_a(T)$ and $\rho_c(T)$ for our oxygenated single crystals with x = 0.42, 0.70 and 0.75 exhibit essentially identical magnitudes and temperature dependences, which rules out an insulating state induced by oxygen deficiency. Indeed, X-ray refinements confirm no discernible oxygen deficiency is present in the single crystals studied.

The effective moment μ_{eff} essentially tracks the change of the ratio of $\rho(2K)/\rho(300K)$, which must associate localized states with the magnetic degrees of freedom (see inserts in **Figs. 8.7d and 8.8a**). The ratio of $\rho(2K)/\rho(300K)$ for both $\rho_a(T)$ and $\rho_c(T)$ also qualitatively tracks the change of transport properties with Rh concentration x (see Inset in **Fig. 8.8c**). The initial, precipitous drop in the ratio from $\sim 10^6$ at x = 0 to ~ 1 near x = 0.16 signals the rapidly growing metallic state. The ratio rises again at x > 0.24, marking the return into an insulating state, before falling back for x > 0.70.

The temperature dependence of the specific heat C for various Rh concentrations x is shown in **Fig. 8.9a**. Fitting the data to C(T) = γT + βT^3 for 10 < T < 50 K yields the coefficient γ for the electronic contribution to C(T), which systematically increases with x from 7 mJ/mole K^2 at x = 0 to 30 mJ/mole K^2 at x = 1. The increased γ for the insulating region 0.24 < x $\leq$ 0.75 may be a result of the states in the gap that are localized due to disorder, which give rise to a finite density of states (**Fig. 8.4d**). Since Rh is not a dopant in the conventional sense, the simple picture of hydrogen-like impurities does not apply. However, any breaking of the translational invariance of the system (such as Rh ions) necessarily introduces a bound state in the gap of a semiconductor, and leads to a gradual filling of the gap with increasing x. Remarkably, C(T)/T exhibits a pronounced peak near $T_M = 100$ mK and 280 mK for x = 0.42 and 0.70, respectively, which can be completely suppressed by a

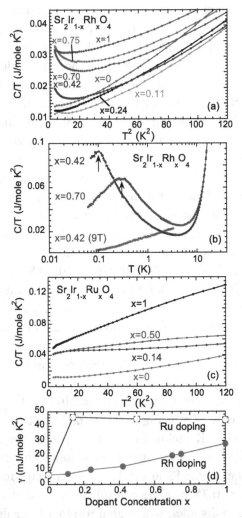

Fig. 8.9. (a) The specific heat divided by temperature T, C(T)/T vs. T^2 for $Sr_2Ir_{1-x}Rh_xO_4$. **(b)** C(T)/T vs. T for 50 mK < T < 20 K at applied magnetic field $\mu_oH = 0$ for x = 0.42 and 0.70, and $\mu_oH = 9$ T applied along the c-axis for x = 0.42. **(c)** C(T)/T vs. T^2 for $Sr_2Ir_{1-x}Ru_xO_4$ for comparison. **(d)** Linear heat capacity coefficient γ vs. x for $Sr_2Ir_{1-x}Rh_xO_4$ and $Sr_2Ir_{1-x}Ru_xO_4$ [Ref. 54].

magnetic field $\mu_oH = 9$ T (**Fig. 8.9b**). This anomaly signals a transition to a low-T spin order from a higher-T frustrated state, characterized by a frustration parameter $f = |\theta_{CW}|/T_M$ (= |-72|/0.1 = 720 for x = 0.42, for example). In contrast, $Sr_2Ir_{1-x}Ru_xO_4$ behaves conventionally (**Fig. 8.9c**), exhibiting γ values considerably larger than those for $Sr_2Ir_{1-x}Rh_xO_4$ (**Fig. 8.9d**), which is consistent with a robust metallic state.

Fig. 8.10 shows a phase diagram for $Sr_2Ir_{1-x}Rh_xO_4$ based on the alloying data presented above, and which illustrates some central results. The initial Rh doping effectively reduces the SOI, or the splitting between the $J_{eff} = 1/2$ and $J_{eff} = 3/2$ bands; that alters the relative strength of the SOI and the tetragonal CEF which dictate the magnetic state, and in turn, affects the band gap near E_F. In addition, the Rh doping also enhances the Hund's rule coupling that competes with the SOI, and prevents the formation of the $J_{eff} = 1/2$ state [25]. These SOI-induced changes account for the simultaneous, precipitate decrease in $\rho(T)$ and T_C, which vanishes at x = 0.16. As x increases further, the Rh/Ir disorder on the transition metal site determines the properties of the system. There is an energy level mismatch for the Rh and Ir sites that makes the hopping of the carriers between an octahedron containing a Rh atom and one with an Ir ion more difficult, and also changes the orientation angles

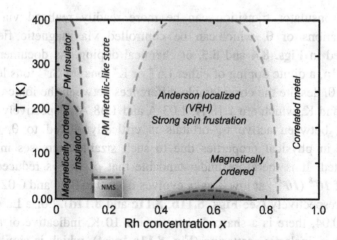

Fig. 8.10. The phase diagram for $Sr_2Ir_{1-x}Rh_xO_4$. Note that PM denotes "paramagnetic state", and NMS a "nonmetallic state" [Ref. 54].

of the octahedra. The random distribution of the Rh and Ir ions gives rise to Anderson localization and an insulating state for $0.24 < x < 0.85$ (± 0.05). In addition, the SOI may no longer be strong enough to support the $J_{eff} = 1/2$ insulating state; and the Hund's rule coupling is enhanced (on the Rh sites), further strengthening the competition between AFM and FM couplings, which generates magnetic frustration at intermediate temperatures. The occurrence of magnetic order below 0.3 K, along with the high θ_{CW}, corroborates the frustrated state. These effects diminish with decreasing disorder as x approaches 1, where the weakened SOI is comparable to other relevant energies, which permits a metallic state. This scenario is qualitatively consistent with recent theoretical studies for Sr_2RhO_4 [5, 6, 26].

In contrast, there is no discernible effect due to disorder in $Sr_2Ir_{1-x}Ru_xO_4$. In the case of isoelectronic Rh substitution the system remains proximate to the Mott condition for an insulator. On the other hand, each Ru atom adds one hole, giving rise to a higher density of states near E_F, which supports a more robust metallic state in Sr_2RuO_4. Under these circumstances disorder in the alloy plays a less dominant role.

8.2.3. *Metal-Insulator Transition via Electron or Hole Doping*

A metal-insulator transition can be more readily realized via slight manipulations of θ, which can be controlled via magnetic field, as illustrated in **Figs. 8.4** and **8.5**, or chemical doping. As documented in **Fig. 8.11a**, a dilute doping of either La^{3+} or K^+ ions for Sr^{2+} ions leads to a larger θ, despite the considerable differences between the ionic radii of Sr, La, and K, which are 1.18 Å, 1.03 Å and 1.38 Å, respectively. Since hopping between active t_{2g} orbitals is critically linked to θ, drastic changes in physical properties due to such sizable increases in θ are anticipated. It is therefore understandable that ρ_a (ρ_c) is reduced by a factor of 10^{-8} (10^{-10}) at low T as x evolves from 0 to 0.04 and 0.02 for La and K, respectively (see **Figs. 8.11b, 8.11c** and **8.11d**). For a La doping of $x = 0.04$, there is a sharp downturn near 10 K, indicative of a rapid decrease in inelastic scattering (**Fig. 8.11c Inset**), which is similar to a low-temperature anomaly observed in oxygen-depleted $Sr_2IrO_{4-\delta}$ with

$\delta = 0.04$ **[15]** (see **Fig. 8.12b** and discussion below). It is noteworthy that T_C decreases with La doping in $(Sr_{1-x}La_x)_2IrO_4$ (not shown), and vanishes at x = 0.04, where the metallic state is fully established. An early study **[32]** also shows a decrease in resistivity at a 2.5% replacement of Sr by La in polycrystalline Sr_2IrO_4. In contrast, magnetic order coexists with a fully metallic state in $(Sr_{0.98}K_{0.02})_2IrO_4$, as shown in **Fig. 8.11d**.

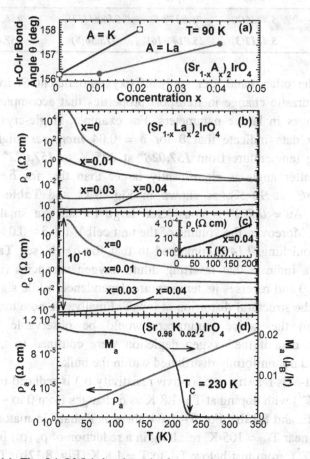

Fig. 8.11. (a) The Ir1-O2-Ir1 bond angle θ as a function of La and K doping concentration x. The temperature dependences of (b) the **a**-axis resistivity ρ_a, and (c) the **c**-axis resistivity ρ_c for $(Sr_{1-x}La_x)_2IrO_4$ with $0 \leq x \leq 0.04$. Inset in (c): Enlarged $\rho_c(T)$ at low T. (d) The temperature dependence of ρ_a and the **a**-axis magnetization M_a at applied field $\mu_oH = 0.1$ T (right scale) for $(Sr_{0.98}K_{0.02})_2IrO_4$. Note that the **c**-axis resistivity for x = 0.04 at 2 K is ten orders of magnitude smaller than that for x = 0 [Ref. 16].

This comparison stresses that the occurrence of a metallic state does not necessarily demand any radical changes in the magnetic state of iridates; this observation is in accord with a conspicuous absence of a resistivity anomaly near T_C = 240 K for Sr_2IrO_4 [7, 15-17].

Table 8.3. *Lattice Parameters at T = 90 K for* δ *= 0 and 0.04*

$\delta(T=90K)$	$a(Å)$	$c(Å)$	$V(Å3)$	Bond angle θ
0	5.4836(8)	25.8270(5)	776.61(22)	156.280°
0.04	5.4812(3)	25.8146(16)	775.56(8)	157.072°

On the other hand, our study of oxygen-deficient $Sr_2IrO_{4-\delta}$ also reveals a drastic change in transport properties that accompany significant changes in lattice parameters. For example, single-crystal X-ray diffraction data indicate that θ for δ = 0.04 *increases* slightly with decreasing temperature from *157.028°* at 295 K to *157.072°* at 90 K, and the latter angle is significantly larger than that for δ = 0, (i.e., $\theta = 156.280°$ at 90 K), as shown in **Fig. 8.12** and **Table 8.3**. The increment $\Delta\theta = 0.792°$ is considered large for such a small oxygen depletion. Moreover, the volume of the unit cell V for $\delta \approx 0.04$ contracts by an astonishing *0.14%* compared to that for δ = 0 (see **Table 8.3**). These data indicate that inserting dilute oxygen vacancies (increasing δ) relaxes θ and reverses its temperature dependence, while significantly reducing the structural distortion at low T. Finally, we note that no such changes in the lattice parameters would be observable in X-ray diffraction data if the oxygen depletion were confined to the crystal surface and not uniformly distributed within the bulk.

The **a**-axis resistivity ρ_a (**c**-axis resistivity ρ_c) is reduced by a factor of 10^{-9} (10^{-7}) with doping at T = 1.8 K as δ changes from 0 to ~ 0.04 (see **Figs. 8.12a** and **8.12b**). For $\delta \approx 0.04$, there is a sharp insulator-to-metal transition near T_{MI} = 105 K, resulting in a reduction of ρ_a (ρ_c) by a factor of 10^{-4} (10^{-1}), from just below T_{MI} to T = 1.8 K (**Fig. 8.12b**). The strong low-T anisotropy reflected in the values of ρ_a and ρ_c and their sensitivity to δ are consistent with a nearly two-dimensional, strongly correlated electron system. Below 20 K, ρ_a has linear-T dependence *without saturation to a residual resistivity limit*; and although there is a plateau in ρ_c for $5 < T < 35$ K, it is followed by a very rapid downturn near $T_a = 5$ K (**Fig. 8.12b** inset), which indicates a sudden, rapid decrease in inelastic scattering.

Oxygen depletion also changes the electronic density of states $g(E_F)$ of $Sr_2IrO_{4-\delta}$, as reflected in the thermoelectric power $S(T)$, as shown in **Fig. 8.12c**. A peak in the c-axis $S_c(T)$ for $\delta = 0.04$ is only $\sim 1/3$ of the

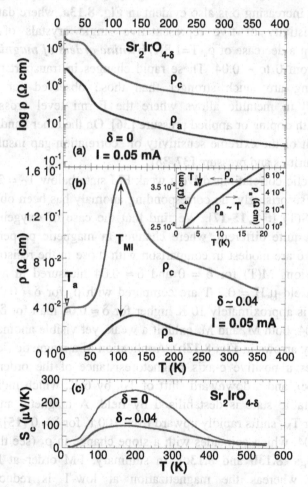

Fig. 8.12. The a- and c-axis resistivities, ρ_a and ρ_c, respectively, as a function of temperature T for (a) $\delta = 0$, and (b) $\delta \approx 0.04$. **Inset** in (b): ρ_a and ρ_c vs. T for 1.7 K < T $\leq$ 20 K; note a downturn near T = 5 K in ρ_c and linear temperature dependence in ρ_a. (c) c-axis thermoelectric power, $S_c(T)$, for $\delta = 0$ and $\delta \approx 0.04$ [Ref. 17].

peak value observed for $\delta = 0$. Since S(T) measures the voltage induced by a ***temperature gradient*** (which cannot be confined to the surface of the crystal), the drastic changes in S(T) shown in **Fig. 8.12c** further reinforces our conclusion that oxygen depletion is a ***bulk effect***, consistent with the observed changes in lattice parameters discussed above. The strong reduction of $S_c(T)$ for $\delta \approx 0.04$ indicates an increase of $g(E_F)$ with increasing δ, since $S \propto 1/g(E_F)$ [35]. The rapid increase of $g(E_F)$ with increasing δ is also evident in **Fig. 8.13a**, where data for the **a**-axis resistivity of five representative single crystals of $Sr_2IrO_{4-\delta}$ document a decrease of $\rho_a(T=1.8K)$ by ***nine orders of magnitude*** as δ changes from 0 to ~ 0.04. These rapid changes in transport properties with doping are much stronger than those observed for "Lifshitz transitions" in metallic alloys where the Fermi level crosses small pockets with doping or applied pressure [36]. On the other hand, they are reminiscent of the extreme sensitivity of "correlation-gap insulators" to dilute impurities and pressure [37, 38].

Magnetic correlations drive a weak FM state below $T_C = 240$ K for $\delta = 0$, but surprisingly, no corresponding anomaly has been observed in $\rho(T)$ and S(T) [5-7, 15-17]. We find that the case of oxygen-depleted crystals is quite different, where changes in magnetic properties with increasing δ are modest in comparison with those in the resistivity. The magnetizations M(T) for $\delta = 0$ and $\delta \approx 0.04$ measured in an applied magnetic field $\mu_oH = 0.2$ T are compared with ρ_a for $\delta \approx 0.04$ in **Fig. 8.13b**. T_C is approximately 10 K higher for $\delta \approx 0.04$ than for $\delta = 0$; and for $\delta \approx 0.04$, both M_a and M_c exhibit a weak, yet visible anomaly at T_{MI} (marked by arrows in **Fig. 8.13b**). Application of a magnetic field $\mu_oH = 7$ T causes a positive **c**-axis magnetoresistance of the order of 10% (**Fig. 8.13c**), and a downward shift of T_{MI} by 6 K, which indicates the low-T metallic state is destabilized by field. A magnetic anomaly in $M_c(T)$ near T_{MI} shifts rapidly upward from 100 K for $\delta = 0$ [15], to 160 K for $\delta \approx 0.04$, which coincides with a slope change in ρ_a (see the dashed line in **Figs. 8.13b** and **8.13c**). In summary, FM order at high-T is stabilized, whereas the magnetization at low-T is reduced, with increasing δ.

The bandwidth W is quite sensitive to structural alterations according to a first-principles calculation [24] that predicts that an

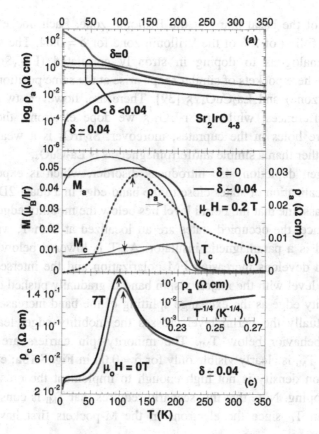

Fig. 8.13. Temperature dependence of (a) log ρ_a for several representative values of δ; (b) the a- and c-axis magnetization, M_a and M_c, for $\delta = 0$ (thin lines) $\delta \simeq 0.04$ (thick lines) and ρ_a (right scale) for $\delta \approx 0.04$; and (c) ρ_c at 0 and 7 T. **Inset in (c):** log ρ_a vs. $T^{-1/4}$ [Ref. 17].

increased θ should cause a broadening of the $J_{eff} = 1/2$ band and a concomitant decrease of the Mott gap by 0.13 eV, if θ increases from 157° to 170°. The observed increment Δθ = 0.792° does not appear nearly sufficient to produce the dramatic changes we have observed in ρ, and we conclude that another mechanism must be responsible for T_{MI}.

Removal of oxygen from $Sr_2IrO_{4-\delta}$ is expected to result in electron doping of the insulating state that is observed to be stable for δ = 0. According to (LDA+SO+U) band structure calculations [24] additional electrons will occupy states in four symmetric pockets located near the

M-points of the basal plane of the Brillouin zone. Each pocket has an estimated filling of 2% of the Brillouin zone for $\delta = 0.04$. The situation appears analogous to doping in strongly correlated $(La_{1-x}Sr_x)_2CuO_4$ (LSCO) (where pockets of similar shape arise at the same positions in the Brillouin zone) and $La_2CuO_{4+\delta}$ [39]. There are, however, two fundamental differences: while in $Sr_2IrO_{4-\delta}$ we dope electrons, the added carriers are holes in the cuprates; moreover, Sr_2IrO_4 is a weak ferromagnet rather than a simple antiferromagnet, as is La_2CuO_4.

Oxygen depletion also introduces disorder, which is expected to lead to localization of states close to the band-edge in a quasi-2D system [40]. We assume that the Fermi level lies below the mobility edge for $\delta \neq 0$, and hence, the occupied states are all localized at high T, where the compound is a paramagnetic insulator. As T is lowered below T_C, the compound develops increasing FM polarization and the intersection of the Fermi level with the majority spin band is gradually pushed closer to the mobility edge as the exchange splitting of the band increases below T_C. Eventually the Fermi level crosses the mobility edge, leading to metallic behavior below T_{MI}. The minority-spin carriers are always localized. T_{MI} is clearly visible only for $\delta \approx 0.04$ in **Fig. 8.13a**; evidently the electron density is not high enough to implement the crossing for smaller doping. Note that this scenario requires that T_{MI} is considerably lower than T_C since the electrons in the M-pockets first have to be polarized.

The reversed trend of the temperature dependence of θ and the increase of θ with doping could be consequences of increased screening in the metallic state. Furthermore, we expect the metallic state to be inhomogeneous and the conductivity to increase due to the growth and percolation of metallic patches; i.e., the metallic state develops out of a phase separation of competing states. The importance of disorder in the physical properties of $Sr_2IrO_{4-\delta}$ is corroborated by fits (**Fig. 8.13c** inset) using a variable range hopping (VRH) relation $\rho_a(T) = A\exp(T_o/T)^\upsilon$ with $\upsilon = \frac{1}{4}$ for $187 < T < 350$ K, and with T_o a characteristic temperature; such behavior is also observed in nonmetallic samples [7].

The radical changes in transport properties of Sr_2IrO_4 with dilute doping strongly suggest that the insulating state driven by a strong SOI is proximate to a metallic state. The inducement of a robust metallic state by either dilute electron or hole doping for Sr^{2+} further reinforces the notion that transport properties in iridates such as Sr_2IrO_4 are mainly dictated by the lattice degrees of freedom. The large and uniquely anisotropic magnetoresistivity in Sr_2IrO_4, and the robust metallic state in doped Sr_2IrO_4, have no apparent correlation with the magnetization, as conventionally anticipated. We attribute these unusual effects to the subtle unbuckling of the IrO_6-octahedra. Furthermore, the lattice distortions that dictate the underlying electronic properties in Sr_2IrO_4 can be controlled via either magnetic field or chemical doping.

8.2.4. *Non-Ohmic Behavior*

The exotic nature of the metallic state of $Sr_2IrO_{4-\delta}$ is revealed in striking non-Ohmic behavior, as exhibited by $\rho_c(T)$ for various applied currents I (see **Fig. 8.14a**; ρ_a behaves similarly, and is not shown). ρ_c changes slightly when $I \leq 1$ mA, but more dramatically when $I \geq 5$ mA (~ 10 A/cm^2 current density). Moreover, there is a characteristic temperature $T^* = 52$ K at which ρ_c sharply drops for $I = 5$ mA, but rises for $I = 14$ mA (see **Figs. 8.14a** and **8.14c**), indicating a current-induced ("dynamic phase") transition. It is particularly noteworthy that the abrupt downturn in ρ_c below $T_a \approx 5$ K disappears for $I \geq 1$ mA, although it is insensitive to applied magnetic field (see **Fig. 8.13c**).

The non-linear I-V characteristic shown in **Fig. 8.14b** shows switching occurs at multiple threshold potentials as I varies from 0.1 μA to 50 mA. We infer a temperature T^* that separates two different regions: For $T < T^*$, there are three threshold potentials, V_{th1}, V_{th2} and V_{th3}. The initial linearity in the I-V curve persists up to $I = 4$ mA for $V < V_{th1} = 0.011$ V at 5 K; between V_{th1} and V_{th2} linearity is briefly restored with a reduced slope. With further increases in I, the I-V response exhibits a third threshold V_{th3}, which marks the onset of ***current-controlled negative differential resistivity*** (NDR), where V across the crystal decreases as I increases. The qualitative difference between the two regions separated by T^* is clearly revealed in the temperature dependences of V_{th1}, V_{th2} and V_{th3}, as well as ρ_c, at $I = 5$ and 14 mA, as shown in **Fig. 8.14c**. Note that V_{th2} and V_{th3} increase with increasing T below T^*. ***For $T > T^*$, the trend is reversed:*** V_{th3} and V_{th2} shift to lower

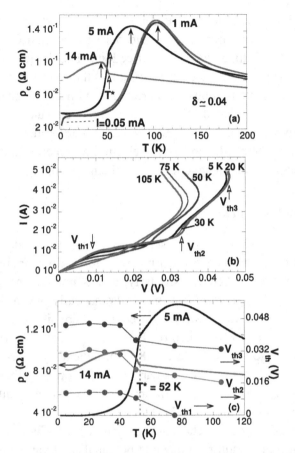

Fig. 8.14. (a) Temperature dependence of ρ_c for $Sr_2IrO_{4-\delta}$ at several representative values of current I. (b) I-V curves at several representative temperatures. (c) Temperature dependence of ρ_c at I = 5 and 14 mA and threshold V_{th} (right scale). V_{th1}, V_2 and V_3 represent three V thresholds [Ref. 17].

values, and V_{th1} tends to zero with increasing T. Note that T* remains sharply defined in $\rho_c(T)$ at 52 K, independent of I (**Fig. 8.14c**). This completely rules out the possibility that self-heating plays a role in the non-Ohmic behavior.

It is noteworthy that non-Ohmic behavior or NDR, which is not commonplace for bulk materials, has been observed in the *insulating state of layered iridates*, such as stoichiometric Sr_2IrO_4 [7] and $BaIrO_3$ [9]. These materials exhibit switching behavior at *a single threshold* V_{th}

that depends on temperature, and is much higher than the upper threshold V_{th3} in $Sr_2IrO_{4-\delta}$ for $\delta \approx 0.04$. The single-threshold effect was attributed to collective charge density wave (CDW) dynamics in the presence of disorder commonly seen in the CDW state [7, 9]. Indeed, density waves pinned by oxygen vacancies and then depinned by applied potential is one possible mechanism for the non-Ohmic behavior. The pockets at the M-points are predicted to have an elongated ellipsoidal shape in the plane, and due to the quasi-2D nature of the compound, a nesting condition between the pockets cannot be ruled out. Such nesting could give rise to either spin or charge density waves that would have to coexist in the presence of ferromagnetism.

The observed non-Ohmic behavior of the *metallic state* therefore poses intriguing questions concerning its origin: Does non-Ohmic behavior observed in both the insulating and metallic states of iridates arise from the same mechanism — i.e., a CDW state? Note that the voltage thresholds ($\sim 10^{-2}$ V) observed for $\delta = 0.04$ are two orders of magnitude smaller than those (~ 1 V) observed in CDW depinning experiments [19], which would suggest an extremely weak CDW pinning by defects that has not been reported before. Otherwise, the observed non-Ohmic behavior of the metallic state must signal a novel metallic state that does not follow Ohm's law.

The novel behavior of $Sr_2IrO_{4-\delta}$ clearly illustrates an intrinsically unstable ground state that readily swings between highly insulating (10^5 Ω cm) and metallic (10^{-5} Ω cm) states with only very slight changes in oxygen content. The non-Ohmic behavior, whose origin is yet to be fully understood, constitutes a new paradigm for device structures in which resistivity can be manipulated with modest applied currents rather than large magnetic fields or much larger voltages.

8.2.5. *Giant Magnetoelectric Effect*

The vast majority of known magneto-electrics and multiferroics are 3d-based compounds [42-51], whereas there are no known examples of ferroelectric 5d materials. Moreover, the traditional view is that the giant magneto-electric effect (GME) depends only on the magnitude and spatial dependence of the magnetization, whereas a recently proposed mechanism points to the critical role of an effective spin-orbit gap Δ_s [52]. Our results indicate this new mechanism is realized in Sr_2IrO_4.

Spin canting, the T-dependent Ir-O-Ir bond angle, and loss of inversion symmetry can also influence the dielectric behavior, which is strongly dependent on crystal symmetry. Indeed, the magnetic anomaly at T_M is closely linked to the dielectric response, as shown in **Figs. 8.9a** and **8.9b**, where M(T) is compared to the real parts $\varepsilon_c'(T)$ and $\varepsilon_a'(T)$ of the **c**-axis and **a**-axis dielectric constants, respectively. Two major features emerge: (1) both $\varepsilon_c'(T)$ and $\varepsilon_a'(T)$ rise by up to ***one order of magnitude*** and peak near T_M, similar to La_2CuO_4 [42]. ($\varepsilon_a(T)$ is loss-dominated above T_M, therefore we focus only on $\varepsilon_c'(T)$ in the discussion that follows.) This strong enhancement of $\varepsilon_c'(T)$ is much larger than that exhibited by well-known magneto-electrics such as $BaMnF_4$ [43], $BiMnO_3$ [44], $HoMnO_3$ and $YMnO_3$ [45]. (2) The peak in $\varepsilon_c'(T \approx T_M)$ separates two regions, I and II, as marked in **Fig. 8.15a**. The weak frequency dependence of $\varepsilon_c'(T,\omega)$ in low-T Region I is typical of a ferroelectric, whereas the stronger frequency dependence of $\varepsilon_c'(T,\omega)$ in the higher-T Region II suggests a relaxor mechanism [46], which is traditionally attributed to disorder and impurities. Alternatively, the sharp peak accompanied by strong frequency dispersion could signal a novel frustrated or disordered magnetic state corresponding to the shaded area in **Figs. 8.15a** and **8.15b**.

It is interesting that electric polarization hysteresis is certainly observed in Sr_2IrO_4 as shown in **Fig. 8.15c**, suggesting a possible existence of some type of ferroelectric state at low T. Furthermore, a magneto-dielectric shift $\Delta\varepsilon_c'(H)/\varepsilon_c'(0)$ is also anticipated and observed in $M_c(H)$ near the metamagnetic transition field H_c (**Fig. 8.16**). (The negligible magnetoresistance in Sr_2IrO_4 at applied fields up to 12 T [7] suggests that $\Delta\varepsilon_c'(H)/\varepsilon_c'(0)$ is an intrinsic effect.) However, ***we do not observe*** $\Delta\varepsilon_c'(H) \propto M^2$ as conventionally expected [44], and M (< 0.1 μ_B/Ir) is exceptionally weak compared to known multiferroics (e.g., $M \approx 6$ μ_B/f.u. for $TbMnO_3$ [51]).

Although the GME in Sr_2IrO_4 is unconventional, it can be understood as a unique manifestation of a recently formulated microscopic mechanism for magneto-electrics with strong spin-orbit coupling; this novel approach yields polarization P proportional to an effective spin-orbit gap Δ_s rather than the magnitude and spatial dependence of the magnetization [52].

In light of all results presented above, it is suggested that T_M defines a drastic change in the coupling between the magnetic and dielectric response, according to the following scenario: In Region II, the strong competition between FM and AFM exchange couplings promotes frustrated or incommensurate magnetic order. A T-dependent magneto-

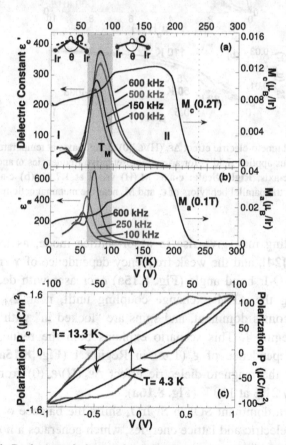

Fig. 8.15. (a) Real part of the **c**-axis dielectric constant $\varepsilon_c{'}(T)$ for representative frequencies ω (left scale), and **c**-axis DC magnetization $M_c(T)$ (right scale); **Fig. 8.15(a) inset:** Schematic change of O-Ir-O bond angle from Region I to Region II. (b) Real part of the **a**-axis dielectric constant $\varepsilon_a{'}(T)$ for representative ω (left scale), and **a**-axis $M_a(T)$ (right scale). (c) Electric polarization P vs voltage V for temperature T = 13.3 K (low V, left scale) and 4.3 K (high V, right scale) [Ref. 15].

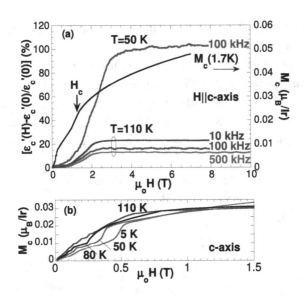

Fig. 8.16. (a) Magneto-electric effect $\Delta\varepsilon_c'(H)/\varepsilon_c'(0)$ along **c**-axis at temperatures T = 50 K and 110 K versus applied field H for a few representative frequencies ω and $\mu_oH \leq 10$ T applied along **c**-axis. **Right Scale: c**-axis $M_c(H)$ vs. H at 1.7 K. **(b) c**-axis $M_c(H)$ at various T. Note the parallel behaviors of ε_c and M_c near the metamagnetic transition field, H_c [Ref. 15].

elastic coupling may give rise to a soft lattice mode, as indicated by optical data [**24**], and the weak frequency dependence of χ' near 135 K [**15**]. The Ir-O-Ir bond angle (**Fig. 8.15a**) decreases with decreasing T, strengthening the AFM exchange coupling until, near T_M, the AFM coupling becomes dominant, and spins are "locked in" with a stiffened lattice in Region I. This scenario explains both the reduction of the frequency dependence of $\varepsilon_c'(T,\omega)$ in Region I (**Fig. 8.15a**), and the reduction of the magneto-dielectric effect $\Delta\varepsilon_c'(H)/\varepsilon_c'(0)$ from 100% at 50 K, to only 21% at 110 K (**Fig. 8.16a**).

Indeed, a dominant SOI in Sr_2IrO_4 shifts the balance of competing magnetic, dielectric and lattice energies, which generates a novel type of GME that is not dependent on the magnetization, but nevertheless is intimately linked with the complex magnetic order that emerges from an exotic Mott insulating state. We expect further examples of exciting new GME, and perhaps novel multiferroics, to be found in other 5d Mott insulators.

8.3. $Sr_3Ir_2O_7$

$Sr_3Ir_2O_7$ has strongly-coupled, double Ir-O layers with adjacent double layers offset along the **c**-axis and separated by Sr-O interlayers, as shown in **Fig. 8.2**. The previous crystal structure study [17] of $Sr_3Ir_2O_7$ reported a tetragonal cell (a = 3.896 and c = 20.879 Å) with space group I4/mmm. A refinement of our single-crystal X-ray diffraction reveals an orthorhombic cell with a = 5.5221 Å, b = 5.5214 Å, and c = 20.9174 Å with a Bbca symmetry [8]. The refinement shows that the IrO_6 octahedra are elongated along the crystallographic **c**-axis; i.e., the Ir-O apical bond distances along the **c**-axis are, on average, 2.035 Å and 1.989 Å in the IrO planes. The octahedra are rotated about the **c**-axis by 11° at room temperature. It is found that within a layer the rotations of the octahedra alternate in sign, forming a staggered structure, with the two layers of a double-layer being out of phase with one another.

8.3.1. *Unusual Magnetism and Transport Properties*

The **a**-axis magnetization M_a of $Sr_3Ir_2O_7$ as a function of temperature T is shown in **Fig. 8.17**. The zero-field-cooled (ZFC) M_a was measured on heating at an applied field $\mu_oH = 0.01$ T after cooling in zero magnetic field (ZFC) to 1.7 K from room temperature, whereas the field-cooled (FC) data were obtained on cooling at $\mu_oH = 0.01$ T from room temperature (or other temperatures specified in **Fig. 8.17b**). It is clear that the FC magnetization exhibits an abrupt FM transition at $T_C = 285$ K that is closely followed by a noticeable change of slope near T* = 260 K. Most strikingly, the FC M_a undergoes a rapid drop below $T_D = 50$ K, and becomes negative at T = 20 K, which clearly signals a magnetization reversal or a rotation of the magnetic moment that is now in opposition to the applied magnetic field. This behavior is robust and not observed in the ZFC magnetization, which instead remains positive and does not exhibit anomalies that are seen in the FC M (see **Fig. 8.17**).

In the entire temperature range measured, the ZFC magnetization reflects a severely reduced moment and a nearly straight line with an extremely weak temperature dependence that only becomes slightly stronger below 20 K. While the irreversibility is characteristic of a

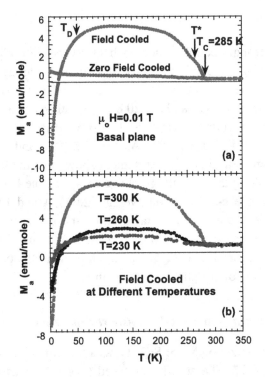

Fig. 8.17. (a) The a-axis magnetization M_a for the basal plane as a function of temperature at 0.01 T; (b) M_a measured using the FC sequence that starts at T=300 K, 260 K, and 230 K, respectively; note that FC sequence starting at different temperatures affects the magnitude of M_a.

ferromagnet, it is unusual that all magnetic anomalies are absent in the case of ZFC measurements, implying a strong spin disordering, or a random orientation of magnetic domains that is persistent through T_C. This strong irreversibility is tightly correlated with both T_C and T*. When $Sr_3Ir_2O_7$ is field cooled from just below T* rather than room temperature, all anomalies become weakened. Data for M vs T obtained in FC measurements that started at T = 300, 260, and 230 K, respectively, are shown in **Fig. 8.17b**. It is apparent that both FM and diamagnetic responses diminish considerably as the temperature from which the FC sequence starts is lowered. Interestingly, when $Sr_3Ir_2O_7$ is field cooled from the temperature range T* < T < T_C, M shows an essentially unchanged diamagnetic response but a significantly reduced ferromagnetic response. While the nature of T* is unclear, it cannot be ruled out that magnetostriction may occur near T* and "lock" up a certain spin configuration that may facilitate the magnetization reversal.

Nevertheless, these observations suggest that any spin disordering below T* has to be eliminated in order to preserve the diamagnetic state that occurs at lower temperatures. The FC M_a for various magnetic fields shown in **Fig. 8.18** suggests the diamagnetic state is metastable; as the magnetic field increases, the FM response becomes stronger, whereas the diamagnetic state diminishes and eventually vanishes at $\mu_oH = 0.25$ T. Although the FM moment is dominant, the downturn of M below T_D still remains. At $\mu_oH = 0.5$ T, the downturn of M_a develops into a sharp upturn below T_D, which suggests that all diamagnetic components that were active at modest applied fields are now aligned with the stronger applied field. Remarkably, M_a can flip back to the direction opposite to the magnetic field without hysteresis when the applied field is reduced back to $\mu_oH = 0.25$ T, so long as T < T*. This full reversibility of the spin reorientation once again implies that the diamagnetic state or the irreversibility seen in **Fig. 8.17** is more likely to be associated with an effect from the crystal structure or a magnetostrictive distortion that stabilizes the metastable diamagnetic state.

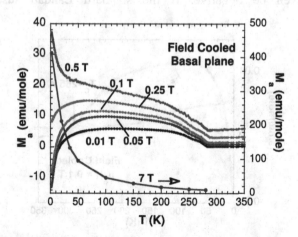

Fig. 8.18. Magnetization M for the basal plane as a function of temperature for $\mu_oH =$ 0.01, 0.05, 0.1, 0.25, 0.5 (all left scale), and 7 T (right scale) [Ref. 8].

We note that T_C remains sharp without shifting with increasing H, and T* stays well defined, unlike FM transitions that are broadened, or AFM transitions that are suppressed to lower temperatures, when a high magnetic field is applied, as shown in **Fig. 8.18.** Applying a Curie-Weiss fit to magnetic susceptibility data for $\mu_o H = 7$ T yields an effective moment $\mu_{eff} = 0.69$ μ_B (compared to 1.73 μ_B expected for S = 1/2), and a Weiss temperature $\theta_{CW} = -17$ K, suggesting an AFM coupling. The variation of the magnetic susceptibility is much weaker for the **c**-axis than for the basal plane; and the signature in the data of the magnetic transition at T_C for the **c**-axis is relatively weak (see **Fig. 8.19**).

The mechanism of such field-induced magnetic behavior is still open to debate. Recent studies of the magnetic structure of $Sr_3Ir_2O_7$ suggest a collinear AFM state along the **c**-axis, but there may exist a nearly degenerate magnetic state with canted spins in the basal plane [12]. In such circumstances, any slight perturbation such as magnetic field could induce a canted spin structure, as indicated in a recent study of magnetic X-ray scattering [14].

The observed anisotropic diamagnetism (see **Fig. 8.19**) can be attributed to the large SOI that breaks spin conservation. The net moment M can be strongly reduced with respect to the usual Pauli susceptibility, and can even be negative. If the standard Landau susceptibility

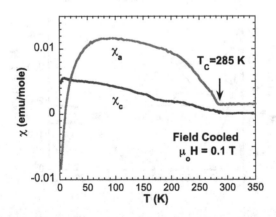

Fig. 8.19. Temperature dependence of the **a**-axis magnetic susceptibility χ_a and the **c**-axis magnetic susceptibility χ_c measured using a field cooled (FC) sequence at $\mu_o H = 0.1$ T. Note that T_C in χ_c is not as sharply defined as for the **a**-axis data [Ref. 8].

(quantization of orbits) and the core diamagnetism are added to the spin susceptibility, it is then likely to result in overall diamagnetism. Since this effect is only observed for FC but not for ZFC, it is conceivable that the bond angle θ depends on the applied magnetic field and has a memory of the recent field history.

The insulating state is revealed in the electrical resistivity ρ(T) over the range 1.7 < T < 1000 K, as shown in **Fig. 8.20**. Both $ρ_a$ and $ρ_c$ increase slowly with temperature decreasing from 1000 to 300 K, but then both rise rapidly in the vicinity of T_C and T_D, demonstrating a coupling between magnetic and transport properties. This behavior sharply contrasts that for Sr_2IrO_4, where such a correlations is absent. Both $ρ_a$ and $ρ_c$ can be fit to a 1/T-dependence for the high-temperature range 300 < T < 1000 K. $ρ_a$ is well fitted by a $1/T^{1/4}$ form for 15 < T < 100 K, suggesting variable range hopping is active. However $ρ_c$ does not obey any known power laws or exponential forms over the same temperature interval. We emphasize that the insulating behavior persists up to 1000 K in $Sr_3Ir_2O_7$.

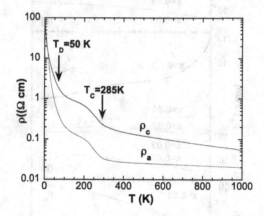

Fig. 8.20. Temperature dependence of the **a**-axis resistivity $ρ_a$ and the **c**-axis resistivity $ρ_c$ for 1.7 < T ≤ 1000 K [Ref. 8].

8.3.2. *Metal-Insulator Transition via Electron Doping and Pressure*

As discussed above, hopping between active t_{2g} orbitals is critically linked to the Ir-O-Ir bond angle θ, and such a strong coupling is manifested in the electrical resistivity ρ of single-crystal $(Sr_{1-x}La_x)_3Ir_2O_7$, as shown in **Fig. 8.21**. The a-axis resistivity ρ_a (the c-axis resistivity ρ_c) is reduced by as much as a factor of 10^{-6} (10^{-5}) at low T as x evolves from 0 to 0.05, (see **Figs. 8.21a** and **8.21b**). For x = 0.05, there is a sharp downturn in ρ_a near 10 K, indicative of a rapid decrease in inelastic scattering (**Fig. 8.21c**). Such low-T behavior is observed in slightly oxygen depleted $Sr_2IrO_{4-\delta}$ with δ = 0.04 and La-doped Sr_2IrO_4 [16, 17]. The radical changes in the transport properties of $Sr_3Ir_2O_7$ with slight doping strongly suggest that the insulating state driven by a strong SOI is

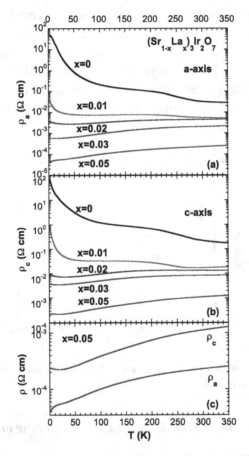

Fig. 8.21. Temperature dependence of **(a)** the a-axis resistivity ρ_a; and **(b)** the c-axis resistivity ρ_c for $(Sr_{1-x}La_x)_3Ir_2O_7$. **(c)** ρ_a and ρ_c for x = 0.05.

proximate to a metallic state. The inducement of such a robust metallic state by impurity doping further reinforces the central finding of this work: that is, transport properties in iridates such as $Sr_3Ir_2O_7$ can be mainly dictated by the lattice degrees of freedom since electron hopping sensitively depends on the bond angle θ. Indeed, It is theoretically anticipated that hopping occurs through two active t_{2g} orbitals, the d_{xy} and d_{xz} when $\theta = 180°$ and the d_{xz} and d_{yz} orbitals when $\theta = 90°$. The data shown in **Fig. 8.21** underline that the larger θ, the more energetically favorable it is for electron hopping and superexchange interactions. Values of the representative bond angle θ for $(Sr_{1-x}La_x)_3Ir_2O_7$ are listed in **Table 8.4**; the data were obtained from our single-crystal X-ray diffraction.

Table 8.4. *The Ir-O-Ir Bond Angle θ at 90 K and 295K*

x	θ (deg) at 90K	θ (deg) at 295K	$\Delta\theta$ (deg)
0	154.922	156.500	1.578
0.05	156.034	156.523	0.489

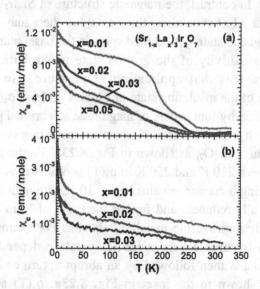

Fig. 8.22. Temperature dependence of (a) the **a**-axis magnetic susceptibilty χ_a; and (b) the **c**-axis magnetic susceptibility χ_c for $(Sr_{1-x}La_x)_3Ir_2O_7$. Note that no magnetic anomalies are observed in χ_c.

It is noteworthy that the magnetically ordered state considerably weakens and T_C decreases, with La doping in $(Sr_{1-x}La_x)_3Ir_2O_7$, but does not vanish at x = 0.05, where the metallic state is fully established. The transition at T_C is largely broadened in the **a**-axis magnetic susceptibility data $\chi_a(T)$, but is hardly visible the **c**-axis magnetic susceptibility data $\chi_c(T)$, as shown in **Fig. 8.22**. This behavior sharply contrasts that in La-doped Sr_2IrO_4, where the occurrence of a fully metallic state is accompanied by a disappearance of magnetic order [16]. This difference could be attributed, in part, to the fact that $Sr_3Ir_2O_7$ with U ~ 0.5 eV is much closer to the borderline of the metal-insulator transition, and this proximity makes it less likely for the magnetic ordering to cause or support an insulating state. Nevertheless, the differences between the two systems indicate that magnetic ordering plays different roles in determining the electronic state of either $Sr_3Ir_2O_7$ or Sr_2IrO_4. Indeed, a recent neutron diffraction study on single-crystal Sr_2IrO_4 reveals a canted spin structure within the basal plane at temperatures below T_C, and forbidden nuclear reflections of space group *I41/acd* appear over a wide temperature range from 4 K to 600 K, which suggests a reduced crystal symmetry [21]. In contrast, the magnetic structure of $Sr_3Ir_2O_7$ is believed to be a collinear AFM state along the **c**-axis; but there may exist a nearly degenerate magnetic state with canted spins in the basal plane [12].

The high sensitivity of the ground state to the lattice degrees of freedom guarantees that application of pressure also provides an effective probe of the insulating state of iridates. Indeed, $\rho_a(T)$ undergoes a drastic reduction by four orders of magnitude at a critical pressure, $P_C = 13.2$ GPa, which marks a transition from an insulating state to a nearly metallic state in $Sr_3Ir_2O_7$, as shown in **Fig. 8.23a**. Furthermore, a broad transition between 210 K and 250 K in $\rho_a(T)$ is observed at P_C. It is noted that the insulating behavior remains below 10 K, although the magnitude of ρ_a is drastically reduced; and further increases of P up to 35 GPa do not significantly improve the metallic behavior. For example, ρ_a at P = 25 GPa (>P_C) features a very weak temperature dependence at high temperatures that is then followed by an abrupt upturn or a transition in ρ_a at 5 K, as shown in the inset in **Fig. 8.23a**. $\rho_a(T)$ approximately follows an activation law, $\rho_a(T) \sim \exp(\Delta/2k_BT)$ (where Δ is the energy gap and k_B the Boltzmann's constant) in the temperature interval

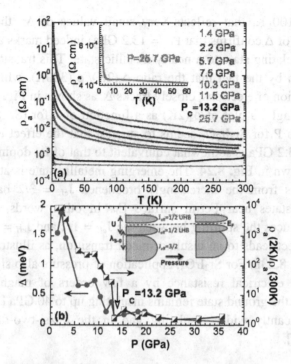

Fig. 8.23. (a) Temperature dependence of the **a**-axis resistivity ρ_a at representative pressures P for Sr$_3$Ir$_2$O$_7$; **inset:** ρ_a at P=25.7 GPa vs. temperature. (b) The activation gap Δ (left scale) and ratio ρ_a (2 K)/ρ_a (300K) (right scale) versus P for Sr$_3$Ir$_2$O$_7$; **inset:** schematic for pressure effect on the bands.

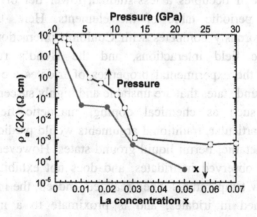

Fig. 8.24. The **a**-axis resistivity ρ_a at 2 K as a function of x for (Sr$_{1-x}$La$_x$)$_3$Ir$_2$O$_7$ and pressure (upper axis) for Sr$_3$Ir$_2$O$_7$.

$10 < T < 100$ K, that reflects very small values of Δ; the pressure dependence of Δ confirms that P_C (= 13.2 GPa) indeed marks a transition from an insulating state to a nearly metallic state. This transition is also corroborated by the fact that the ratio $\rho_a(2K)/\rho_a(300K)$, which reflects the localization of electrons, closely tracks Δ, as shown in **Fig. 8.23b**.

Interestingly, a plot of $\rho_a(2K)$ as a function of x for $(Sr_{1-x}La_x)_3Ir_2O_7$ and pressure P for $Sr_3Ir_2O_7$ seems to suggest that the effect of pressure near P_C = 13.2 GPa is somewhat equivalent to that of La doping near x = 0.03, as shown in **Fig. 8.24**. The emerging metallic state is attributed to contributions from the increasingly broadened J_{eff} = 3/2 band to the density of states near Fermi surface E_F. In other words, increasing pressure reduces the splitting between the J_{eff} = 1/2 and J_{eff} = 3/2 bands, and this effect leads to an insulator-metal transition, as illustrated in the inset in **Fig. 8.23b**. For Sr_2IrO_4, application of pressure also significantly reduces the electrical resistance by a few orders of magnitude near 20 GPa, but the ground state remains insulating up to 40 GPa [10], owing to a significantly wider insulating gap in the quasi-two-dimensional system [1, 2].

8.4. Conclusions

Iridium oxides present a fertile new field of condensed matter physics, mainly because Ir occupies a less-studied, lower tier of the transition series in the periodic table of the elements. Here, the spin-orbit interaction vigorously competes with Coulomb interactions, non-cubic crystal electric field interactions, and the Hund's rule coupling. Consequently, the experimental properties of iridates are often exotic, and reflect ground states that are unstable and highly susceptible to small perturbations such as chemical doping, magnetic field and high pressures. In particular, traditional arguments would predict that iridates should have metallic, Fermi liquid ground states. However, the metallic state is rarely observed in iridates, and does not exhibit Fermi-liquid behavior at low temperatures when realized. Indeed, the insulating state is often attained in iridates, and is proximate to a metal-insulator transition.

A number of unexpected properties of iridates have been explained in terms of a recently predicted, $J_{eff} = 1/2$ insulating state that is a profound manifestation of the SOI. Because of the strong SOI, spin is no longer a good quantum number, which is manifest in the unusual magnetic properties of Sr_2IrO_4 and $Sr_3Ir_2O_7$. It is particularly noteworthy that magnetic properties do not seem to be as closely associated with electric transport properties, as is the case in 3d transition metal oxides. Chemical doping is the most effective way to achieve a metallic state in iridates via increasing the Ir-O-Ir bond angle in IrO_6 octahedra, which points to the importance of electron-lattice couplings. It is therefore surprising that, in contrast to magnetic materials based on 3d elements, high pressure is not very effective in inducing a metallic state in iridates.

Acknowledgments

We wish to acknowledge the support of our research on transition metal oxides at the University of Kentucky by the U. S. National Science Foundation under grants DMR-0856234 and EPS-0814194, and the U. S. Department of Energy Office of Science, Basic Energy Sciences Grant No. DE-FG02-97ER45653. GC would like to thank Professor Pedro Schlottmann for important input for this work and Dr. Shalinee Chikara, Miss Panpan Kong, Dr. Changqing Jin, Dr. Oleksandr Korneta, Dr. Li Li, Dr. Ge Min and Dr. Tongfei Qi and Dr. Shujuan Yuan for their contributions to the experimental work reviewed in this Chapter.

References

1. B. J. Kim, Hosub Jin, S. J. Moon, J.-Y. Kim, B.-G. Park, C. S. Leem, Jaejun Yu, T. W. Noh, C. Kim, S.-J. Oh, V. Durairai, G. Cao, and J.-H. Park, Phys. Rev. Lett. **101**, 076402 (2008)
2. S.J. Moon, H. Jin, K.W. Kim, W.S. Choi, Y.S. Lee, J. Yu, G. Cao, A. Sumi, H. Funakubo, C. Bernhard, and T.W. Noh, Phys. Rev. Lett. **101**, 226401 (2008)
3. B.J. Kim, H. Ohsumi, T. Komesu, S. Sakai, T. Morita, H. Takagi, T. Arima, Science **323**, 1329 (2009)
4. Q. Huang, J. L. Soubeyroux, O. Chmaissen, I. Natali Sora, A. Santoro, R. J. Cava, J.J. Krajewski, and W.F. Peck, Jr., J. Solid State Chem. **112**, 355 (1994)
5. R.J. Cava, B. Batlogg, K. Kiyono, H.Takagi, J.J. Krajewski, W. F. Peck, Jr., L.W. Rupp, Jr., and C.H. Chen, Phys. Rev. B **49**, 11890 (1994)

6. M.K. Crawford, M.A. Subramanian, R.L. Harlow, J.A. Fernandez-Baca, Z.R. Wang, and D.C. Johnston, Phys. Rev. B **49**, 9198 (1994)
7. G. Cao, J. Bolivar, S. McCall, J.E. Crow, and R. P. Guertin, Phys. Rev. B **57**, R11039 (1998)
8. G. Cao, Y. Xin, C. S. Alexander, J.E. Crow and P. Schlottmann, *Phys. Rev. B* **66**, 214412 (2002)
9. G. Cao, J.E. Crow, R.P. Guertin, P. Henning, C.C. Homes, M. Strongin, D.N. Basov, and E. Lochner, *Solid State Comm.* **113**, 657 (2000)
10. D. Haskel, G. Fabbris, Mikhail Zhernenkov, P. P. Kong, C. Jin, G. Cao and M. van Veenendaal, *Phys. Rev Lett.* **109**, 027204 (2012)
11. Q. Wang, Y. Cao, J. A. Waugh, T. F. Qi, O. B. Korneta, G. Cao, and D. S. Dessau, submitted to Phys. Rev Lett., 2012
12. Jean-Michel Carter and Hae-Young Kee, arXiv: 1207.2183v2 (2012)
13. G. Jackeli and G. Khaliulin, Phys. Rev. Lett. **102**, 017205 (2009)
14. J. P. Clancy, K. W. Plumb, C. S. Nelson, Z. Islam, G. Cao, T. Qi, and Young-June Kim, 2012
15. S. Chikara, O. Korneta, W. P. Crummett, L.E. DeLong, P. Schlottmann, and G. Cao, Phys. Rev. B **80**, 140407 (**R**) (2009)
16. M. Ge, T. F. Qi, O.B. Korneta, L.E. De Long, P. Schlottmann and G. Cao, Phys. Rev. B **84**, 100402(R), (2011)
17. O.B. Korneta, Tongfei Qi, S. Chikara, S. Parkin, L. E. DeLong, P. Schlottmann and G. Cao, Phys. Rev. B **82**, 115117 (2010)
18. D. Hsieh, F. Mahmood, D. Torchinsky, G. Cao and N. Gedik, *Phys. Rev. B* **86**, 035128 (2012)
19. G. M. Sheldrick, Acta Crystallogr A **64**, 112 (2008)
20. S. J. Zhang et al., Europhys. Lett. **88**, 47 008 (2009)
21. Feng Ye, Songxue Chi, Bryan C. Chakoumakos, Jaime A. Fernandez-Baca, Tongfei Qi, and G. Cao, *Phys. Rev. B* **87**, 140406(R) 2013
22. D. Haskel, N. Souza-Neto, J. C. Lang, V. V. Krishnamurthy, S. Chikara, G. Cao, and Michel van Veenendaal, *Phys. Rev. Lett.* 105, 216407 (2010)
23. I. Franke, P.J. Baker, S.J. Blundell, T. Lancaster, W. Hayes, F.L. Pratt, and G. Cao, Phys. Rev. B **83**, 094416 (2011)
24. S. J. Moon, Hosub Jin, W. S. Choi, J. S. Lee, S. S. A. Seo, J. Yu, G. Cao, T. W. Noh, and Y. S. Lee, Phys. Rev. B **80**, 195110 (2009)
25. Hiroshi Watanabe, Tomonori Shirakawa, and Seiji Yunoki, Phys. Rev. Lett. **105**, 216410 (2010)
26. Cyril Martins, Markus Aichhorn, Loig Vaugier, and Silke Biermann, Phys. Rev. Lett. **107**, 266404 (2011)
27. M.A. Subramanian, M.K. Crawford, R. L. Harlow, T. Ami, J. A. Fernandez-Baca, Z.R. Wang and D.C. Johnston, Physica C **235**, 743 (1994)
28. R. S. Perry, F. Baumberger, L. Balicas, N. Kikugawa, N.J. Ingle, A. Rost, J. F. Mercure, Y. Maeno, Z.X. Shen and A. P. Mackenzie, New Journal of Physics **8**, 175 (2006)

29. S. J. Moon, M. W. Kim, K. W. Kim, Y. S. Lee, J.-Y. Kim, J.-H. Park, B. J. Kim, S.-J. Oh, S. Nakatsuji, Y. Maeno, I. Nagai, S. I. Ikeda, G. Cao, and T. W. Noh, Phys. Rev. B **74**, 113104 (2006)

30. Y. Maeno, H. Hashimoto, K. Yoshida, S. Nishizaki, T. Fujita, J.G. Bednorz, and F. Lichtenberg, Nature **372**, 532 (1994)

31. M.W. Haverkort, I.S. Elfimov, L.H. Tjeng, G.A. Sawatzky and A. Damascelli, Phys. Rev. Lett. **101**, 026406 (2008)

32. Guo-Qiang Liu, V.N. Antonov, O. Jepsen, and O. K. Andersen, Phys. Rev. Lett. **101**, 26408 (2008)

33. J. S. Lee, Y. Krockenberger, K. S. Takahashi, M. Kawasaki, and Y. Tokura, Phys. Rev. B **85**, 035101 (2012)

34. Nevill Mott, *Metal-Insulator Transition*, Taylaor & Francis, London, 1990; p.52.

35. P. A. Cox, *Transition Metal Oxides*, p.163 (Clarendon Press, Oxford, 1995)

36. T. F. Smith, J. Low Temp. Phys. **11**, 581 (1973)

37. P. Schlottmann, Phys. Rev. B **46**, 998 (1992)

38. J. Beille, M. B. Maple, J. Wittig, Z. Fisk and L. E. De Long, Phys. Rev. B **28**, 7397 (1983)

39. J. D. Jorgensen, B. Dobrowski, S. Pei, D. G. Hinks, L. Soderholm, M. Morosin, J. E. Schirber, E. L. Venturini and D. S. Ginley, Phys. Rev. B **38**, 11337 (1988).

40. N.F. Mott, *Metal-Insulator Transitions*, p. 36 (Taylor and Francis, London, 1990)

41. G. Gruner, *Density Waves in Solids* (Addison-Wesley, New York, 1994)

42. G. Cao, J. W. O'Reilly, J. E. Crow and L. R. Testardi, Phys. Rev. B **47**, 11510 (1993)

43. J.F. Scott, Phys. Rev. B **16** 2329 (1977)

44. T. Kimura, S. Kawamoto, I. Yamda, M. Azuma, M.Takano and Tokura, Phys. Rev B **67**, 180401 (R) (2003)

45. B. Lorenz, Y. Q. Wang, Y.Y. Sun and C.W. Chu, Phys. Rev. B **70**, 212412 (2004)

46. George Samara, in: H. Ehrenreich and F. Spaepen, Eds., *Solid State Physics* Vol. 56 (Academic Press, New York, 2001), page 270.

47. N. A. Hill J. Phys. Chem. B **104**, 6694 (2000)

48. N. Hur, S. Park, P. A. Sharma, J. S. Ahn, S. Guha and S-W. Cheong, Nature **429**, 392 (2004)

49. W. Eerenstein, N. D. Mathur and J. F. Scott, Nature **442**, 759 (2006)

50. T. Kimura, Annu. Rev. Mater. Res. **37**, 387 (2007)

51. R. Ranjith, A. K. Kunu, M. Filippi, B. Kundys. W. Prellier, B. Ravaeu J. Lavediere, M.P. Singh and S. Jandl, Appl. Phys. Lett. **92**, 062909 (2008)

52. Jiangping Hu, Phys. Rev. Lett. **100**, 077202 (2008)

53. M. A. Subramanian, M. K. Crawford, and R. L. Harlow, Mater. Res. Bull. 29, 645 (1994)

54. T.F. Qi, O. B. Korneta, L. Li, K. Butrouna, V.S. Cao, Xiangang Wan, R. Kaul and G. Cao, *Phys. Rev. B* **86**, 125105 (2012)

Index